Adapting to Climate Change in Agriculture-Theories and Practices

Syed Sheraz Mahdi · Rajbir Singh ·
Bhagyashree Dhekale
Editors

Adapting to Climate Change in Agriculture-Theories and Practices

Approaches for Adapting to Climate Change in Agriculture in India

Springer

Editors
Syed Sheraz Mahdi
Division of Agronomy
Sher-e-Kashmir University of Agricultural Sciences and Technology of Kashmir (SKUAST-K)
Srinagar, Jammu and Kashmir, India

Rajbir Singh
Department of Plant Pathology
Gochar Mahavidyalaya
Rampur Maniharan, Saharanpur, Uttar Pradesh, India

Bhagyashree Dhekale
Division of Agricultural Statistics
Sher-e-Kashmir University of Agricultural Sciences and Technology of Kashmir (SKUAST-K)
Srinagar, Jammu and Kashmir, India

ISBN 978-3-031-28141-9 ISBN 978-3-031-28142-6 (eBook)
https://doi.org/10.1007/978-3-031-28142-6

This Springer imprint is published by the registered company Springer Nature Switzerland AG
The registered company address is: Gewerbestrasse 11, 6330 Cham, Switzerland

Paper in this product is recyclable.

Contents

Chapter 1
Conservation of Carnivorous Plants in Odisha: A Key Challenge for the Policy Makers

Prasad Kumar Dash, Subhrakanta Jena, Rakesh Kumar Mohalik, and Hemanta Kumar Sahu

1.1 Introduction

Nature has always been a matter of inquisitiveness to human beings due to its stunning and vibrant life forms. Many of its creations have created wave within the scientific communities and scientists are still trying to know and disclose explanations for several such outstanding life forms. Generally, insects eat upon plants and being preyed by the higher animal in the food web operating in nature. However, when it comes to the role of nature the plants can be serve as predators in the food web. Amongst numerous such unclear mysteries, there lies a group of angiosperms that attract, catch, kill, eat, digest and absorb the body juice of animals are called Carnivorous Plants behaviour we normally associate with the animal kingdom. In these plants, the leaves are modified into trap leaves to catch, kill and digest insects. To the plant world, they are also known as "Insectivorous" or 'Hunter plants'. Even though the studies on these menacing predators have started in fourteenth Century, it was Charles Darwin, who was mesmerized by these plants in 1857 and through his experiments and research proved the carnivorous syndrome in these plants. It is called as the Venus flytrap (*Dionaea muscipula*) as "The most wonderful plant in the world." Apart from the angiosperms, there are about 50 species of carnivorous fungi. In the

P. K. Dash (✉)
Odisha Biodiversity Board, Forest and Environmental Department, Government of Odisha, Regional Plant Research Centre Campus, Nayapalli, Bhubaneswar, Odisha 751015, India
e-mail: prasad.dash2008@gmail.com

S. Jena · H. K. Sahu
P.G. Department of Zoology, North Orissa University, Takatpur, Baripada, Mayurbhanj, Odisha 757003, India

R. K. Mohalik
Keonjhar Wildlife Division, Anandapur, Keonjhar, Odisha 758021, India

S. Sheraz Mahdi et al. (eds.), *Adapting to Climate Change in Agriculture-Theories and Practices*, https://doi.org/10.1007/978-3-031-28142-6_1

evolution process, the insectivorous plants have evolved from 6 distinct lineages. All these plants arisen with a special type of modified leaves to trap different arthropod species and have potential to digest the captured animal (arthropods) to absorb the nutrients. The technique of trapping and utilizing the prey (all arthropod species) makes these plants unique specialized in the nature besides that these plants are also adopted to persist in the marshy places with poor nutrient soil.

1.2 Significance of Carnivorous Plants

Apart from their peculiar behavior and unique food habit in the plant world, these plants also provide various ecosystem services to human beings. These plants also play significant role in human society by eradicating the diseases transmitted through different dipterans vectors as they preferably feed on the dipteran species like mosquitos, midges, deerflies and horse flies (Jennings & Rohr, 2011). Human schistosomes, a disease caused by flatworms inhabiting in the contaminated water along with other parasites. This disease is wide spread over the earth and past record revels that 207 million people are infected and approximately 700 million are at risk. The death records suggest globally, 20 lakh/year death are occurring due to schistosomiasis. The *Utricularia* belong to the carnivorous plant category usually found in the aquatic bodies feeds on mosquito eggs and larvae and even depredates human schistosome miricidia and cercariae, which indicate these plants, can be a source to control and reduce the protozoans parasite diseases there by reducing their vectors through which they are transmitted (Jennings & Rohr, 2011). The carnivorous plants are fatal to those animal species, which fits in their leave traps. Some arthropods are evolved according with carnivorous plant and adapted to survive against the predation of these plants. An Australian bug called as Cryptopcltis bug, which can walk on the sticky traps of these plants and feeds on the prey species of these plants. Several other species insects like some mosquito larvae are adopted to live in the water bodies where the pitcher plants are present. Similarly, some spider species are able to construct their web over the leave of these plants, which prevents the prey fall into the traps and feeds on them.

All most all Droseras are important medicinally. The traditionally ethnobotanical claims suggests that brulsed leave with salt applied to treat blisters corns have been protected by applying leaf extracts of *D. indica*. The past records suggests that the Ayurvedic traditional healers prepare 'Swarnabhasma' (gold-bhasma/Golden ash) from these plants and used to treat different ailments like bronchial asthma, rheumatoid arthritis, diabetes mellitus and nervous disorders, besides that it was also used for poor memory, eye sight, infertility and physical weakness problems. The leaf juice of *D. indica* is used as the antisyphilitic tonic by the tribal healers. In home textile industry, the yellow crystalline substances are obtained from *D. peltata* to make yellow color silk. It has been observed that the extracts of the *D. peltata* has the antimicrobial activity against human pathogenic microbes (Didry et al., 1998). *D. burmanii* is used by the traditional healer inhabiting in Tripura as a antisipetic for snakebites and also used as effective medicine in thought infection. It is locally famous in Tripura as Bishkatali as it reduces the effect of any oision bite in the body (Majumdar et al., 2011). *Drosera burmannii* possess medicinally active compounds like quinones (plumbagin), hydroplumbagin glucoside, flavonoids (kaempferol, myricetin, quercetin, and hyperoside), rossoliside (7-methyl-hydrojuglone-4-glucoside), as a result this species could be a source of making effective medicine (Raju & Christina, 2013). Similarly, Bladderworts like *Utricularia aurea* and *U. reticulata* are ornamental, species used in aquaria and rockeries. *U. stellaris* can cure cough; *U. caerulae* has wound healing property; *U. bifida* is mostly useful for urinary infection. Anti-tumor activity has been reported in *Utricularia aurea. Utricularia bifida* is used in Indian medicine for urinary disorders (Choosawad et al., 2005).

The changes in the environmental parameters in the habitat influences the diversity and distribution of these plants, as the plant species are easily identified so these species can serve as biological indicator in the assessment of ecosystem for health benefits.

1.3 Morphology, Biology and Ecology of Carnivorous Plants

Except few genera, generally carnivorous plants are annual to perennial herbs with rhizomes, fibrous roots, or tubers with a vertical stolon below ground (*Drosera*) or herbaceous, terrestrial (Drosera and Utricularia), epiphytic and saprophytic (Utricularia) or aquatic (*Utricularia*) or climbing shrubs (*Nepanthes*). The leaves are whorled in rosettes. The inflorescence varies from a scapose raceme in Lentibulariaceae to cymose helicoid or dichasial in Droseraceae and racemes or panicles in *Nepanthes.* Flowers bisexual, actinomorphic with brilliant colors, secrete nectars and other ways to attract their prey.

In the ecosystem the carnivorous plants stablish a peculiar type of prey-predator relationship that instead of being eaten by herbivorous their emerge with dual mode of nutritional method i.e. photo autotrophic and heterotrophic nutrition specifically carnivorous nature of eating insects and other animal species, which allow these plants are adopted in such a way that they can able to grow upon poor soil quality, where most of the plants are unable do so. These plants utilize insect for two main purposes i.e., to get nourishment by capturing them and pollinated to continue sexual reproduction. Some of the adaptive feature make these species more efficient to persist in the poor soil i.e., the capability to capture insect, early growth of flower and seed dispersal. The reproduction in carnivorous plant may be of different type some are entomophiles, some reproduce through self-pollination and show vegetative propagation.

The victims of these plants also fall in a broad range comprising of almost 150 species of pray like mosquitoes, housefly, small butterflies and beetles, arachnids including mainly spiders and mites, mollusks like snails and slugs, sometimes earthworms. It has observed that the *D. burmanii* species capture 6 insect orders like Lepidoptera, Isoptera, Diptera, Orthoptera, Hymenoptera, and Coleoptera (Majumdar et al., 2011), whereas *Daphnia*, nematodes, mosquito larvae and tadpoles are the major pray species for Utricularias. Zooplankton like rotifers, cladocerans, copepods, annelids, rhizopodeans, and insects; and the phytoplankton like Bacillariophyta, Chlorophyta, Cyanophyta, and Euglenophyta are other victims of aquatic *U. gibba* and *U. inflata.*

Carnivorous plant can maintain aquatic, semiaquatic and terrestrial mode of life. Most of the carnivorous plant species are seasonal as they are inhabiting in marshy, submerged, muddy and sandy areas, usually the soil of their habitat constitute low concentration of N, P and K. In these type of habitats, usually anaerobic conditions are found due to the partial decomposition of organic matters, which generates other acid forms and some unfavorable conditions for the plant.

1.4 Diversity and Distribution

Carnivorous plants are distributed over the world except Antarctica (Barthlott et al., 2007). About 133, 86 and 65 carnivorous plant species are flourishing in the Malaysia, Australia and Brazil region respectively. They represent members of five orders: Poales, Oxalidales, Caryophyllales, Ericales and Scrophulariales. Seven hundred carnivorous plant species are widespread over the world, described under 15 genera and 9 families of dicotylidonous type (Fleischmann, 2012). Very interestingly in India 44 species under five genera are reported so far viz. *Utricularia* (38 species), *Drosera* (3 species), *Nepenthes* (1 species), *Pinguicula* (1 species), and *Aldrovanda* (1 species) belong mainly to three families: Droseraceae, Nepenthaceae and Lentibulariaceae (Kamble et al., 2012). The comparison of carnivorous plant diversity has given in Table 1.1.

1.5 Carnivorous Plants of Odisha Past and Present

Odisha with its unique phytogeographic location in India having varied topography, frequent climatic change and its presence above 5000 ft MSL biogeographic provinces of India—the Eastern Ghats and Chhotanagpur Plateau makes Odisha a unique habitat flourishing flora and fauna in a better way starting from microbes to mammals.

Initially Dr. Saxena, 1965 reported the presence of *Utricularia pubescens* from Odisha. Latter Saxena and Brahmam (1994) reported the occurrence of 14 species of carnivorous plants under 2 families and 2 genera from different districts of Odisha. This includes 3 species *Drosera* under family of Droseraceae and 11 species of *Utricularia* under family Lentibuleriaceae. The present report which is a result of

Table 1.1 Comparison of carnivorous plant diversity

Order	Family	Genus	Number of species		
			World	India	Odisha
Oxalidales	Cephalotaceae	*Cephalotus*	1		
Nepenthales (Caryophyllales)	Droseraceae	*Drosera*	168	3	3
		Aldrovanda	1	1	
		Dionaea	1		
	Drosophyllaceae	*Drosophyllum*	1		
	Nepenthaceae	*Nepenthes*	127	1	
	Dioncophyllaceae	*Triphyophyllum*	1		
Ericales	Sarraceniaceae	*Darlingtonia*	1		
		Heliamphora	20		
		Sarracenia	8		
	Roridulaceae	*Roridula*	2		
Scrophulariales (Lamiales)	Byblidaceae	*Byblis*	7		
	Lentibulariaceae	*Pinguicula*	94	1	
		Genlisea	27		
		Utricularia	234	38	14
Total: 4 orders	9 families	15 genera	693	44	17

Source http://www.omnisterra.com/bot/cp_home.cgi

the survey made by the authors for the period from 2009 to 2013 has include 2 more species of *Utricularia* i.e. *Utricularia reticulata* and *Utricularia praeterita* from the state enhancing the total no. of species in to 17. However, we could not locate the presence of *Utricularia pubescence* during our survey. However, based on the report of Saxena (1965), we have kept this species in the final list of carnivorous plants of Odisha. Odisha hosts 17 species of Carnivorous Plants including 3 species of *Drosera* under family Droseraceae and 14 species of *Utricularia* under family Lentibulariaceae.

1.5.1 The Sundews of Odisha and Its Distribution

Each species of the Droseraceae family popularly called as Sundew. These species have small deep red colour leave covered with small hairy bristles, these leave produces shiny sticky secretions to encounter insects. Insects are trapped or captured when they encounter this leave and, in the mean, while the hairy out growth bend over the insect, the insect dies with suffocation and digested, out of the 4 genera placed under Droseraceae, 2 of the genera are commonly found in India i.e. *Drosera* and *Aldrovanda*. In Odisha, so far only one genus is reported to occur under Droseraceae, i.e., *Drosera*. It is the largest carnivorous plant having more than 170 species distributed over the world, Indian flora represents 3 species of *Drosera* such as *D. burmannii*, *D. indica* and *D. peltata* and all of them are found in Odisha.

Drosera burmannii occurs in moist or exposed sandy soils, in shaded places, nearby rice fields, lowland (Barren land at foothills of Sikharchandi hill) to mountain areas up to 4000 ft. It is distributed in Barkot block of Deogarh, Krishnamali and Khandualmali hills of Karlapat wildlife sanctuary, above 1000 m on laterite plateau in Kalahandi district, Gandhamardan hills of Bolangir district and Mahendragiri hills of Gajapati district and Hemgiri range of Sundargarh district etc. *Drosera indica* occurs in laterite plateaus of Karlapat wildlife sanctuary, above 3000 ft associated with *Eriocaulon* sp., grasses and other carnivorous plants such as *Drosera burmanii*, *Utricularia scandens*, *Utricularia bifida* and *Utricularia minutissima* of Krishnamali and Khandualmali hills, Kalahandi district. Laterite plateau above 3000 ft in Gandhamardan hills of Bolangir district and above 4000 ft in grasslands of Mahendragiri hills, Gajapati district are other critical habitat for *D. indica*. So far, *Drosera peltata* is reported only from Deomali hills of Koraput district above 5000 ft associated with grasses, *Eriocaulon* sp., and as epiphytes on old tree trunks covered with mosses.

1.5.2 The Bladderworts of Odisha and Its Distribution

The members of the genus *Utricularia* are commonly called as Bladderworts. They are not only the largest, but also the most cosmopolitan plant carnivorous genus, with a distribution ranging from high latitudes and boreal environments to the tropics. Their ecological success is further underscored by the great variety of life forms which include aquatic (lentic and lotic), terrestrial, epilithic, and epiphytic species. Utricularias are remarkable for their traps. Though the species show a wide morphologic and biologic diversity, a single type of trap is observed all over the genus. This family of insectivore's plants have 4 genera and more than 230 species are found worldwide of which 50 species are submerged or amphibious aquatics. Two genera such as *Utricularia* and *Pinguicula*, occurs in India. Only one genus i.e. *Utricularia* occurs in Odisha. Fifteen species of *Utricularia* have so far been reported from Odisha, Out of which 2 species (*U. reticulata* and *U. praeterita*) are new distributional record for the state. Other species of bladderworts found in Odisha are *U. aurea, U. bifida, U. caerula, U. exoleta, U. hirta, U. minutissima, U. polygaloides, U. pubescens, U. scandens, U. stellaris, U. striatula* and *U. uliginosa*. Out of them *U. aurea, U. exoleta, U. stellaris* are aquatic and the rest are terrestrial in habitat. The aquatic Utricularias are found to occur in roadside wetlands and swamps from Cuttack to Paradeep, Bhitarkanika mangrove ecosystems, Kanjia lake and RPRC Lake.

The terrestrial Utricularias are common in marshy places, along grasses in rice fields associated with *Eriocaulon* sp., such as *Utricularia hirta*, *Utricularia polygaloides*, *Utricularia scandens* and *Utricularia bifida* occurs in ditches and banks of small perennial water holes in Kamakhyanagar, Dhenkanal district, Paniganda wetlands in Ganjam, Similipal, Talcher, Sambalpur, Sunabedha wl sanctuary etc. Species like *Utricularia minutissima, Utricularia caerula, Utricularia praeterita, Utricularia uliginosa, Utricularia reticulata and Utricularia striatula* occurs in associated with grasses, in low land areas and swamps of Ranpur, Nayagarh district, Swamps of Bhitarkanika, Kendrapada district, small water wholes of Kamakhya Nagar, Dhenkanal district to high altitudinal plateaus of mahendragiri hills of Gajapati district, Deomali hills of Koraput district, Krishnamali and Khandualmali hills of Karlapat wildlife sanctuary in Kalahandi district, Laterite plateau above 3000 ft in Gandhamardan hills of Bolangir district and above 4000 ft, above 2500 ft in 2400 ft, Malayagiri hills of Angul district and in Phulbani forest divisions of Kandhamal district.

1.6 Carnivorous Plants in Biotechnology

The carnivorous plants are rich source of a number phytochemicals. The *Drosero* genus contains phytochemicals like Plumbaginin, 7-methyljuglone and flavonoids, quercetins and quinines. These phytochemicals have immunomodulatory, antispasmodics, cytotoxic activity against cancer cell, antibacterial, antifungal and antiviral, anti-inflammatory and antioxidant activities. It has observed that this phytochemical also promotes the phagocytic activities of WBC (granulocytes) and boost the immune response against different pathogens of human body. *D. indica* extracts have potential bioactivities like antioxidant and anticancer activity. The phytochemicals present in the *D. indica* also have nutritional and have property to prevent free radicals to reduce toxic level in the body. These plants do have organic acids and enzyme, which can help in curdling the milk. *Drosera burmannii* contains quinones, hydroplumbagin glucoside, flavonoids (kaempferol, myricetin, quercetin, and hyperoside), rossoliside (7-methyl-hydrojuglone-4-glucoside), and useful for several disorders.

The genus *Utricularia* is also very interesting at the molecular level. Some species are known to have minute genomes: *U. gibba* has the smallest known plant genome, which has 80 mega bases, comprises the half of *Arabidopsis*. The molecular genomic studies also focused on the evolution rate of chloroplast, mitochondria and nuclear ribosomal sequence (Greilhuber et al., 2006).

1.7 Threats to Carnivorous Plants of Odisha

All most 50% of the carnivorous plant species have assessed through IUCN with respect to the threaten level but still not all the species have not been quantified systematically and not been placed in any IUCN threat level. Few records published in last few years suggests that IUCN listed 102 species of 7 genera have been placed under critically endangered 11 species under endangered category and 39 species are in vulnerable rank. Consequently, of these IUCN-evaluated species, 56% are considered to be threatened, which is lower than the whole angiosperm present in the earth i.e., 70%. IUCN, records suggest that diversity is depleted due to sampling of wild plants for search, pollution, and natural modifications. Some species are at the verge of extinction over the world. In India 3 species of carnivorous plants are categorized under red data book as endangered plants.

The species *D. indica* and *D. burmanii* are commercially imported from Asia to European countries for the production of "Herba dorserae" a Ayurvedic herbal cosmetic products instead of focusing their own localized carnivorous plant i.e., *D. rotundifilia.* The exporter are frequently exploiting these plants in a huge quantity annoyingly about the abundance of these plant species in their own locality (Jayaram & Prasad, 2006). In our neighboring state like Andhra Pradesh, *D. indica* and *D. burmanii* have categorized under vulnerable category by IUCN. All these threatened species are in the vulnerable category now they are going to enlist under the endangered category if the Govt. will not take any initiative conservation strategy. These plants are declined accordingly due to their popularity of being used medicinal herbs and their commercial values in local as well as in the abroad market. These species are used for making cosmetic herbal products in foreign countries. The problem of conversion of land into either agricultural or residential areas is threatening the natural habitat of these carnivorous plants. In addition, mining and pollution of water bodies highly affects the growth of these plants. Currently the anthropogenic activities as urban development and industrialization are putting impact on the carnivorous plants diversity as a result their number is declining in the wetland habitats. Now days due to the continuous exposure of influents or pollutants, the wetland is declining day by day and putting on the impact on the diversity and distribution of the carnivorous plant. As they are adopted to poor nutrient environment, so they also could not grow up on the high nutrient, containing soils. Currently the contaminated water bodies are populated with weeds, which indirectly wipe out carnivorous plants from their niche.

1.8 Conservation Strategies

Public awareness is essential for the conservation of these plants and should be circulated among the student communities, researchers, scientists, foresters, policy makers. The biodiversity assessment should be periodically conducted regarding the diversity, distribution and other ecological parameters in relation to these plants. The early botanical survey regarding these represents that besides the natural causes the anthropogenic activities are vulnerable to this carnivorous plant diversity. All the facts and problem should be recorded through documentation and store in cataloged format so that more research will be conducted in a better way to conserve these plant species in their own natural habitat. The research should be carried out in a sustainable manner as these plant species are not only a good medicinal source but also play vital role in the ecosystem and serving as an indicator species in the nature.

The state legislation should establish some conservation strategy special meant for carnivorous plants, which should be operated by the forest department. It is imperative to locate areas such as plateaus, bordering areas of rice fields, wetlands etc. that have a high probability of *Drosera* and associated occurrence of terrestrial *Utricularia* species and aquatic Utricularias, so in the natural areas in-situ conservation strategy should be build up in a better way. The above mention suggestions not only helpful to conserve the carnivorous plants but also brings ecological integrity among the co-existence species in their habitats. Several other methods like field transfer of these plants from hazardous region to the conservation territory, seedbed preparation in the in-vivo condition, seed preservation and their maintenance through standardize methods. The molecular genomic information should be encoded and kept in the research database of different institutes to make the future these plants in safe side. The first step towards preserving these plants is to preserve at least some parts of their natural habitats. Along with it, it is also necessary to develop horticultural techniques for the species and promote ex-situ conservation. Otherwise, we may end up having only photographic evidence proving that the plant predators existed in Odisha.

1.9 Future Prospects in Research on Carnivorous Plants

The Carnivorous Plants are these days coming into limelight of modern research because of their characteristic enzyme complexes, absent in other plants. Unfortunately, these plants are on the decline when they have the potential to open up new vistas in the field of medicine. Given such a significant body of phytochemical research to date, future research prospects for the genus Drosera appear promising. Apart from micro propagation, It is hypothesize that three particularly novel lines may include the discovery of medicinal agents against multi-drug resistant (MDR) TB, applications in green chemistry, and anthocyanin-mediated nitrogen transport in these plants.

References

Choosawad, D., Leggat, U., Dechusukhum, C., Phongdara, A., & Chotigeat, W. (2005). Anti-tumor activities of fucoidan from the aquatic plant *Utricularia aurea* Lour. *Songklanakarin Journal of Science and Technology, 27*(3), 799–807.

Didry, N., Dubreuil, L., Trotin, F., & Pinkas, M. (1998). Antimicrobial activity of aerial parts of *Drosera peltata* Smith on oral bacteria. *Journal of Ethnopharmacology, 60*(1), 91–96.

Fleischmann, A. (2012). The new *Utricularia* species described since Peter Taylor's monograph. *Carnivorous Plant Newsletter, 41*(2), 67–76.

Greilhuber, J., Borsch, T., Müller, K., Worberg, A., Porembski, S., & Barthlott, W. (2006). Smallest angiosperm genomes found in Lentibulariaceae, with chromosomes of bacterial size. *Plant Biology, 8*(6), 770–777.

Jayaram, K., & Prasad, M. N. V. (2006). Drosera indica L. and D. burmanii Vahl., medicinally important insectivorous plants in Andhra Pradesh-regional threats and conservation. *Current Science, 91*, 943–946.
Jennings, D. E., & Rohr, J. R. (2011). A review of the conservation threats to carnivorous plants. *Biological Conservation, 144*(5), 1356–1363.
Kamble, M. V., Harikrishnan, S., & Balakumar, P. (2012). *Utricularia caerulea* (Lentibulariaceae): A new record to flora of Andaman & Nicobar islands. *Rheedea, 22*(2), 116–118.
Majumdar, K., Datta, B. K., & Shankar, U. (2011). Community structure and population status of Drosera burmanii Vahl. with new distributional record in Tripura, India. *Journal of Ecology and the Natural Environment, 3*(13), 410–414.
Raju, A., & Christina, A. J. M. (2013). Drosera burmannii Vahl: Antioxidant potential in Dalton's Ascites Lymphoma (DAL) bearing mice. *Journal of Medicinal Plants, 1*(4), 152–159.
Saxena, H. O. (1965). Utricularia pubescens Sm.-a new record for India. *Indian Forester, 91*(1), 73–75.
Saxena, H. O., & Brahmam, M. (1994). The flora of Orissa (Vol. 1, pp. 437–439). Regional Research Laboratory; Orissa Forest Development Corporation.

Chapter 2
Climate-Smart Millets Production in Future for Food and Nutritional Security

Sagar Maitra, Tanmoy Shankar, Akbar Hossain, Masina Sairam, Lalichetti Sagar, Upasana Sahoo, Dinkar Jagannath Gaikwad, Biswajit Pramanick, Tanuj Kumar Mandal, Sukamal Sarkar, Harun I. Gitari, and Esmaeil Rezaei-Chiyaneh

2.1 Introduction

The millet crops belong to the family of grasses which show tolerance to soil moisture stress and different adverse weather conditions. They are mostly annuals with small grains and warm weather coarse cereals which are often used as food and

S. Maitra (✉) · T. Shankar · M. Sairam · L. Sagar · U. Sahoo · D. J. Gaikwad
M.S. Swaminathan School of Agriculture, Centurion University of Technology and Management, Paralakhemundi, India
e-mail: sagar.maitra@cutm.ac.in

A. Hossain
Bangladesh Wheat and Maize Research Institute, Dinajpur, Bangladesh

B. Pramanick
Department of Agronomy, Dr. Rajendra Prasad Central Agricultural University, Samastipur, Bihar, India

T. K. Mandal
School of Agriculture and Allied Sciences, The Neotia University, Jhinga, Sarisha, West Bengal, India

S. Sarkar
School of Agriculture and Rural Development, Ramakrishna Mission Vivekananda Educational and Research Institute, Narendrapur, West Bengal, India

H. I. Gitari
School of Agriculture and Enterprise Development, Kenyatta University, Nairobi, Kenya

E. Rezaei-Chiyaneh
Department of Plant Production and Genetics, Faculty of Agriculture, Urmia University, Urmia, Iran

S. Sheraz Mahdi et al. (eds.), *Adapting to Climate Change in Agriculture-Theories and Practices*, https://doi.org/10.1007/978-3-031-28142-6_2

fooder (Fahad et al., 2017; Maitra et al., 2023a, b). During last few decades when major emphasis was given to fine cereals, namely, rice, wheat and maize, millets were neglected and treated as 'orphan cereals.' But over time millets have been re-evaluated and considering their nutritional value these are further treated as 'nutri-cereal'. Still millets are grown by the tribal and small farmers under the drought and rainfed conditions of mainly in arid and semi-arid regions (Saxena et al., 2018). Millets cultivation is predominantly confined in Africa, Asia and few regions of Europe. Worldwide, millets are grown in 33.56 million ha with an output of 31 million t of grains (FAOSTAT, 2020). Generally, millets are grouped into two categories, such as major and small millets. Pearl millet (*Pennisetum glaucum* L.) and sorghum (*Sorghum bicolor* L.) and fit into major millets, whereas, minor millets are barnyard millet (*Echinochloa frumentacea* L.), brown-top millet (*Brachiaria ramose* L. Stapf; *Panicum ramosum* L.) finger millet (*Eleusine coracana* L. Gaertn), foxtail millet (*Setaria italica* L.), kodomillet (*Paspalum scrobiculatum* L.), proso millet (*Panicum miliaceum* L.), little millet (*Panicum sumatrens* L.) and so on (Maitra, 2020a).

Presently, climatic aberration is appeared as a menace to agriculture and the modification of normal climate is very common. The anthropogenic intervention leading to climate change resulted production of greenhouse gases (GHGs) and aerosol which adversely impacted primarily on rainfall and temperature. As per the estimation of the Intergovernmental Panel on Climate Change (IPCC, 2007), if the anthropogenic activities go on in the same manner there will be a possibility of enhancement of earth's temperature by 1.1 to 5.4 °C by 2100. Moreover, global warming may trigger the occurrence of natural calamities such as excess rain, inundation and floods, scanty rain, soil moisture stress and drought, and cyclonic storms as resultant of increase of temperature and improper distribution of rain. The cumulative effect of climatic aberrations change in rainfall, temperature and elevated CO_2 ultimately causes hindrances to normal farming activities. Climate change hampers crop productivity with qualitative changes (Aryal et al., 2020). Mitigation of the adverse impacts of climatic abnormalities and global warming on farming and quality agricultural output are tremendous jobs (Ergon et al., 2018; Nuttall et al., 2017). However, to combat with the situation, adaptation options have already been taken into consideration in different regions of the world. There are several thermo-tolerant cultivars which have been developed and already are under cultivation (Ishimaru et al., 2016; Morita et al., 2016). The adverse influence of climate change has already been reflected in the performance of major food crops (Gaikwad et al., 2022), namely, rice (Bhatt et al., 2019; Rahman et al., 2017; Soora et al., 2012), wheat (Chakraborty et al., 2019; Hossain et al., 2021; Mukherjee et al., 2019; Xiao et al., 2018) and maize (Ureta et al., 2020). In the present consequence of climatic abnormalities, millets can be considered as climate-smart crops as they are drought and thermo-tolerant, rich in nutrients, can ensure bio-diversity, check soil erosion in marginal lands, as C_4 plants enable to use elevated atmospheric CO_2 and suitable to grown in wider ecological conditions (Banerjee & Maitra, 2020; Brahmachari et al., 2018; Srinivasarao et al., 2014). However, millets can be stored better than other food grains under normal condition for the quality of resistance from damage of insect attack (Adekunle, 2012; Li & Brutnell, 2011; Sage & Zhu, 2011; Sage et al., 2011).

In the present world, the country leaders and policy makers including leading international organizations implemented various initiatives to eradicate hunger towards achieving sustainable development goal (SDG), but hunger is still prevailing particularly in some corners of the developing countries (Rimas & Fraser, 2010).

The prime issues faced by the all concerned are population growth, urbanization and change in food demand and enough need for agricultural produces with dwindling and degrading natural resources (Gladek et al., 2016). Further, climate change imposed an interruption in achieving the targetted food security. The food security emphasizes availability, accessibility and proper use of food (with nutrition security) (Gross et al., 2000). A large portion of small and marginal farmers of dryland areas aground the world grows millets in the subsistence farming. Millets consumption can fulfill the food as well as nutritional security to the undernourished populace residing in the under-developped countries. In the chapter, effects of climate change on agricultural productivity, suitability of millets under the circumstances and climate-smart technologies for millets cultivation have been discussed thoroughly.

2.2 Climate Change Impact on Agriculture

During recent years, disasters occurring very frequently and climate change is responsible for occurrence of disasters like flood, drought, and cyclonic storms and so on. The developing countries are mostly affected by the climatic aberration (Maitra & Shankar, 2019). The production of important cereals such as rice, wheat and maize has declined drastically by ill effects of temperature rise and erratic rainfall (Lesk et al., 2016). The projected prediction has indicated that population growth in the developing countries; especially, the Sub-Saharan Africa and South Asia will be nourishing added population of 2.4 billion by the middle of the present century. The population living in the above-mentioned geographical locality lives on farming and allied activities and about one-fifth of the human population residing in the developing part of the world are suffering from starvation and malnutrition (Saxena et al., 2018). In future food crisis may be more crucial to the under changed climatic conditions. A general recommendation mentioned that there will be need for further enhancement of agricultural production by 60% in 2050 to fulfill the foodneeds of the future population. The present context demands for enhanced agricultural productivity and revenue in the developing countries (Lipper et al., 2014).

The collective effect of climatic abnormalities results in disturbance in normal agricultural activities. Agriculture is an anthropogenic activity and dependent on climatic parameters, namely, humidity, temperature, rainfall and so on (Gornall et al., 2010; Yohannes, 2016). Climatic aberration affects qualitative and quantitative fluctuation on agricultural productivity. Alteration in agro-ecosystem may also decline intensity of cropping and drought or water stagnation led to degradation of natural resources and biodiversity. Agriculture has enough importance in the economy and livelihood in the developing countries (Ackerman & Stanton, 2013). A projection has indicated that there will be the need for around 14,886 million tonnes of cereal equivalent food in the world in 2050 (Islam & Karim, 2019) to feed 9.7 billion people. As per present concept, food security is synonymous to food and nutritional security. To meet the target, latest and proven technologies are adopted considering the cropping as well as farming systems of various agro-ecological regions. As the productivity of fine cereals are adversely affected by climate change, millets can be chosen targeting uninterrupted production of food grains, because millets are hardy crops with wider adaptability to diverse agro-climatic conditions and cropping systems (Arendt & Dal Bello, 2011; Upadhyaya et al., 2008). Further, millets can easily be stored under normal storage conditions and so can be treated as famine food under contingency situations (Michaelraj & Shanmugam, 2013).

2.3 Adaptation Options Against Climate Change

Climate change denotes aberrations in the normal system and adaptation options are essential (Iizumi, 2019). To combat with the adverse impacts of climatic aberration, crop management options are considered which include changes in crop cultivation methods, cultivation of existing crops and cultivars with modified agronomic management, giving preference to the varieties of the same crop with abiotic stress tolerance, substitution to the crops with abiotic stress tolerance, availing crop insurance facilities, providing more emphasis to weather forecast and agrometeorological advisories, and crop insurance. Further, manipulation of sowing date, nutrient management, irrigation, drainage and water management, conservation agriculture inclusive of tillage, mulching and cover cropping are common agronomic management practices generally adopted against climate change (Fujibe et al., 2006). During present times, prediction of climate variability has become easier and considering the climate extremes suitable agronomic measures are adopted (IPCC, 2013; Vrieling et al., 2016). In this regard, different agro-meteorological tools are useful in weather related decision support system as farmers can adopt suitable measures (Hayashi et al., 2018). Further, there is the need for more precise information (Iizumi, 2019) with proper communication network to the farmers, particularly, smallholders in their local vernacular. Crop insurance is another adaptive measure to safeguard the smallholders from crop failure due to climatic issues. During present time, elimination of hunger and food security can be achieved by combined application biotechnology and information technology with ecofriendly adoption of agronomic management (Swaminathan & Kesavan, 2012). Development of climatic stress tolerant cultivar is genetic and biotechnological approaches suitable as adaptation measure. Recent advancement in the front of science and technology provided sophisticated tools for precision agriculture.

2.4 Millets: The Climate-Smart Crops

The climate change for agriculture is becoming a major challenge. The different factors which act as important issues like scanty rainfall and temperature raise directly increase the rate of evapotranspiration and reduction of water table in poor and marginal soil. Further, increased level of CO_2 and other GHGs are major issues influence crop production. So, to combat with the environmental issues, smart crops such as millets cultivation may be considered because millets come under C_4 plants category which are acquiescent with climate change. The C_4 mechanism can fight against drought and some other environmental stresses and these are of short to medium in duration with requirement of less number of inputs like labour, irrigation and nutrients. Generally, C_4 plants (millets) show greater nitrogen use efficiency than C_3 plants. As millets are C_4 plants, produce more phosynthate with enhanced temperature iclusive of increased level of flexible distribution arrays of dry matter and reduced hydraulic conductivity per unit leaf area (Sage & Zhu, 2011). Moreover, millets register more water use efficiency (WUE) than prime cereal crops. To produce 1 g dry matter foxtail millet uses 257 g water, whereas maize and wheat requires 470 g and 510 g of water respectively to yield the same dry matter (Bandyopadhyay et al., 2017). In future, when water scarcity will be more crucial, millets will be preferred to fine cerals to manage the food grain production target. Millets are hardy in nature and the crops show less susceptibility against pest and disease attack. Millets are C_4 plants which can use more of CO_2 and register less carbon footprint in agriculture (Aubry et al., 2011; Bandyopadhyay et al., 2017; Li & Brutnell, 2011). Agronomic measures are important in contribution of GHGs emission. Production of maize, wheat and rice contributes carbon equivalent emission of 935, 1000 and 956 kg C ha^{-1}, respectively (Jain et al., 2016). However, millets register less compared to above fine cereals and cultivation of millets is known to minimize C footprint in agriculture (Saxena et al., 2018). Further, chemical fertilizers are generally applied to crop field to supply the nutrients need of the crops and chemical N fertilizer is a very common input in agriculture. The production process of chemical N fertilizer produces CO_2. An estimate mentioned that the quantity of chemical N fertilizers produced worldwide generates 300 Tg of CO_2 to the atmosphere (Jensen et al., 2012).

Millets need a smaller amount nutrirnts than other fine cereals and hence, application chemical inputs are less which is environment friendly. In developing conutries, over-dependence on major cerals caused erosion of genetic diversity during last five decades. In contrast, diversified millets have enough potential to create diversity in agroecosystem ensuring superior ecosystem services. Moreover, millets play multifaceted role in food production system and sustainability of rural livelihood (Fig. 2.1) by providing food as well as nutritional and environmental security.

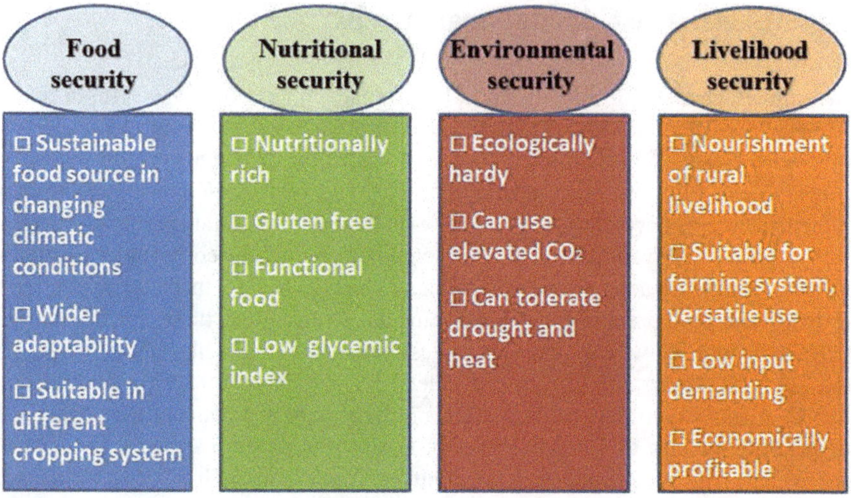

Fig. 2.1 Versatile role of millets in climate smart and sustainable agriculture

2.5 Nutritional Importance of Millets

The millets are also known as 'nutri-cereals' as they contain protein, fats, carbohydrates, vitamins, minerals and some micronutrients and phytochemicals (Table 2.1) (Banerjee & Maitra, 2020; Saleh et al., 2013). Further, millets are treated as functional food. During recent years, health-conscious people started consuming millets in their diet.

The richness in the nutritional quality has elevated millets as healthy foods for proper nutritional requirements. Millets are primarily used as food, however, they are also used as animal feed. Millets are generally gluten free and so preferred by

Table 2.1 Nutritional quality of millets (per 100 g of edible portion)

Crops	Crude fibre (g)	Protein (g)	Fat (g)	Mineral (g)	Carbohydrate (g)
Barnyard millet	14.7	11.6	5.8	4.7	74.3
Brown-top millet	–	9.0	1.9	3.9	71.3
Finger millet	3.6	7.3	1.3	2.7	72.0
Foxtail millet	8.0	12.3	4.3	3.3	60.0
Kodo millet	9.0	8.3	1.4	2.6	65.9
Little millet	8.6	8.7	5.3	1.7	75.7
Pearl millet	1.2	11.6	5.0	2.3	67.5
Proso millet	2.2	12.5	1.1	1.9	70.4
Sorghum	1.6	10.4	1.9	1.6	72.6

Source GOI (2020), Maitra et al. (2022), Banerjee and Maitra (2020)

the people suffering from gluten allergy and celiac disease. Besides, millets are comprised of enough of fibre content, vitamins and essential mineral matters which are vital to fulfill the nutritional security of undernourished people.

Millets are comprised of healthy phytochemicals like polyphenols, lignans, phytosterols, phyto-oestrogens andphytocyanins. The millets are treated as functional food because of presence of antioxidants, detoxifying agents and immune modulators that can potentially benefit against hyperglycemia, cardiovascular diseases, tumour, respiratory diseases, Parkinson's diseases and so on (Chandrasekara et al., 2012; Rao et al., 2011, 2012). The antioxidants present in millets can protect the DNA, proteins molecules and lipids membranes (Banerjee & Maitra, 2020).

2.6 Demand of Foods in Future and Role of Millets

For achieving the security of food and nutrition, it is very important to acquire the yield enhancement for the increasing population and to manage the distribution of the food grains. There is limitation in the world to provide healthy and nutritious food to all. In the developing countries of Africa and Asia, the problem is more crucial. Due to climate change the available resource and their limited utilization is raising the problem for food and nutritional security (Committee on World Food Security, 2012). About 815 million people of Africa and Asia are facing malnutrition (El Bilali, 2018; El Bilali et al., 2019). The potential of millets with rich nutritive and healthy benefit which is consumed as staple food and due to the high nutritional value of these crops is called as nutri-cereals. Further, consumption of millets crops is better in comparison to fine cereals because it contents more fibre content and easy digestive food (Banerjee & Maitra, 2020). The estimated population in the world will be 9.7 and 11.2 billion by 2050 and 2100 (FAO, 2017). There are already shrinkage and decline of land and water resources. On the other hand, urbanization is taking place rapidly with change food habits. In urban areas, demand for value added and animal source foods are more which need more energy to produce. The change in food demand is combined effect of increased population and income growth (Valina et al., 2014). Now there is the urgent need for sustainable intensification of farm productivity (Garnett, 2014). Earlier, farm output in the developing countries has been enhanced by adoption input driven technologies and over a period of few decades environmental degradation has been noticed. In the present context, the agricultural productivity has to be increased by about 50 per cent to meet the demand in 2050 (World Bank Group, 2016). Mueller et al. (2012) estimated that the enhancement of crop production should be 45 to 70% more than the present level. In 2050, the cereal equivalent food demand will be around 14,886 million tonnes (Islam & Karim, 2019). Food waste is another important factor to be considered while addressing the food requirement for the future and wastage of food is observed in several corners of the world. The climate change impacts created an additional burden in this regard. All these factors clearly indicate the requirement of more food production targeting food and nutritional security (FAO, 2018). Considering above constraints, targets and the huge potential of millets in terms of ecological soundness and nutritional value, it can be mentioned that millets will be one of the suitable options to ensure food and nutritional security of a considerable number of world population.

2.7 Value Added Food from Millets

Millets are multipurpose grains used as food and feed because of nutritional composition (Devi et al., 2014). The straw of millet crops is valuable as livestock feed and livestock is an essential component for smallholders in their integrated farming system. Sorghum is used as pet feed preparation (Aruna & Visarada, 2019). In northern India, different traditional festival this crop used from ancient time during fasting period for making sweet dishes. Finger millet grains are used for traditional food preparation in different countries including alcoholic and non-alcoholic beverages (Ramashia et al., 2019). Different products like *rawa*, flour, sweet, cake, pasta, biscuits, cookies, chocolates are made by using millets as ingredients. Value added products developed from sorghum in India are nutritional enriched (Table 2.2).

Use of sorghum mill feed and pellets are very common as fish and shrimp feed. Different value-added food products and health drinks are prepared from millets and the course cereals are of high demand in food and health industry. The phenolic

Table 2.2 Nutritional composition of sorghum based valued added products (per 100 g)

Parameters	Grain	Flour	Fine rawa	Medium rawa	Flakes	Vermicelli	Pasta	Pops	Biscuits
Moisture (%)	11.9	13.8	10.2	9.0	13.8	8.4	11.5	5.9	5.7
Ash (%)	1.6	1.6	0.7	2.0	0.6	0.8	0.8	0.6	2.0
Protein (%)	10.4	6.2	6.7	7.2	5.1	8.4	8.4	5.0	4.6
Fat (%)	1.9	2.8	1.7	1.2	2.4	1.4	1.4	2.6	24.5
Carbohydrate (%)	72.6	76.2	77.8	77.7	75.0	76.2	76.2	83.1	60.3
Iron (mg)	4.1	8.4	10.6	5.1	87.8	64.5	64.5	2.4	2.3
Calcium (mg)	25.0	10.0	7.6	5.8	93.2	54.5	64.5	10.3	68.8
Chromium (mg)	0.0	0.0	1.3	1.5	0.9	0.2	0.2	1.4	0.2
Zinc (mg)	1.6	1.3	1.2	1.4	8.8	7.5	5.7	4.5	BDL
Magnesium (mg)	171.0	171.0	76.5	86.0	80.5	67.5	67.5	86.8	56.1
Riboflavin (mg)	0.1	0.4	0.1	1.1	0.0	1.3	1.3	0.2	2.3
Energy (Kcal/100 g)	349.0	355.0	350.0	350.0	342.0	355.0	355.0	376.0	481.0

Source Dayakar et al. (2017), Kumar and Maitra (2020), Abah et al. (2020)

compounds present in sorghum (Dykes & Rooney, 2006; Dykes et al., 2013) are beneficial against non-communicable diseases and widely used for pasta making by substituting wheat (Khan et al., 2015). The gluten free millet-based products ultimately lower blood sugar and energy intake and increase antioxidant status (Cardoso et al., 2017). Further, sorghum is known in treatment of sickle cell disease and orthopedic treatment (Aruna & Visarada, 2019) and tablet preparation (Alebiowu & Itiola, 2002; Zhu, 2014). The edible cutlery and syrup are also produced from millets. Bioindustrial products like ethanol (Corredor et al., 2006), biodegradable and edible films for packaging (Kaur et al., 2014), food colourants (Clifford, 2000) are other industrial products derived from millets. In paper and construction industries also stover of sorghum, pearl millets and other millets is used (House et al., 2000; Saeed et al., 2017).

2.8 Climate-Smart Technologies in Millets Cultivation

The climate change impacts imposed a question mark before the enhancement of production and yield of major cereals and automatically millets could be considered as climate-smart crops because of their resilience against climatic aberrations. To fulfill the present requirement as well as sustainable production of food grains, millets production should be directed in a climate-smart way where all suitable technologies of Good Agricultural Practice (GAP) should be adopted. Moreover, technology enabled precision crop management should also be taken into consideration for maximization of input use efficiency. Following are the climate-smart technologies for sustainable millets cultivation.

2.8.1 Integrated Nutrient Management

Integrated nutrient management (INM) shows the positive impact on yield by applying with integration of different nutrient sources such as organic manures, biofertilizers and inorganic fertilizers which enhance the soil health (Kumara et al., 2007). The nonjudicious supply of chemical nutrients inputs is not properly utilized by the plants. As a result, applied chemical fertilizers register very poor nutrient use efficiency (NUE) for different crops (Parkinson, 2013; Zhang et al., 2012) as well as in millets. In the world, sustaining agricultural productivity is a hugetask under the present threat of climaticfactors. Production of chemical fertilizers consumes energy and causes emission of GHGs. By substituting chemical nutrients with biofertilizers and organic manures in crop production, atmospheric pollution can be checked. The INM targets sustainability in crop production along with enhancement of productivity and economically viability (Chen et al., 2011; Jagathjothi et al., 2010, 2011; Pallavi et al., 2017; Wu & Ma, 2015;). Generally, organic manures are having low analytical value and huge quantity of bulky organic manures is required to fulfill the demand of the crops. But millets are less nutrient demanding crops. Hence, a portion of chemical nutrients can easily be substituted by organic sources and biofertilizers. Research evidences indicated better performance of INM practices in different millets (Table 2.3).

Moreover, nano materials are presently in use as nutrients. A study revealed that foliar application of nano-urea supplement along with the recommended dose of nitrogen increased growth and yield of finger millet (Samanta et al., 2022).

2.8.2 *Nutrient Management Based on Soil Test Crop Response (STCR)*

The soil test crop response (STCR) is an approach of nutrient management that aims for precision supply of fertilizers based on the nutrient status of the soil and its response for a target yield. Among different nutrient management practices adopted in crop production, the STCR method quantifies nutrients from applied inputs and soil for a target yield (Maitra et al., 2020a; Regar & Singh, 2014). The focus of the STCR approach is to ensure fertilization application in a balanced manner considering the role of soil and nutrients provided (Choudhary et al., 2019). As per STCR method fertilizers can be recommended based on regression analysis of certain percent of maximum yield. The STCR considers the three factors, namely, nutrient requirement of the crop, percentage contribution from soil available nutrients and percentage contribution from added fertilizers. For achieving a target yield of crop in a given location, the STCR approach may be considered as aprecision decision making tool where the right amount of nutrient application in the soil is prescribed depending upon soil value to maintain soil fertility. The STCR approach enhances profitability with more yield in an environmentally friendly way (Das et al., 2015) and it further increases the NUE (Jemila et al., 2017; Lal, 2015; Sekaran et al., 2018a; Santhi et al., 2011a, b). As per the STCR, finger millet responded well to the application of 200% N, 100% P, 100% K, 25% Zn, 25% S, 25% B and 5 t ha^{-1} FYM (for a target productivity of 4 t ha^{-1}) against RDF (Sandhya Rani et al., 2017). Shetty and Kumar (2018) also mentioned that STCR-based NPK along with compost 10 t ha^{-1} performed better compared to other nutrient doses in alfisols of Karnataka, India. The STCR method clearly indicated that it was the suitable method to maintain nutrient balance and soil health. A long-termtrial conducted at Indian Agricultural Research Institute, India on pearl millet–wheat cropping system clearly revealed that STCR based nutrient arrangement was better for a target yield of cereals (Sharma et al., 2016). Researches carried out on STCR based integrated plant nutrition system (STCR-IPNS) for nutrient recommendation in pearl millet under Inceptisol of Tamilnadu, India and revealed that for a yield target of 4 t ha^{-1}, STCR-IPNS expressed its superiority over other practices. Further, STCR recorded more grain yield of pearl millet than blanket application of nutrients, blanket supply of chemical fertilizers along with FYM and farmer's practice of the locality (Sekaran et al., 2018a, b).

Table 2.3 Studies on INM in different millets

Crop	Study area	Salient feature of the research	Reference
Sorghum	India	In Sorghum—wheat cropping system, INM enhanced grain and fodder productivity of sorghum by 18.5 and 9.4%	Patidar and Mali (2001)
	India	Application of 75% recommended dose of fertilizer (RDF) + 3 t ha^{-1} of farmyard manure (FYM) along with either *Azospirillum* or phosphate solubilizing bacteria (PSB) yielded at par with RDF (chemicals) in winter sorghum	Srinivas et al. (2008)
Pearl millet	India	In pearl millet—wheat cropping system, pearl millet with 50% RDF + 50% N FYM yielded at par with 100% RDF	Kumar et al. (2005)
	Niger	In the Sahelian zone, combined application 0.9 t ha^{-1} millet stover and 2.7 t ha^{-1} organic manure along with 15 kg N and 4 kg P ha^{-1} through chemicals resulted in 132% grain productivity	Akponikpe et al. (2008)
	India	INM with chemical ferilizers, FYM and biofertilizer combination enhanced growth, yield attributes and yield ofpearlmillet during rainy season	Rajput (2008a, b)
	Zimbabwe	Cattle manure with a combination of either biomass transfer and/or ammonium nitrate showed increases of soil fertility parameters, panicle length and grain yields	Kokerai et al. (2019)
	India	The RDF (120 kg N + 45 kg P + 45 kg K + 20 kg $ZnSO_4$ ha^{-1}) produced more grain and fodder yield of pearl millet and the treatment remained at par with 50% RDF + 5 t ha^{-1} FYM + bio-fertilizer	Kadam et al. (2019)
Finger millet	Nepal	Organi manures with chemical fertilizers produced more grain yield over chemical nutrients	Pilbeam et al. (2002)
	India	Phosphorus-enriched FYM and recommended dose of N (RDN) increased grain yield over application of RDF	Jagathjothi et al. (2011)
	India	A combination of 10 t ha^{-1} FYM + 100% NPK and 5 t ha^{-1} maize residue incorporation + 100% NPK yielded more grains in semi-arid tropical Alfisol	Sankar et al. (2011)
	India	Combined application of 10 t ha^{-1} FYM + biofertilizer consortia (*Azospirillum brasilense* + *Bacillus* spp. + *Psuedomonasflurosence* @ 20 g kg^{-1} seed each) + $ZnSO_4$ (12.5 kg ha^{-1}) + Borax (kg ha^{-1}) + 100% RDF (50:30:25) yielded more than RDF in sandy loam soil	Roy et al. (2018)

(continued)

Table 2.3 (continued)

Crop	Study area	Salient feature of the research	Reference
Foxtail millet	India	Application of 50% RDF (through chemicals) + 25% N (through neem cake) + microbial fertilizer registered more yield than only RDF	Monisha et al. (2019)
	India	Integrated application of FYM + RDF + 3% *Panchagavya* produced more grains than RDF	Kumaran and Parasuraman (2019)
	India	Application of 75% RDN (chemical) + 25% N (poultry manure) + seed inoculation with *Azospirillum* increased grain yield	Selectstar Marwein et al. (2019)
Little millet	India	Application of 7.5 t ha^{-1} of FYM, NPK (40:20:10 kg ha^{-1}) calcium carbonate, zinc sulphate, and borax produced higher grain yield than RDF	Parihar et al. (2010)
	India	A combination of 100% RDF + 1 t ha^{-1} neem ensured higher yield than RDF	Sandhya Rani et al. (2017)
		Combined application of 75% RDN (chemical) + 25% RDN (vermicompost) yielded more than RDF	Thesiya et al. (2019)
Kodo millet	India	Integrated application of of 125% RDF + soil *Azospirillum* @ 2 kg ha^{-1} (soil) + vermicompost @ 2 t ha^{-1} + foliar application of 1% nutrient supplement increased grain yield of kodo millet	Prabudoss et al. (2014)

2.8.3 *Site Specific Nutrient Management (SSNM)*

The different nutrients which are deficient worldwide in the soil are mainly six elements, namely, N, P, K, S, Zn and B. Presently, precision management of essential nutrients can be adopted as different tools and decision support systems are available for the purpose. These tools fine tune the fertilizer application to the crop fields. The identification and management of variability and site-specific management is one of the best ways of crop nutrients management. During last six decades, enough of chemical inputs have been applied to crop field and unbalanced application of chemical fertilizers in intensive agriculture caused wastage, pollution and deficiency of some specific nutrients. In this regard site specific nutrient management (SSNM) can be adopted for judicious use of fertilizers. The SSNM works in such condition where deficient nutrients can be reclaimed by this methodology. The primary thing needs to be done under SSNM is initial soil test and based on the soil test results, a yield target can be fixed and nutrients are applied accordingly to the soil (Rathod et al., 2012). The research on SSNM for major cereals has been carried out, but limited research has been carried out on millets. Ramachandrappa et al. (2015) noted that the impact of SSNM on finger millet under intercropping with red gram performed well in Bangalore, India for a target productivity of 4 t ha^{-1} and SSNM resulted in higher yield of finger millet and profitability with soil health improvement. Singh and Bharadwaj (2017) studied on multi-locational trial and mentioned that SSNM practice gave more grain yield of pearl millet than the recommended practice and farmer's practice. The result of an experiment conducted in Uttar Pradesh, India clearly indicated that the effect of SSNM on pearl millet—wheat cropping system yielded more than farmer's practice and state recommendation (Kumar & Singh, 2019). In soybean–sorghum cropping system, the SSNM practice resulted in better productivity for both the crops at Raichur, Karnataka, India (Ravi et al., 2020).

2.8.4 Resource Conservation Technology (RCT)

Resource conservation technologies (RCTs) are important as mitigation and adaptation options to combat climate change because of numerous benefits. RCTs focus on conservation agriculture (CA) practices that include soil cover, minimum tillage, crop diversification and application of organic inputs (FAO, 2020). CA is a farming system approach which promotes minimum use of high energy inputs in agriculture with a goal of resource conservation, enhancement of nutrient and water use efficiency leading to agricultural sustainability. As regard of soil health and its management, minimum or zero tillage is a wonder technique for the different millet crops which expressed better results on growth and productivity under resource poor conditions (Verma et al., 2017, 2018; Wilson et al., 2008). Besides, zero tillage or conservation tillage is economically beneficial because of less energy involvement in farming compared to conventional tillage. Further in conventional tillage, farm machineries are operated by fossil fuel burning which causes emission of more GHGs (Martin-Gorriza et al., 2020). Reduced tillage also save labour input in agriculture compared to conventional tillage (Choudhary et al., 2018; Malviya et al., 2019). Millets are diversified grains of various nature and thus millets cultivation creates on-farm biodiversity suitable to drylands facilitation a new green revolution (Goron & Raizada, 2015; Michaelraj & Shanmugam, 2013). In erosion prone areas, residue incorporation and mulching are beneficial for soil conservation (Mgolozeli et al., 2020). In drylands, soil moisture and fertility are two major contraints for a good harvest (Choudhary et al., 2018; Schlegel et al., 2017) and CA has enough potential to overcome these issues because cover cropping and mulching can enable higher soil moisture content and residue incorporation in soil can ensure higher organic C and other nutrients (Chehade et al., 2019; Prasad et al., 2016; Srinivasarao et al., 2013). Intensive tillage causes loss of soil organic carbon (SOC) and global loss of SOC in this operation has been quantified as 60–90 Pg (Lal, 1999). Not only loss of SOC, but also conventional tillage impacts negatively on soil physical, chemical and biological properties (Lal, 2004). In contrast, CA facilitates gain in SOC inclusive of improvement of soil properties. Studies conducted at different locations clearly indicated positive impacts of RCTs on soil health improvement. In pearl millet—wheat cropping system, zero tillage resulted in a greater SOC and available nutrients than conventional tillage (Kaushik et al., 2018). Inclusion of crop residue was advantageous in pearl millet and sorghum cultivation in West Africa as it decreased top-soil temperature, increased water availability and improved soil physico-chemical properties (Buerkert et al., 2000). Sankar et al. (2011) observed from multi-locational trials carried out in Inceptisol, Vertisol and Aridisol of India and mentioned that reduced tillage was more productive and economic for production of pearl millet under arid and semi-arid conditions. Finger millet yield was increased by substituting 50% of the RDN with organic manures in Alfisol. Further, a conservation tillage enhanced SOC (Prasad et al., 2016). Malviya et al. (2019) concluded that farmers should adopt reduced tillage as well as inclusion of residue of previous crop as mulch material for kodo millet cultivation in Rewa, Madhya Pradesh of India.

2.8.5 Inoculation of Growth Promoting Microorganisms

Under the present context of climate change, plants are supposed to fetch weather abnormalities and stress due biotic and abiotic factors. Different plant growth promoting microorganisms are capable to provide support to the plants to overcome these abnormalities (Ojuederie et al., 2019). Research conducted on microbes mediated abiotic stress tolerance revealed that plant growth promoting rhizobacteria (PGPR), namely, *Bacillus atrophaeus, B. sphaericus, B. subtilis, Pseudomonas* spp. and *Staphylococcus kloosii* are capable to reduce stress in finger millet by enhancing root and shoot growth (Chandra et al., 2018; Shultana et al., 2020). The endophytic bacteria *Bacillus amyloliquefaciens* EPP90 played versatile role in stress tolerance in pearl millet (Kushwaha et al., 2019). Niu et al. (2018) showed that the isolates of bacterial strains of *Pseudomonas fluorescens, Entero bacterhormaechei*, and *Pseudomonas migulae* enhanced seed germination and seedling vigour of foxtail millet under drought conditions because of ability to produced exopolysaccharide and 1-aminocyclopropane-1-carboxylic acid (ACC) deaminase. Abiotic factors are responsible for different biotic stress also and PGPR can be used for recovery of abiotic stress. The study indicated that the rhizobacterial strain of *Pseudomonas* sp. MSSRFD41 was effective against blast disease of finger millet (*Pyriculariagrisea*) and growth enhancement (Sekar et al., 2018). Further, biopriming of finger millet seeds with *Pseudomonas* sp. MSSRFD41 was beneficial in terms of increase in germination and plant growth. The strains of *Pseudomonas* spp., UOM ISR 17 and UOM ISR 23 were able to check the spread of downy mildew (*Sclerospora graminicola* (Sacc.) Schroet) effectively in pearl millet (Jogaiah et al., 2010).

2.8.6 Application of Growth Promoters

The phytohormones and growth promoting substances play vital role in plant growth and stress mitigation. Seed treatment and foliar application of hormones and growth promoters resulted in better growth and development of different crops including millets (Appu & Senthilmurugan, 2014). Hydro-priming in pearl millet expressed vigorous growth of finger millet in drylands (Kumar et al., 2002). Earlier Maitra et al. (1997) reported that pre sowing seed 100 ppm Na_2HPO_4 and KH_2PO_4 registered better yield of finger millet. Similarly, overnight seed soaking with water or 0.25% $CaCl_2$ resulted in increasing yield of finger millet over no treatment (Maitra et al., 1998). In pearl millet, seed priming with NaCl impaired the ill effect of salinity and later showed more growth (Ashraf & Iram, 2002). Pearl millet seed priming with solution of GA (at the rate of 50 mg per litre of water) showed more growth (Vijayaraghavan, 1999). Seeds of pearl millet treated with chlormequat chloride (CCC) or 0.15% succinic acid also recorded higher germination percentage than untreated seeds (Shanmugasundaram & Kannaiyan, 1989). Effect of seed treatment in sorghum was studied by Kadiri and Hussaini (1999) and they noted seed soaking with $CaCl_2$ or KNO_3 solution resulted in better germination, growth and chlorophyll content. In barnyard millet (*Echinochloa frumentacea*), seed treatment with 100 ppm IAA or 1% KH_2PO4 expressed more germination, plant height, seedling vigour index and drymatter accumulation (Sujatha et al., 2013). Foliar spray of plant growth regulators (PGRs) is also beneficial in millets. Prabha et al. (2016) recommended that foliar spray of nutrients and PGR as consortia could be used for enhancement of growth, productivity and economic return of finger millet and they used brassinosteroids, mepiquat chloride and chlormequat chloride as growth promoters. Growth and yield attributes of finger millet were enhanced by the foliar application of salicylic acid showed enhanced growth and yield attributing characters of finger millet (Sathishkumar et al., 2018). The PGRs are potential to combat abiotic stress in enhancing the growth parameters of pearl millet as reported by Suresh et al. (2018) when NAA application (40 ppm) at 20 and 40 days after sowing was proved beneficial. Earlier, Sivakumar et al. (2002) reported that foliar application of brassinosteroid (0.1 ppm) and triacontanol (10 ppm) expressed more pearl millet grains as well as enhanced grain protein and sugar in pearl millet. The studies clearly indicated that growth promoting substances played a vital role in enhancement of growth and productivity of different millets even under abiotic stress conditions.

2.8.7 Terminal Drought and Agronomic Manipulation

In arid and semi-arid conditions, erratic rainfall is a very common during monsson season and rainfed crops may face the consequences of drought. In dryland regions of Africa and Asia, millets are grown mostly as rainfed crop in low fertile soils and occurrence of terminal drought stress results in yield loss. Terminal drought appears in the end season that is in the later part of the reproductive stage. Though millets are drought tolerant and hardy crops, terminal drought creates yield loss drastically in some millet. Finger millet, suffers a lot because of terminal drought and little millet (80.1%), prosomillit (34.6%) and pearl millet (60.1%) also show yield reduction (Bidinger et al., 1987; Goron & Raizada, 2015; Tadele, 2016). However, foxtail millet does not exhibit terminal drought stress. Cultivation of drought tolerant and short duration cultivars is one of the suitable options to overcome such stress (Vadez et al., 2012). Further, taking advantage of pre-rainy season shower millets can be sown early to avoid terminal drought stress.

2.8.8 Integrated Weed Management (IWM)

The crop weed competition may cause insufficient share by the crops for resources, namely, soil moisture, light and nutrient which ultimately results in reduction of yield and quality of millets (Mishra et al., 2018). The weed management can be done through integrated manner by using physical, chemical and mechanical methods to manage weed infestation in millets. The millets mainly belong to the grass family, but during the early growth stages weeds can do harm. The growth period from seeding to45 days after sowing is considered as the critical period when crop-weed competition can reduce an average yield between 20 and 60% in different millets. The yield losses due to various weed species in different millets such as sorghum (15–83%), pearl millet (16–94%) and finger millet (55–61%) are more compared to other millets. Striga (*Striga hermonthica*) is a prominent parasitic weed mainly for sorghum and pearl milletthat may cause 50% of yield loss (Oduori, 2007; Wanyera, 2007). Use of pre- and post-emergence herbicides such as oxyfluorfen, atrazine and 2,4-D are effective (Mishra et al., 2018). The different millets affected by various weeds are grass, broad-leaved and sedges. Common weeds of different millets are *Ageratum conyzoides, Commelina benghalensis, Cynodon dactylon, Cyperus rotundus, Dactyloctenium aegyptium, Echinochloa colona, Eleusine indica, Euphorbia hirta, Solanum nigrum, Sorghum halepense, Striga litura* and *Trianthema portulacastrum.* In gerenal, cultural management and mechanical practices like off-season tillage, deep summer ploughing and hand weeding are very common among smallholders of drylands. But chemically weeds can be managed by application of herbicides (Table 2.4) (Mishra et al., 2018). However, integrated weed management (IWM) is considered as the best option for weed management as well as sustaining crop productivity.

2.8.9 Integrated Pest and Disease Management

Though millets are ecologically hardy crops, pest disease incidence is found in millets also (Table 2.5). Different insects cause damage to millets and loss due to insect pest attack was ranged between 10 and 20% in India and about 50% in Ghana (Bekoye & Dadie, 2015; Gahukar & Jotwani, 1980; Kumar & Channaveerswami, 2015; Tanzubil & Yakubu, 1997). Similarly, under favourable conditions diseases may also cause yield loss.

To protect the crops from pests and diseases integrated pest management (IPM) should be adopted. The choice of pest-disease tolerant cultivars, use of quality and

Table 2.4 Herbicide recommendation for millets

Millets	Herbicide	Dose (kg or litre per ha)	Time of application (PE = pre-emergence; PoE = post-emergence	Target weeds	Remarks
Sorghum	Atrazine	0.75	PE and PoE	*Trianthema pottulacastrum* and *Echinochloacolona*	Effective under sole cropping
	2,4-D	0.50–0.75	PoE	Broad-leaved weeds	Applied between 4 and 6 weeks after planting
	2,4-D	1.0	PoE	Parasitic weeds	Sole cropping
	Metolachlor	1.0–1.5	PE	Grasses	Suitable for intercropping also
	Atrazine + pendimethalin	0.75 + 0.75	PE	Broad spectrum weeds	Suitable for intercropping
	Atrazine + pendimethalin	0.75 + 0.50	PE	Broad spectrum weeds	Suitable for intercropping
Pearl millet	Atrazine	0.50	PE and PoE	Grassy weeds	Effective under sole cropping
Finger millet	Isoproturon	0.50	PE	Grassy weeds	Rainfed areas
	Oxyflurofen	0.1	PoE	Grassy weeds	Irrigated areas
	2, 4-D sodium salt	0.75	PoE	Broad leaved weeds	Sprayed after three weeks of sowing
Foxtail/ Little/ Proso/ Kodo/ Barnyard millet	Isoproturon	0.50	PE		Rainfed areas
	2, 4-D sodium salt	0.75	PoE	Broad leaved weeds	Sprayed after three weeks of sowing

Source Chapke et al. (2018), Mishra et al. (2018)

certified seeds, adoption of summer and deep tillage, soil solarization, management of nutrients and establishment of optimum plant stand are some cultural methods which can be adopted for a healthy crop. Mechanical measures like collection and destruction of pests and disease infected plants, use of different traps and erection of bird perches should be taken into consideration. Use of biopesticidesand organic formulations like *Trichoderma* spp. and *Pseudonomas* sp. for disease management and application of *Bacillus thuringiensis* and neem-based products are beneficial to

Table 2.5 Common pests and diseases of different millets

Crop	Insect	Disease
Sorghum	Shoot fly (*Atherigonasoccata*), stem borer (*Chilopartellus*), aphids (*Melanaphissacchari*)	Grain mould (*Fusariumsemitectum, F. moniliforme, Curvularialunata*), downy mildew (*Peronosclerosporasorghi*), charcoal rot (*Macrophominaphaseolina*)
Pearl millet	White grub (*Holotrichiaconsanguinea*), shoot fly (*Atherigona approximate, A. hyalinipennis*), grass hoppers (*Aiolopussavignyi, Aiolopussimulatrix, Oedaleus senegalensis, Kraussariaangulifera*), stem borer (*Acigonaignefusalis, Sesamiainferens*)	Downy mildew (*Sclerosporagraminicola*), rust (*Puccinia substriata*), smut (*Pennisetum glaucum*), blast (*Pyricularia grisea*)
Finger millet	Cut worm (*Spodoptera exigua*), leaf aphid (*Rhopalosiphummaidis*), stem borer (*Chilozonellus, Saluriainficita*), earhead caterpillars (*Cryptoblabesangustipennella, C. gnidiella*)	Blast (*Pyricularia grisea*), leaf spot (*Cercosporaeleusine*), brown spot (*Drechsleranodulosum*), Smut (*Melanopsichiumeleusinis*), rust (*Uromyces eragrostidis*)
Foxtail millet	Shoot fly (*Atherigonaatripalpis*)	Downy mildew (*Sclerophthoramacrospora*)
Little millet	Shoot fly (*Atherigonaatripalpis*), stem borer (*Acigonaignefusalis*), termites (*Microtemes* spp., *Odontotermes* spp.)	Blast (*Pyricularia grisea*)
Proso millet	Shoot fly (*Atherigonaatripalpis*)	Smut (*Melanopsichiumeleusinis*)
Kodo millet	Shoot fly (*Atherigonaatripalpis*), termites (*Microtemes* spp., *Odontotermes* spp.), stem borer (*Atherigonaatripalpis*)	Head smut (*Ustilagotrichophora*)
Brown-top millet	Armyworms (*Syrphisunipuncta, Mythimnaunipuncta*), Grasshopper (*Oedaleus senegalensis, Aiolopussimulatrix*)	Greengram yellow mosaic virus
Barnyard millet	Pink borer (*Sesamiainferens*)	Head smut (*Ustilagotrichophora*)

Source Chapke et al. (2018), El-Shafie (2020), Gahukar and Reddy (2019), ICRISAT (1989), Kaurav et al. (2018), Reddy (1991), Tetreault et al. (2019)

manage pest population. Further, need based chemicals can be applied when the pest-disease population dynamics will exceed the economic threshold level. Ultimately, good crop management (GAP) practices should be adopted for ensuring sustainability in millets cultivation.

2.8.10 *System of Millet Intensification (SMI)*

System of crop intensification (SCI) is a new approach of sustainable intensification (Adhikari et al., 2018). During last two decades, System of Rice Intensification (SRI) was adopted by rice farmers of different countries. The main aspects of SRI are transplanting of early aged seedlings, square planting width wide spacing, incorporation of organic manures and judicious water use. Research evidences and farmers experiences revealed that SRI crop yielded more (Adhikari et al., 2018; Kassam et al., 2009; Pradan/SDTT, 2012; Uphoff, 2017). In the present context of gradual shrinkage of frees water availability, SRI has become more relevant because stagnation of water in rice field as per the conventional system causes low water use efficiency. In arid and semi-arid conditions, adverse impact of climate change is more prominent and sustainable agricultural production is a great challenge. The situation warrants adoption of more innovative and environment-friendly approaches in drylands (Gurjeet et al., 2011). In millets cultivation also SCI technologies adopted and research conducted on system of finger millet intensification (SFMI) showed that finger millet yielded more with transplanting seedlings of less than two weeds and planting with row x hill spacing of 25 cm × 25 cm compared to conventional practices with closer spacing and planting of aged seedlings (Bhatta et al., 2017). There is urgent need of resource conservation with production enhancement and SCI focuses to that direction; thus, system of millets intensification (SMI) may be a boon for drylands, if the technologies for different millets are standardized. However, SCI is typically suitable for transplanted crops and the scope of direct seeded crops has not been demonstrated.

2.8.11 *Adoption of Smart Technologies in Millet Cultivation in the Future*

Under the present scenario of climatic aberrations, agriculture should be smart enough to combat the situation and accordingly climate smart technologies should be adopted. Millets-based intercropping systems can be considered as a climate-smart technology as it can provide a natural insurance against failure of a component crop (Maitra, 2020b; Maitra et al., 2000). There is enough advancement of science and technologies which has been reflected in human civilization. But in agriculture, there is very limited reflectance of latest innovation of science and that too in neglected crops cultivation like millets. Precision agriculture (PA) is a new concept in the developing world where smart technologies are adopted to take appropriate decision and make farming operations easier. The effiecient use of inputs in agriculture can be increased by adoption of PA and thus more agricultural production can be obtained with the precise quantity of inputs. PA is information and technology (IT) based crop production technology which identifies, analyzs and manages the onfarm variability for a target yield and it ensures profit in crop production and efficient management of resources and thus, makes agriculture economically viable and sustainable (Saiz-Rubio & Rovira-Más, 2020). In this regard, remote sensing, different sensors and internet of things (IoT) for irrigation and water management (Maitra & Pine, 2020), artificial intelligence (AI) and machine learning (ML), easily applicable decision support tools like different apps will be in practice in the future. Further, hyperspectral imagery and image processing and use of drones and unmanned air vehicles (UAV) are important tools for decision making and crop health and soil monitoring and crop management. Site specific nutrient management (SSNM) can be an ideal option for nutrient management in millets. Presently, millets are still neglected crops and raising of millets is confined among the smallholders in drylands with less investment in farming. The nutritional importance of millets has been re-evaluated recently which created optimism that such climate resilient crops will be the future foods under the threat of climate change. But the situation warrants research needs for flourish of millets in daily food basket and PA can play a great role in the direction. All the new technologies are future of farming and can make application of valuable inputs more precisely for a target yield. Moreover, with the pace of technological innovation, it may be stated that in the future smart farms will be equipped with sensing technology, applications with IoT-based crop and precision water management, data analytics, technology enabled input delivery as well as smart crop management. In the dawn of technological embellishment, it may be anticipated that the conventional millets cultivation may be switched over into the direction of fast-growing concept of Agriculture 5.0 that infers precise management with automation in which robots, UAVs and application of AI and ML will be more prominent (Saiz-Rubio & Rovira-Más, 2020) for a target productivity on sustainable basis.

2.9 Conclusion

Millets fulfill some desired qualities in terms of health and nutrition benefits and ecological soundness which are very much important in present context. These are ancient grains, but remained neglected due to more instutional focus on fine cereals. Recently under climate change scenario, millets regained their old pride and became a suitable option amongst policymakers and health-conscious consumers recognized the nutritional composition. As millets remained neglected for last few decades, sufficient research works have not been conducted. There is enough scope for enhancement of productivity eco-friendly millets by intensifying research on climate-smart technologies as well as good agricultural practices. The focus areas for research are nutrients management options like INM, SSNM and STCR, RCT and organic agriculture as well as CA, other climate-smart agronomic manipulations inclusive of change of sowing time and planting geometry and water management. Further, PA technologies can further refine the hurdles of millets farming and thus can ensure food and nutritional security along with agricultural sustainability in the future.

References

Abah, C. R., Ishiwu, C. N., Obiegbuna, J. E., & Oladejo, A. A. (2020). Sorghum grains: Nutritional composition, functional properties and its food applications. *European Journal of Nutrition and Food Safety, 12*(5), 101–111. https://doi.org/10.9734/ejnfs/2020/v12i530232

Ackerman, F., & Stanton, E. A. (2013). Climate impacts on agriculture: A challenge to complacency. Global Development Environment Institute, Working Paper No. 13-01. http://citeseerx.ist.psu.edu/viewdoc/download?doi=10.1.1.394.2273&rep=rep1&type=pdf. Accessed August 29, 2021.

Adekunle, A. A. (2012). Agricultural innovation in sub-Saharan Africa: Experiences from multiple stakeholder approaches. Forum for Agricultural Research in Africa, Ghana. ISBN 978-9988-8373-2-4.

Adhikari, P., Araya, H., Aruna, G., Balamatti, A., Banerjee, S., Baskaran, P., Barah, B. C., Behera, D., et al. (2018). System of crop intensification for more productive, resource-conserving, climate-resilient, and sustainable agriculture: Experience with diverse crops in varying agroecologies. *International Journal of Agricultural Sustainability, 16*(1), 1–28. https://doi.org/10.1080/14735903.2017.1402504

Akponikpe, P., Michels, K., & Bielders, C. (2008). Integrated nutrient management of pearlmillet in the sahel combining cattle manure, crop residue and mineral fertilizer. *Experimental Agriculture, 44*, 453–472.

Alebiowu, G., & Itiola, O. A. (2002). Compressional characteristics of native and pregelatinized forms of sorghum, plantain, and corn starches and the mechanical properties of their tablets. *Drug Development and Industrial Pharmacy, 28*, 663–672.

Appu, M., & Senthilmurugan, M. (2014). Foliar application of salicylic acid stimulates flowering and induce defense related proteins in finger millet plants. *Universal Journal of Plant Science, 2*, 14–18.

Arendt, E., & Dal Bello, F. (2011). Gluten-free cereal products and beverages. Academic Press.

Aruna, C., & Visarada, K. B. R. S. (2019). Other industrial uses of sorghum. In *Breeding sorghum for diverse end uses* (pp. 271–292). https://doi.org/10.1016/b978-0-08-101879-8.00017-6

Aryal, J. P., Sapkota, T. B., Khurana, R., Khatri-Chhetri, A., Bahadur Rahut, D. B., & Jat, M. L. (2020). Climate change and agriculture in South Asia: Adaptation options in smallholder production systems. *Environment, Development and Sustainability, 22*, 5045–5075. https://doi.org/10.1007/s10668-019-00414-4

Ashraf, M., & Iram, A. (2002). Optimization and influence of seed priming with salts of potassium or calcium in two spring wheat cultivars diVering in salt tolerance at the initial growth stages. *Agrochimica, 46*, 47–55.

Aubry, S., Brown, N. J., & Hibberd, J. M. (2011). The role of proteins in C_3 plants prior to their recruitment into the C_4 pathway. *Journal of Experimental Botany, 62*, 3049–3059. https://doi.org/10.1093/jxb/err012

Bandyopadhyay, T., Muthamilarasan, M., & Prasad, M. (2017). Millets for next generation climate-smart agriculture. *Frontiers in Plant Science, 8*, 1266. https://doi.org/10.3389/fpls.2017.01266

Banerjee, P., & Maitra, S. (2020). The role of small millets as functional food to combat malnutrition in developing countries. *Indian Journal of Natural Sciences, 10*(60), 20412–20417.

Bekoye, N. B. M., & Dadie, A. (2015). Evaluation des pertesen grains de mil dues aux insectes. *European Scientific Journal, 11*(21), 266–275.

Bhatt, D., Sonkar, G., & Mall, R. K. (2019). Impact of climate variability on the rice yield in Uttar Pradesh: An agro-climatic zone-based study. *Environmental Processes, 6*, 135–153. https://doi.org/10.1007/s40710-019-00360-3

Bhatta, L. R., Subedi, R., Joshi, P., & Gurung, S. B. (2017). Effect of crop establishment methods and varieties on tillering habit, growth rate and yield of finger-millet. *Agricultural Research and Technology Journal, 11*(5), 555826. https://doi.org/10.19080/ARTOAJ.2017.11.555826

Bidinger, F. R., Mahalakshmi, V., & Rao, G. D. P. (1987). Assessment of drought resistance in Pearl-Millet (Pennisetum-Americanum). 2. Estimation of genotype response to stress. *Australian Journal of Agricultural Research, 38*(1), 49–59.

Brahmachari, K., Sarkar, S., Santra, D. K., & Maitra, S. (2018). Millet for food and nutritional security in drought prone and red laterite region of eastern India. *International Journal of Plant Soil and Science, 26*(6), 1–7.

Buerkert, A., Bationo, A., & Dossa, K. (2000). Mechanisms of residue mulch-induced cereal growth increases in West Africa. *Soil Science Society of America Journal, 64*, 346–358.

Cardoso, L. D. M., Pinheiro, S. S., Martino, H. S. D., & SantAna, H. M. P. (2017). Sorghum (Sorghum bicolor L.): nutrients, bioactive compounds, and potential impact on human health. *Critical Reviews in Food Science and Nutrition, 57*, 372–390.

Chakraborty, D., Sehgal, V. K., Dhakar, R., Ray, M., & Das, D. K. (2019). Spatio-temporal trend in heat waves over India and its impact assessment on wheat crop. *Theoretical and Applied Climatology*. https://doi.org/10.1007/s00704-019-02939-0. Accessed July 6, 2020.

Chandra, D., Srivastava, R., Glick, B. R., & Sharma, A. K. (2018). Drought-tolerant Pseudomonas spp. improves the growth performance of finger millet (Eleusine coracana (L.) Gaertn.)under non-stressed and drought-stressed conditions. *Pedosphere, 28*, 227–240.

Chandrasekara, A., Naczk, M., & Shahidi, F. (2012). Effect of processing on the antioxidant activity of millet grains. *Food Chemistry, 133*(1), 1–9. https://doi.org/10.1016/j.foodchem.2011.09.043

Chapke, R. R., Prabhakar, Shyamprasad, G., Das, I. K., & Tonapi, V. A. (2018). Improved millets production technologies and their impact. In *Technology bulletin* (84p). ICAR-Indian Institute of Millets Research. ISBN: 81-89335-69-3.

Chehade, L. A., Antichi, D., Martelloni, L., Frasconi, C., Sbrana, M., Mazzoncini, M., & Peruzzi, A. (2019). Evaluation of the agronomic performance of organic processing of tomato as affected by different cover crop residues management. *Agronnmy, 9*, 504. https://doi.org/10.3390/agronomy9090504

Chen, X.-P., Cui, Z.-L., Vitousek, P. M., Cassman, K.G., Matson, P.A., Bai, J.-S., Meng, Q. F., Hou, P., Yue, S.-C., Römheld, V., & Zhang, F. S. (2011). Integrated soil-crop system management for food security. *Proceedings of National Academy of Sciences of USA, 108*, 6399–6404.

Choudhary, M., Datta, A., Jat, H. S., Yadav, A., Gathala, M., Sapkota, T., Das, A., Sharma, P., Jat, M. L., Singh, R., & Ladha, J. (2018). Changes in soil biology under conservation agriculture based sustainable intensification of cereal systems in Indo-Gangetic Plains. *Geoderma, 313*, 193–204. https://doi.org/10.1016/j.geoderma.2017.10.041

Choudhary, S., Baghel, S. S., Upadhyay, A. K., & Singh, A. (2019). STCR-based manure and fertilizers application effect on performance of rice and chemical properties of vertisol. *International Journal of Current Microbiology and Applied Sciences, 8*(3), 2080–2086.

Clifford, M. N. (2000). Anthocyanins-nature, occurrence and dietary burden. *Journal of Agricultural and Food Chemistry, 80*, 1063–1072.

Committee on World Food Security. (2012). *Coming to terms with terminology: Food security, nutrition security, food security and nutrition, food and nutrition security*. http://www.fao.org/fsnforum/sites/default/files/file/Terminology/MD776(CFS_Coming_to_terms_with_Terminology).pdf. Accessed May 16, 2020.

Corredor, D. Y., Bean, S. R., Schober, T., & Wang, D. (2006). Effect of decorticating sorghum on ethanol production and composition of DDGS. *Cereal Chemistry, 83*, 17–21.

Das, D., Dwivedi, B., Meena, M., Singh, V. K., & Tiwari, K. N. (2015). Integrated nutrient management for improving soil health and crop productivity. *Indian Journal of Fertilisers, 11*, 64–83.

Dayakar, R. B., Bhaskarachary, K., Arlene Christina, G. D., Sudha Devi, G., & Tonapi, V. A. (2017). *Nutritional and health benefits of millets* (pp. 112). ICAR_Indian Institute of Millets Research (IIMR).

Devi, P. B., Vijayabharathi, R., Sathyabama, S., Malleshi, N. G., & Priyadarisini, V. B. (2014). Health benefits of finger millet (Eleusine coracana L.) polyphenols and dietary fiber: A review.

Journal of Food Science and Technology, 51, 1021–1104. https://doi.org/10.1007/s13197-011-0584-9

Dykes, L., & Rooney, L. W. (2006). Sorghum and millet phenols and antioxidants. *Journal of Cereal Science, 44*(3), 236–251.

Dykes, L., Rooney, W. L., & Rooney, L. W. (2013). Evaluation of phenolics and antioxidant activity of black sorghum hybrids. *Journal of Cereal Science, 58*, 278–283.

El Bilali, H. (2018). Relation between innovation and sustainability in the agro-food system. *Italian Journal of Food Science, 30*, 200–225. https://doi.org/10.14674/ijfs-1096

El Bilali, H., Callenius, C., Strassner, C., & Probst, L. (2019). Food and nutrition security and sustainability transitions in food systems. *Food and Energy Security, 8*, e00154. https://doi.org/10.1002/fes3.154

El-Shafie, H. A. F. (2020). Integrated insect pest management. In *Pests control and acarology*. https://doi.org/10.5772/intechopen.81827

Ergon, Å., Seddaiu, G., Korhonenc, P., Virkajärvic, P., Bellocchid, G., Jørgensene, M., Østremf, L., Reheulg, D., & Volaire, F. (2018). How can forage production in Nordic and Mediterranean Europe adapt to the challenges and opportunities arising from climate change. *European Journal of Agronomy, 92*, 97–106. https://doi.org/10.1016/j.eja.2017.09.016

Fahad, S., Bajwa, A. A., Nazir, U., Anjum, S. A., Farooq, A., Zohaib, A., Sadia, S., Nasim, W., Adkins, S., Saud, S., Ihsan, M. Z., Alharby, H., Wu, C., Wang, D., & Huang, J. (2017). Crop production under drought and heat stress: Plant responses and management options. *Frontiers in Plant Science, 8*, 1147. https://doi.org/10.3389/fpls.2017.01147

FAO. (2017). *HLPE reports. High level panel of experts on food security and nutrition (HLPE)*. Rome, Italy. http://www.fao.org/cfs/cfs-hlpe/reports/en, Accessed 29 August 2021

FAO. (2018). *The future of food and agriculture, alternative pathways to 2050*. Rome, Italy. http://www.fao.org/3/CA1553EN/ca1553en.pdf. Accessed August 29, 2021.

FAO. (2020). *Conservation agriculture*. http://www.fao.org/3/a-i6169e.pdf. Accessed September 7, 2020.

FAOSTAT. (2020). *Crop statistics*. http://www.fao.org/faostat/en/#data/QC. Accessed June 15, 2020.

Fujibe, F., Yamazaki, N., & Kobayashi, K. (2006). Long-term changes of heavy precipitation and dry weather in Japan (1901–2004). *Journal of the Meteorological Society of Japan, 84*, 1033–1046.

Gahukar, R. T., & Jotwani, M. G. (1980). Present status of field pests of sorghum and millets in India. *Tropical Pest Management, 26*(2), 138–151. https://doi.org/10.1080/09670878009414385

Gahukar, R. T., & Reddy, G. V. P. (2019). Management of economically important insect pests of millet. *Journal of Integrated Pest Management, 10*(1), 28. https://doi.org/10.1093/jipm/pmz026

Gaikwad, D. J., Ubale, N. B., Pal, A., Singh, S., Ali, M. A., & Maitra, S. (2022). Abiotic stresses impact on major cereals and adaptation options—A review. *Research on Crops, 23*(4), 896–915.

Garnett, T. (2014). Three perspectives on sustainable food security: Efficiency, demand restraint, food system transformation. What role for life cycle assessment. *Journal of Cleaner Production, 73*, 10–18.

Gladek, E., Fraser, M., Roemers, G., Sabag Munoz, O., Hirsch, P., & Kennedy, E. (2016). *The global food system: An analysis. Amsterdam*. https://www.metabolic.nl/publications/global-food-system-analysis. Accessed August 29, 2020.

GOI. (2020). Government of India, Directorate of Millet Research. Importance of coarse cereals. http://millets.dacfw.nic.in/Importance_CoarseCereals.html. Accessed August 29, 2020.

Gornall, J., Betts, R., & Burke, E. (2010). Implications of climate change for agricultural productivity in the early twenty-first century. *Philosophical Transactions of the Royal Society of London Series B, Biological Science, 365*, 2973–2989.

Goron, T. L., & Raizada, M. N. (2015). Genetic diversity and genomic resources available for the small millet crops to accelerate a New Green Revolution. *Frontiers in Plant Science, 6*, 157. https://doi.org/10.3389/fpls.2015.00157

Gross, R., Schoeneberger, H., Pfeifer, H., & Hans-Joachim, A. P. (2000). The four dimensions of food and nutrition security: Definitions and concepts. http://www.fao.org/elearning/course/FA/en/pdf/P-01_RG_Concept.pdf. Accessed August 29, 2020.

Gurjeet, S. Y., Humphreys, G., Kukal, E., & Walia, S. S. (2011). U.S Effect of water management on dry seeded and puddled transplanted rice. Part 1. Crop performance. *Field Crops Research, 120*, 112–122.

Hayashi, K., Llorca, L., Rustini, S., Setyanto, P., & Zaini, Z. (2018). Reducing vulnerability of rainfed agriculture through seasonal climate predictions: A case study on the rainfed rice production in Southeast Asia. *Agricultural Systems, 162*, 66–76.

Hossain, A., Skalicky, M., Brestic, M., Maitra, S., Ashraful Alam, M., Syed, M. A., Hossain, J., Sarkar, S., Saha, S., Bhadra, P., Shankar, T., Chaki, A. K., El Sabagh, A., & Islam. (2021). Consequences and mitigation strategies of abiotic stresses in wheat (Triticum aestivum L.) under the changing climate. *Agronomy, 11*, 241. https://doi.org/10.3390/agronomy11020241

House, L. R., Gomez, M., Murty, D. S., Sun, Y., Verma, & B. N. (2000). Development of some agricultural industries in several African and Asian countries. In C. W. Smith, & R. A. Frederiksen (Eds.), *Sorghum: Origin, history, technology, and production* (pp. 131–190). Wiley.

ICRISAT (International Crops Research Institute for the Semi-Arid Tropics). (1989). *International workshop on sorghum stem borers, 17–20, lCRISAT Center, Patancheru, India.*

Iizumi, T. (2019). Emerging adaptation to climate change in agriculture. In T. Iizumi, R. Hirata, & R. Matsuda (Eds.), *Emerging adaptation to climate change in agriculture, research and practices* (pp. 1–16). Springer Nature Singapore. https://doi.org/10.1007/978-981-13-9235-1

IPCC. (2007). Intergovernmental panel on climate change, climate change: Impacts, adaptation and vulnerability. In *Contribution of working group II to the fourth assessment report of the Intergovernmental Panel on Climate Change.* Cambridge University Press.

IPCC. (2013). Intergovernmental panel on climate change, climate change: The physical science basis. In *Contribution of working group I to the fifth assessment report of the IPCC*. Cambridge University Press.

Ishimaru, T., Hirabayashi, H., Sasaki, K., Ye, C., & Kobayashi, A. (2016). Breeding efforts to mitigate damage by heat stress to spikelet sterility and grain quality. *Plant Production Science, 19*, 12–21. https://doi.org/10.1080/1343943X.2015.1128113

Islam, S. M. F., & Karim, Z. (2019). World's demand for food and water: The consequences of climate change. In *Desalination—Challenges and opportunities*. https://doi.org/10.5772/intechopen.85919. Online First. https://www.intechopen.com. Accessed August 29, 2020,

Jagathjothi, N., Ramamoorthy, K., & Kuttimani, R. (2011). Integrated nutrient management ongrowth and yield of rainfed direct sown finger millet. *Research on Crops, 12*, 79–81.

Jagathjothi, N., Ramamoorthy, K., & Priya, R. S. (2010). Influence of enriched FYM with inorganic fertilizers on nutrient uptake, soil available nutrients and productivity of rainfed finger millet. *Madras Agricultural Journal, 97*(10–12), 385–387.

Jain, N., Arora, P., Tomer, R., Mishra, S. V., Bhatia, A., Pathak, H., Chakraborty, D., Kumar, V., Dubey, D., & Harit, R. C. (2016). Greenhouse gases emission from soils under major crops in northwest India. *Science of the Total Environment, 542*, 551–561. https://doi.org/10.1016/j.scitotenv.2015.10.073

Jemila, C., Saliha, B. B., & Udayakumar, S. (2017). Evaluating the effect of phosphatic fertilizers on soil and plant P availability and maximising rice crop yield. *Oryza, 54*, 305–313.

Jensen, E. S., Peoples, M. B., Boddey, R. M., Gresshoff, P. M., Hauggaard-Nielsen, H., Alves, B. J., & Morrison, M. J. (2012). Legumes for mitigation of climate change and the provision of feedstock for biofuels and biorefineries. A review. *Agronomy for Sustainable Development, 32*, 329–364.

Jogaiah, S., Shivanna, R. K., Gnanaprakash, P. H., & Hunthrike, S. S. (2010). Evaluation of plant growth-promoting Rhizobacteria for their efficiency to promote growth and induce systemic resistance in pearl millet against downy mildew disease. *Archives of Phytopathology and Plant Protection, 43*(4), 368–378. https://doi.org/10.1080/03235400701806377

Kadam, S., Pawar, S. B., & Rajeshwar, J. (2019). Impact of integrated nutrient management on growth, yield of summer pearl millet. *Trends in Bioscience, 12*(4). ISSN 0974-8431.

Kadiri, M., & Hussaini, M. A. (1999). EVect of hardening pretreatments on vegetative growth, enzyme activities and yield of Pennisetum americanum and Sorghum bicolor. *Global Journal of Pure and Applied Sciences, 5*, 179–183.

Kassam, A., Friedrich, T., Shaxson, F., & Pretty, J. (2009). The spread of conservation agriculture: Justification, sustainability, and uptake. *International Journal of Agricultural Sustainability, 7*, 292–320.

Kaur, K. D., Jha, A., Sabikhi, L., & Singh, A. K. (2014). Significance of coarse cereals in health and nutrition: a review. *Journal of Food Science and Technology, 51*, 1429–1441.

Kaurav, A., Pandya, R. K., & Singh, B. (2018). Performance of botanicals and fungicides against blast of pearl millet (Pennisetum glaucum). *Annals of Plant and Soil Research, 20*(3), 258–262.

Kaushik, U., Raj, D., Rani, P., & Antil, R. S. (2018). Impact of zero tillage on available nutrients status on pearlmillet wheat cropping system. *International Journal of Chemical Studies, 6*(3), 2997–3000.

Khan, I., Yousif, A. M., Johnson, S. K., & Gamlath, S. (2015). Acute effect of sorghum flour-containing pasta on plasma total polyphenols, antioxidant capacity and oxidative stress markers in healthy subjects: A randomised controlled trial. *Clinical Nutrition, 34*(3), 415–421

Kokerai, L. K., Kugedera, A. T., & Chimbwanda, F. (2019). Evaluation of economic returns of in-situ rainwater harvesting and integrated nutrient management options on sorghum production. *Amity Journal of Agribusiness, 3*(2), 36–46.

Kumar, A., Gangwar, J. S., Prasad, S. C., & Harris, D. (2002). On-farm seed priming increases yield of direct-sown finger millet in India. *International Sorghum and Millets Newsletter, 43*, 90–92.

Kumar, P., Nanwal, R. K., & Yadav, S. K. (2005). Integrated nutrient management in pearl millet (Pennisetum glaucum)–wheat (Triticum aestivum) cropping system. *Indian Journal of Agricultural Sciences, 75*, 640–643.

Kumar, S. D., & Maitra, S. (2020). Sorghum-based intercropping system for agricultural sustainability. *Indian Journal of National Sciences, 10*, 20306–20313.

Kumar, V., & Singh, V. (2019). Effect of site-specific nutrient management on yield and nutrient uptake in pearl millet (Pennisetum glaucum)–wheat (Triticum aestivum) cropping sequence. *Annals of Plant and Soil Research, 21*(1), 1–6.

Kumar, A. D., & Channaveerswami, A. S. (2015). Pre and post emergencecontrol measures for shoot fly incidence and its influence on seed yield oflittle millet (Panicum sumatrense). *Journal of Experimental Zoology, India, 18*, 811–814.

Kumara, O., Naik, T. B., & Palaiah, P. (2007). Effect of weed management practices and fertility levels on growth and yield parameters in finger millet. *Karnataka Journal of Agricultural Sciences, 20*, 230–233.

Kumaran, G., & Parasuraman, P. (2019). Effect of enriched FYM and Panchagavya spray on foxtail millet (Setariaitalica) under rainfed conditions. *International Journal of Chemical Studies, 7*(2), 2121–2123.

Kushwaha, P., Kashyap, P. L., Kuppusamy, P., Srivastava, A. K., & Tiwari, R. K. (2019). Functional characterization of endophytic bacilli from pearl millet (Pennisetum glaucum) and their possible role in multiple stress tolerance. *Plant Biosystems,* 1–12. https://doi.org/10.1080/11263504.2019.1651773

Lal, R. (1999). Soil management and restoration for C sequestration to mitigate the accelerated greenhouse effect. *Progress in Environmental Science, 1*, 307–326.

Lal, R. (2004). Soil carbon sequestration impacts on global climate change and food security. *Science, 304*, 1623–1627.

Lal, R. (2015). Restoring soil quality to mitigate soil degradation. *Sustainability, 7*, 5875–5895.

Lesk, C., Rowhani, P., & Ramankutty, N. (2016). Influence of extreme weather disasters on global crop production. *Nature, 529*, 84–87.

Li, P., & Brutnell, T. P. (2011). Setaria viridis and Setariaitalica, model genetic systems for the panicoid grasses. *Journal of Experimental Botany, 62*, 3031–3037. https://doi.org/10.1093/jxb/err096

Lipper, L., Thornton, P., Campbell, B. M., Baedeker, T., Braimoh, A., Bwalya, M., Caron, P., Cattaneo, A., Garrity, D., & Henry, K. (2014). Climate-smart agriculture for food security. *Nature Clinical Practice Endocrinology & Metabolism, 4*, 1068–1072.

Maitra, S. (2020a). Potential horizon of brown-top millet cultivation in drylands: A review. *Crop Research, 55*(1–2), 57–63. https://doi.org/10.31830/2454-1761.2020.012

Maitra, S. (2020b). Intercropping of small millets for agricultural sustainability in drylands: A review. *Crop Research, 55*(3–4), 162–171. https://doi.org/10.31830/2454-1761.2020.025

Maitra, S., Ghosh, D., Sounda, G., & Jana, P. K. (1998). Effect of seed treatment on growth and productivity of finger millet under rained lateritic belt of West Bengal. *Indian Agriculture, 42*, 37–43.

Maitra, S., Patro, T. S. S. K., Reddy, A., Hossain, A., Pramanick, B., Brahmachari, K., Prasad, K., Santosh, D. T., Mandal, M., Shankar, T., Sagar, L., Banerjee, M., Palai, J. B., Praharaj, S., & Sairam, M. (2023a). Brown top millet (Brachiaria ramosa L. Stapf; Panicum ramosum L.)—A neglected and smart crop in fighting against hunger and malnutrition. In M. Farooq, & K. H. M. Siddique (Eds.), *Neglected and underutilized crops, future smart food* (pp. 221–245). Academic Press. https://doi.org/10.1016/B978-0-323-90537-4.00012-0

Maitra, S., & Pine, S. (2020). Smart irrigation for food security and agricultural sustainability. *Indian Journal of Natural Sciences, 10*, 20435–20439.

Maitra, S., Reddy, M. D., & Nanda, S. P. (2020a). Nutrient management in finger millet (Eleusine coracana L. Gaertn) in India. *International Journal of Agriculture Environment and Biotechnology, 13*(1), 13–21.

Maitra, S., Pine, S., Banerjee, P., Pramanick, B., & Shankar, T. (2022). Millets: Robust entrants to functional food sector. In Pirzadah et al. (Ed.), *Bioresource technology: Concept, tools and experience* (pp. 1–27). John Wiley & Sons Ltd.

Maitra, S., Praharaj, S., Hossain, A., Patro, T. S. S. K., Pramanick, B., Shankar, T., Pudake, R. N., Gitari, H. I., Palai, J. B., Sairam, M., Sagar, L., & Sahoo, U. (2023b). Small millets: The next-generation smart crops in the modern era of climate change. In R. N. Pudake et al. (Eds.), *Omics of climate resilient small millets* (pp. 1–25). Springer Nature Singapore Pte Ltd. https://doi.org/10.1007/978-981-19-3907-51

Maitra, S., & Shankar, T. (2019). Agronomic management in little millet (Panicum sumatrense L.) for enhancement of productivity and sustainability. *International Journal of Biosciences, 6*(2), 91–96.

Maitra, S., Sounda, G., Ghosh, D., & Jana, P. (1997). Effect of seed treatment on finger millet (Eleusine coracana) varieties in rainfed upland. *The Indian Journal of Agricultural Sciences, 67*, 478–480.

Maitra, S., Ghosh, D. C., Sounda, G., Jana, P. K., & Roy, D. K. (2000). Productivity, competition and economics of intercropping legumes in finger millet (Eleusine coracana) at different fertility levels. *The Indian Journal of Agricultural Sciences, 70*(12), 824–828.

Malviya, K. S., Bakoriya, L., Kumar, S., Aske, S., Mahajan, G., & Malviya, K. D. (2019). Effect of tillage and cultural practices on growth, yield and economics of kodomillet. *International Journal of Current Microbiology and Applied Sciences, 8*(06), 890–895. https://doi.org/10.20546/ijcmas.2019.806.107

Martin-Gorriza, B., Maestre-Valeroa, J. F., Almagrob, M., Boix-Fayosb, C., & Martínez-Mena, M. (2020). Carbon emissions and economic assessment of farm operations under different tillage practices in organic rainfed almond orchards in semiarid Mediterranean conditions. *Scientia Horticulturae*. https://doi.org/10.1016/j.scienta.2019.108978

Mgolozeli, S., Nciizah, A. D., Wakindiki, I. I. C., & Mudau, F. N. (2020). Innovative pro-smallholder farmers' permanent Mulch for better soil quality and food security under conservation agriculture. *Agronomy, 10*, 605. https://doi.org/10.3390/agronomy10040605

Michaelraj, P. S. J., & Shanmugam, A. (2013). A study on millets based cultivation and consumption in India. *International Journal of Marketing, Financial Services & Management Research, 2*(4), 49–58.

Mishra, J. S., Kumar, R., Upadhyay, P., & Hans, H. (2018). Weed management in millets. *Indian Farming, 68*, 77–79.

Monisha, V., Rathinaswamy, A., Mahendran, P. P., & Kumutha, K. (2019). Influence of integrated nutrient management on growth attributes and yield of foxtail millet in red soil. *International Journal of Chemical Studies, 7*(3), 3536–3539.

Morita, S., Wada, H., & Matsue, Y. (2016). Countermeasures for heat damage in rice grain quality under climate change. *Plant Production Science, 19*, 1–11.

Mueller, N. D., Gerber, J. S., Johnston, M., Ray, D. K., Ramankutty, N., & Foley, J. A. (2012). Closing yield gaps through nutrient and water management. *Nature, 490*, 254–257.

Mukherjee, A., Simon Wang, S. Y., & Promchote, P. (2019). Examination of the climate factors that reduced wheat yield in Northwest India during the 2000s. *Water, 11*, 343. https://doi.org/10.3390/w11020343

Niu, X., Song, L., Xiao, Y., & Ge, W. (2018). Drought-tolerant plant growth-promoting rhizobacteria associated with foxtail millet in a semi-arid agroecosystem and their potential in alleviating drought stress. *Frontiers in Microbiology, 8*, 2580. https://doi.org/10.3389/fmicb.2017.02580

Nuttall, J. G., O'Leary, G. J., Panozzo, J. F., Walker, C. K., Barlow, K. M., & Fitzgerald, G. J. (2017). Models of grain quality in wheat–A review. *Field Crops Research, 202*, 136–145.

Oduori, C. O. A. (2007). Finger millet blast management in East Africa. Creating opportunities for improving production and utilization of finger millet. In M. A. Mgonja, J. M. Lenné, E. Manyasa, & S. Sreenivasaprasad (Eds.), *Proceedings of the first international finger millet stakeholder workshop, projects R8030 and R8445 UK Department for International Development—crop protection programme held 13–14 September 2005 at Nairobi* (196 pp). Patancheru 502 324, Andhra Pradesh, India: International Crops Research Institute for the Semi-Arid Tropics. ISBN: 978-92-9066-505-2

Ojuederie, O. B., Olanrewaju, O. S., & Babalola, O. B. (2019). Plant growth promoting rhizobacterial mitigation of drought stress in crop plants: Implications for sustainable agriculture. *Agronomy, 9*(11), 712. https://doi.org/10.3390/agronomy9110712

Pallavi, C., Joseph, B., Khan, M. A., & Hemalatha, S. (2017). Effect of organic fertilizers and biofertilizers on yield and yield attributing traits of direct sown rainfed finger millet, Eleusine coracana (L). Gaertn. *International Journal of Farm Sciences, 7*(3), 1–5.

Parihar, S. K., Dwivedi, B. S., Khan, I. M., & Tiwari, R. K. (2010). Effect of integrated nutrient management on yield and economics of little millet (Panicurn sumatrense roth exRoem, and Schult). *Journal of Soils and Crops, 20*(2), 211–215.

Parkinson, R. (2013). System based integrated nutrient management. *Soil Use and Management, 29*, 608.

Patidar, M., & Mali, A. L. (2001). Integrated nutrient management in sorghum (Sorghum bicolor) and its residual effect on wheat (Triticum aestivum). *The Indian Journal of Agricultural Sciences, 71*, 587–591.

Pilbeam, C., Gregory, P., Tripathi, B., & Munankarmy, R. (2002). Fate of nitrogen-15-labelled fertilizer applied to maize-millet cropping systems in the mid-hills of Nepal. *Biology and Fertility of Soils, 35*, 27–34.

Prabha, V. V., Senthil, A., Sritharan, N., & Boominathan, P. (2016). Effect of foliar application of plant growth regulators and nutrients on physiological traits of finger millet (Eleusine coracana). *Research on Crops, 17*(3), 483–488.

Prabudoss, V., Jawahar, S., Shanmugaraja, P., & Dhanam, K. (2014). Effect of integrated nutrient management on yield and nutrient uptake of transplanted Kodo millet. *European Journal of Biotechnology and Bioscience, 1*(5), 30–32.

Pradan/SDTT. (2012). *Cultivating finger millet with SRI principles: A training manual*. Mumbai: PRADAN, Ranchi, and Sir Dorabji Tata Trust. http://sri.cals.cornell.edu/aboutsri/othercrops/fingermillet/In_SFMI_Pradan.pdf

Prasad, J. V. N. S., Srinivasa, R. C., Srinivasa, K., Naga Jyothia, C., Venkateswarlub, B., Ramachandrappac, B. K., Dhanapalc, G. N., Ravichandraa, K., & Mishra, P. K. (2016). Effect of ten years of reduced tillage and recycling of organic matter on crop yields, soil organic carbon and its fractions in Alfisols of semi arid tropics of southern India. *Soil and Tillage Research, 156*, 131–139. https://doi.org/10.1016/j.still.2015.10.013

Rahman, M. A., Kang, S., Nagabhatla, N., & Macnee, R. (2017). Impacts of temperature and rainfall variation on rice productivity in major ecosystems of Bangladesh. *Agriculture and Food Security Journal, 6*, 10. https://doi.org/10.1186/s40066-017-0089-5

Rajput, S. C. (2008a). Effect of integrated nutrient management of productivity and monetaryreturns of pearl millet (Pennisetum glaucum). *Research on Crops, 9*, 248–250.

Rajput, S. C. (2008b). Yield and economics of rainfed pearl millet (Pennisetum glaucum) as influenced by integrated nutrient management. *Research on Crops, 9*, 253–254.

Ramachandrappa, B. K., Thimmegowda, M. N., Sathish, A., Devaraja, K., Jagadeesh, B. N., & SandhyaKiranmai, M. (2015). Sustainable dryland technologies for improving productivity and livelihood security in Alfisols of Karnataka. *Indian Journal of Dryland Agricultural Research and Development, 30*(2), 105–112.

Ramashia, S. E., Anyasi, T. A., Gwata, E. T., Meddows-Taylor, S., & Jideani, A. I. O. (2019). Processing, nutritional composition and health benefits of finger millet in sub-saharan Africa. *Food Science and Technology, 39*(2), 253–266.

Rao, B. K. R., Krishnappa, K., Srinivasarao, C., Wani, S. P., Sahrawat, K. L., & Pardhasaradhi, G. (2012). Alleviation of multinutrient deficiency for productivity enhancement of rain-fed soybean and finger millet in the semi-arid region of India. *Communications in Soil Science and Plant Analysis, 43*, 1427–1435.

Rao, B. R., Nagasampige, M. H., & Ravikiran, M. (2011). Evaluation of nutraceutical properties of selected small millets. *The Journal of Pharmacy and Bioallied Sciences, 3*, 277–279. http://www.jpbsonline.org/text.asp?2011/3/2/277/80775

Rathod, D. D., Meena, M. C., & Patel, K. P. (2012). Crop yields and nutrient uptake as affected by different Zn-enriched organics under wheat and fodder maize crop sequence in typicHaplustept of Gujarat. *The Journal of Soil and Water Conservation, 11*, 67–72.

Ravi, S., Narayana Rao, K., Ravi, M. V., Veeresh, H., Dodamani, B. M., & Swamy, M. (2020). Effect of nutrient management approaches on seed yield and nutrient uptake of soybean-sorghum based cropping system. *International Journal of Chemical Studies, 8*(1), 2762–2766. https://doi.org/10.22271/chemi.2020.v8.i1ap.8687

Reddy, K. V. S. (1991). Insect pests of sorghum in Africa. *International Journal of Tropical Insect Science, 12*(5–6), 653–657. https://doi.org/10.1017/s1742758400013151

Regar, K. L., & Singh, Y. V. (2014). Fertilizer recommendation based on soil testing for the targeted yield of rice in eastern plain zone of Uttar Pradesh. *The Bioscan, 9*(2), 531–534.

Rimas, A., & Fraser, E. (2010). *Empires of food: Feast, famine, and the rise and fall of civilizations* (p. 302). Free Press.

Roy, A. K., Ali, N., Lakra, R. K., Alam, P., Mahapatra, P., & Narayan, R. (2018). Effect of integrated nutrient management practices on nutrient uptake, yield of finger millet (Eleusine coracana L. Gaertn.) and post-harvest nutrient availability under rainfed condition of Jharkhand. *International Journal of Current Microbiology and Applied Sciences, 7*(08), 339–347. https://doi.org/10.20546/ijcmas.2018.708.038

Saeed, H. A. M., Liu, Y., Lucia, L. A., & Chen, H. (2017). Evaluation of Sudanese sorghum and bagasse as a pulp and paper feedstock. *Bio Resources, 12*(3), 5212–5222

Sage, R. F., & Zhu, X. G. (2011). Exploiting the engine of C_4 photosynthesis. *Journal of Experimental Botany, 62*, 2989–3000. https://doi.org/10.1093/jxb/err179

Sage, R. F., Pascal-Antoine, C., & Edwards, E. J. (2011). The C_4 plant lineages of planet Earth. *Journal of Experimental Botany, 62*(9), 3155–3169. https://doi.org/10.1093/jxb/err048

Saiz-Rubio, V., & Rovira-Más, F. (2020). From smart farming towards agriculture 5.0: A review on crop data management. *Agronomy, 10*, 207. https://doi.org/10.3390/agronomy10020207

Saleh, A. S. M., Zhang, Q., Chen, J., & Shen, Q. (2013). Millet grains: Nutritional quality, processing, and potential health benefits. *Comprehensive Reviews in Food Science and Food Safety, 12*, 281–295. https://doi.org/10.1111/1541-4337.12012

Samanta, S., Maitra, S., Shankar, T., Gaikwad, D., Sagar, L., Panda, M., & Samui, S. (2022). Comparative performance of foliar application of urea and nano urea on finger millet (Eleusine coracana L. Gaertn). *Crop Research, 57*(3), 166–170. https://doi.org/10.31830/2454-1761.2022.025

Sandhya Rani, Y., Triveni, U., Patro, T. S. S. K., Divya, M., & Anuradha, N. (2017). Revisiting of fertilizer doses in finger millet (Eleusine coracana (L.) Garten.) through targeted yield and soil test crop response (STCR) approach. *International Journal of Current Microbiology and Applied Sciences, 6*(7), 2211–2221.

Sankar, G. R. M., Sharma, K. L., Dhanapal, G. N., Shankar, M. A., Mishra, P. K., Venkateswarlu, B., & Grace, J. K. (2011). Influence of soil and fertilizer nutrients on sustainability of rainfed finger millet yieldand soil fertility in semi-arid Alfisols. *Communications in Soil Science and Plant Analysis, 42*, 1462–1483.

Santhi, R., Bhaskaran, A., & Natesan, R. (2011a). Integrated fertilizer prescriptions for beetroot through inductive-cum-targeted yield model on an alfisol. *Communications in Soil Science and Plant Analysis, 42*, 1905–1912.

Santhi, R., Poongothai, S., & Maragatham, S. (2011b). STCR-IPNS prescription for higher productivity of beetroot and sustained soil fertility. In *Proceedings national symposium on soil health improvement for enhancing crop productivity, Tamil Nadu Agricultural University, Coimbatore and AICRIP on soil test crop response correlation* (pp. 17–18). *Indian Institute of Soil Science, Bhopal*.

Sathishkumar, A., Sakthivel, N., Subramanian, E., Kalpana, R., Janaki, P., & Rajesh, P. (2018). Foliar application of nutrients and plant growth regulators on growth and yield of finger millet. *Journal of Pharmacognosy and Phytochemistry, 7*(3), 3032–3035.

Saxena, R., Vanga, S. K., Wang, J., Orsat, V., & Raghavan, V. (2018). Millets for food security in the context of climate change: A review. *Sustainability, 10*(7), 2228. https://doi.org/10.3390/su10072228

Schlegel, A. J., Assef, Y., Bond, H. D., Haas, L. A., & Stone, L. R. (2017). Changes in soil nutrients after 10 years of cattle manure and swine effluent application. *Soil and Tillage Research, 172*, 48–58.

Sekar, J., Raju, K., Duraisamy, P., & Ramalingam Vaiyapuri, P. (2018). Potential of finger millet indigenous rhizobacterium Pseudomonas sp. MSSRFD41 in blast disease management—Growth promotion and compatibility with the resident rhizomicrobiome. *Frontiers in Microbiology, 9*, 1029. https://doi.org/10.3389/fmicb.2018.01029

Sekaran, U., McCoy, C., Kumar, S., & Subramanian, S. (2018). Soil microbial community structure and enzymatic activity responses to nitrogen management and landscape positions in switchgrass (Panicum virgatum L.). *GCB Bioenergy*. https://doi.org/10.1111/gcbb.12591

Selectstar Marwein, B., Singh, R., & Chhetri, P. (2019). Effect of integrated nitrogen management on yield and economics of foxtail millet genotypes. *International Journal of Current Microbiology and Applied Sciences, 8*(08), 2543–2546.

Shanmugasundaram, V. S., & Kannaiyan, M. (1989). EVect of concentration of seed hardening chemicals on physiological characters of pearl millet (Pennisetum typhoids Stapf and Hubb). *Journal of Agronomy and Crop Science, 163*, 174–176.

Sharma, V. K., Pandey, R. N., Kumar, S., Chobhe, K. A., & Chandra, S. (2016). Soil test crop response-based fertilizer recommendations under integrated nutrient management for higher productivity of pearl millet (Pennisetumglaucum) and wheat (Triticum aestivum) under long term experiment. *The Indian Journal of Agricultural Sciences, 86*(8), 1076–1081.

Shetty, V. S., & Kumar, M. D. (2018). Effect of NPK application through different approaches on yield and secondary nutrient uptake by finger millet under rainfed conditions. *International Journal of Pure and Applied Bioscience, 6*(2), 735–741.

Shultana, R., Zuan, A. T. K., Yusop, M. R., Saud, H. M., & Ayanda, A. F. (2020). Effect of salt-tolerant bacterial inoculations on rice seedlings differing in salt-tolerance under saline soil conditions. *Agronomy, 10*, 1030. https://doi.org/10.3390/agronomy10071030

Singh, J. P., & Bhardwaj, K. K. (2017). Effect of site-specific nutrient management on yield and nutrient uptake by pearl millet and wheat crops. *Indian Journal on Fertilisers, 13*(3), 28–33.

Sivakumar, R., Pathmanaban, G., Kalarani, M. K., Vanangamudi, M., & Srinivasan, P. S. (2002). Effect of foliar application of growth regulators on biochemical attributes and grain yield in pearl millet. *Indian Journal of Plant Physiology, 7*(1), 79–82.

Soora, N. K., Singh, A. K., Agarwal, P. K., Rao, V. U. M., & Venkateswarlu, B. (2012). *Climate change and Indian agriculture: Impact, adaptation and vulnerability—Salient achievements from ICAR network project* (p. 32). IARI Publication.

Srinivas, M., Rao, S. V., & Seetharama, N. (2008). Effect of integrated nutrient managementand different moisture conservation practices on yield and economics of rabisorghum. *Research on Crops, 9*, 557–560.

Srinivasarao, C., Lal, R., Kundu, S., Babu, M. B., Venkateswarlu, B., & Singh, A. K. (2014). Soil carbon sequestration in rainfed production systems in the semiarid tropics of India. *Science of the Total Environment, 487*, 587–603. https://doi.org/10.1016/j.scitotenv.2013.10.006

Srinivasarao, C., Venkateswarlu, B., Lal, R., Singh, A. K., & Kundu, S. (2013). Sustainable management of soils ofdry land ecosystems of India for enhancing agronomic productivity and sequestering carbon. *Advances in Agronomy, 121*, 253329.

Sujatha, K., Selvarani, K., Vijayalakshmi, V., Vanniarajan, C., & Sivasubramaniam, K. (2013). Seed fortification studies in barnyard millet (Echinochloa frumentacea) cv. CO1. *IOSR Journal of Agriculture and Veterinary Science, 5*(4), 22–24.

Suresh, G., Guru, G., & Lokanadan, S. (2018). Effect of nutrient levels and plant growth regulators on growth parameters of pearl millet. *International Journal of Pure and Applied Bioscience, 6*(3), 271–277. https://doi.org/10.18782/2320-7051.6562

Swaminathan, M. S., & Kesavan, P. C. (2012). Agricultural research in an era of climate change. *Agricultural Research, 1*, 3–11.

Tadele, Z. (2016). Drought adaptation in millets. In A. K. Shanker, & C. Shanker (Eds.), *Abiotic and biotic stress in plants—Recent advances and future perspectives* (pp. 639–662). IntechOpen.

Tanzubil, P. B., & Yakubu, E. A. (1997). Insect pets of millet in northern Ghana. Farmers' perceptions and damage potential. *International Journal of Pest Management, 43*, 133–136.

Tetreault, H. M., Grover, S., Scully, E. D., Gries, T., Palmer, N. A., Sarath, G., Louis, J., & Sattler, S. E. (2019). Global Responses of Resistant and Susceptible Sorghum (Sorghum bicolor) to Sugarcane Aphid (Melanaphissacchari). *Frontiers in Plant Science, 10*, 145. https://doi.org/10.3389/fpls.2019.00145

Thesiya, N. M., Dobariya, J. B., & Patel, J. G. (2019). Effect of integrated nutrient management on growth and yield parameters of kharif little millet under little millet-green gram cropping sequence. *International Journal of Pure and Applied Bioscience, 7*(3), 294–298. https://doi.org/10.18782/2320-7051.7392

Upadhyaya, H., Reddy, V.G., & Sastry, D. (2008). Regeneration guidelines finger millet; CGIAR System-Wide Genetic Resource Programme, Rome, Italy, 2008.

Uphoff, N. (2017) SRI: An agroecological strategy to meet multiple objectives with reduced reliance on inputs. *Agroecology and Sustainable Food Systems, 41*, 825–854.https://doi.org/10.1080/21683565.2017.1334738

Ureta, C., González, E. J., Espinosa, A., Trueba, A., Piñeyro-Nelson, A., & Álvarez-Buylla, E. R. (2020). Maize yield in Mexico under climate change. *Agricultural Systems, 177*, 102697. https://doi.org/10.1016/j.agsy.2019.102697

Vadez, V., Hash, T., Bidinger, F. R., Kholova, J. (2012). II.1.5 Phenotyping pearl millet for adaptation to drought. *Frontiers in Physiology, 3*, 386. https://doi.org/10.3389/fphys.2012.00386

Valina, H., Sands, R. D., van der Mensbrugghe, D., Nelsond, G. C., Ahammad, H., Blanc, E., Bodirsky, B., Fujimori, S., Hasegawa, T., Havlik, P., Heyhoe, E., Kyle, P., Mason-D'Croz, D., Paltsev, S., Rolinski, S., Tabeau, A., van Meijl, H., von Lampe, M., & Willenbockelm, D. (2014).

The future of food demand: Understanding differences in global economic models. *Agricultural Economics, 45*, 51–67.

Verma, J. K., Pareek, N. K., Saharan, B., Singh, H., & Vimal Khinchi, V. (2017). Influence of tillage practices on growth and yields of pearl millet (Pennisetum glaucum (L.) R. Br.) cultivars under rainfed conditions of Western Rajasthan. *International Journal of Plant & Soil Science, 16*(1), 1–9.

Verma, M., Singh, B. D., Gupta, P. C., & Kumar, R. (2018). Zero Tillage technique with high yielding variety for wheat cultivation as compared to traditional method of farmers: Evidences of FLD. *International Journal of Current Microbiology and Applied Sciences, 7*(08), 2512–2516. https://doi.org/10.20546/ijcmas.2018.708.255

Vijayaraghavan, H. (1999). EVect of seed treatment with plant growth regulators on bhendi (Abelmoschus esculentus L.) grown under sodic soil conditions. *The Madras Agricultural Journal, 86*, 247–249.

Vrieling, A., Meroni, M., Mude, A. G., Chantarat, S., Ummenhofer, C. C., & de Bie, C. A. J. M. (2016). Early assessment of seasonal forage availability for mitigating the impact of drought on East African pastoralists. *Remote Sensing of Environment, 174*, 44–55.

Wanyera, N. M. W. (2007). Finger millet blast management in East Africa. Creating opportunities for improving production and utilization of finger millet. In *Proceedings of the first international finger millet stakeholder workshop, projects R8030 and R8445 UK Department for international development—Crop protection programme held 13–14 September 2005 at Nairobi* (196 pp). Patancheru 502 324, Andhra Pradesh, India: International Crops Research Institute for the Semi-Arid Tropics. ISBN: 978-92-9066-505-2

Wilson, C. A., Davidson, D. A., & Cresser, M. S. (2008). Multi-element soil analysis: An assessment of its potential as an aid to archaeological interpretation. *Journal of Archaeological Science, 35*(2), 412–424. https://doi.org/10.1016/j.jas.2007.04.006

World Bank Group. (2016). *Global monitoring report 2015–16: Development goals in an era of demographic change*. World Bank

Wu, W., & Ma, B. (2015) Integrated nutrient management (INM) for sustaining crop productivity and reducing environmental impact: A review. *Science of the Total Environment,* 512–513

Xiao, D., Bai, H., & Liu, D. L. (2018). Impact of future climate change on wheat production: A simulated case for China's wheat system. *Sustainability, 10*, 1277. https://doi.org/10.3390/su10041277

Yohannes, H. A. (2016). Review on relationship between climate change and agriculture. *Journal of Earth Science and Climatic Change, 7*, 335. https://doi.org/10.4172/2157-7617.1000335

Zhang, F., Cui, Z., Chen, X., Ju, X., Shen, J., Chen, Q., Liu, X., Zhang, W., Mi, G., Fan, M., & Jiang, R. (2012). Integrated nutrient management for food security and environmental quality in China. *Advances in Agronomy, 116*, 1–40.

Zhu, F. (2014). Structure, physiochemical properties, modifications and uses of sorghum starch. *Comprehensive Reviews in Food Science and Food Safety, 13*, 597–610.

Chapter 3
Applying Genomics Resources to Accelerate the Development of Climate Resilient Crops

Dinesh Kumar Saini, Sourabh Kumar, and Rajvir Kaur

3.1 Introduction

Abiotic stresses, for instances, drought, submergence, salinity, cold/chilling, high temperature or heat, and mineral toxicity, are the major causes of crop yield reductions, worldwide. These abiotic stresses collectively reduce expected average yields of the major staple crops by more than 50% (Tanin et al., 2022). At present, global agriculture is facing a serious threat from climate change which further intensifies the harmful effects of major abiotic stresses such as drought, heat, waterlogging/submergence, chilling, and salinity on crop plants. This reduced productivity may increase the food prices and led to the greater global food insecurity. If this scenario persists, it might lead to social unrest and famines in certain instances. Changes in climate will continue to affect food supply unless efforts are made to increase the resilience of the crops; some projections have had previously showed a drastic decrease in the production of major staple crops in the year i.e., 2020, including 14% for wheat, 11% for rice and 9% for maize (Hisas, 2011). Moreover, it has been predicted by the Intergovernmental Panel on Climate Change (IPCC) that the mean temperatures around the planet may rise between 2 and 5 °C or more by 2050 (www.ipcc.ch). Therefore, increasing the resilience of crops to climate change or production of "climate ready crops" or "climate resilient crops" represents a critical component towards ensuring food security. Overall, this situation presents two potential breeding challenges (a) to modify the selection criteria to focus on adaptation of these staple crops to different abiotic stresses rather than total yield, (b) to ensure the

D. K. Saini (✉) · R. Kaur
Department of Plant Breeding and Genetics, Punjab Agricultural University, Ludhiana, Punjab, India
e-mail: dineshsaini96344@gmail.com

S. Kumar
Department of Genetics and Plant Breeding, Chaudhary Charan Singh University, Meerut, Uttar Pradesh, India

S. Sheraz Mahdi et al. (eds.), *Adapting to Climate Change in Agriculture-Theories and Practices*, https://doi.org/10.1007/978-3-031-28142-6_3

presence of multiple stress tolerance genes and their uses in elite material and wider breeding germplasm of these staple crops. Fortunately, from the "Mendelian era" of nineteenth century we have now entered into the era of "Genomics" in twenty-first century where numerous efficient and cost-effective tools and techniques are available to understand the complex genetics of traits associated with abiotic stress tolerance and accelerate the breeding process.

Climate resilient crops can be developed through molecular breeding approaches or genetic engineering-based strategies. On the important note, one must know that any stress tolerant crop can not be considered as climate resilient crop, as tolerance means that variety will behave well under target stress condition only, while resilient crop is supposed to be resilient to different changes in climate, means it should perform well under different climatic conditions such as drought, heat, chilling, and submergence etc. In general, breeders focus on the development of variety suited to the target condition only whether it may be drought or heat or any other stress. Exploitation of molecular breeding strategies for crop improvement has proved successful in crossbreeding. Advances in plant genomics may provide further ways to improve the understandings of underlying genetics of the traits and offer molecular or DNA based markers to accelerate the pace of genetic improvement. Bi-parental QTL mapping and genome wide association studies are known to provide the information regarding the QTLs/genes and markers associated with the traits. These markers can be then indirectly used for efficient selection of the plants carrying desired genes, and this strategy is called as marker assisted selection (MAS). Tremendous advancements have been made in genomics techniques with the advent of next-generation sequencing (NGS) technologies, which facilitated the movement from low-throughput (LTP) to SNP (Single Nucleotide Polymorphism)-based high-throughput (HTP) systems (Yang et al., 2013a).

Despite these advancements in genomics of crops, a lack of precise phenotypic data has led to poor results in QTL/gene discovery, limiting progress in genomic-assisted breeding of the crops. Thus, now the research bottleneck in plant sciences is shifting from genotyping (genomics) to phenotyping (phenomics). To alleviate this 'phenomics bottlenecks' various sophisticated non-invasive imaging, image analysis, spectroscopy, robotics, high performance computing facilities and phenomics databases have been developed which systematically collect precise phenotypic data. These phenomics platforms and tools can record data on different traits like plant architecture, development, plant photosynthesis, biomass, productivity etc. on thousands of plants in a single day (reviewed in Mir et al., 2019). This high-throughput phenotyping will further assist genomics-assisted breeding approaches such as selection of promising progenies/germplasm for further use in crop breeding aiming at both genetic improvement of the population and cultivar release. An advanced form of MAS, marker assisted recurrent selection (MARS) have also been used in some crop species mainly in maize (Abdulmalik et al., 2017; Beyene et al., 2015; e.g., Bankole et al., 2017), which allows simultaneous identification and improvement of quantitative traits by collecting favorable alleles at many loci. After bi-parental QTL mapping and GWAS, now, we have reached an era from studying a few loci to all loci of the genome. This new technology is termed as genomic selection (GS) or genome wide selection (GWS) which assume that all marker loci of the genome contribute to the trait-expression either negatively or positively, so small-effect marker loci are also taken into consideration.

Other important approach for production of climate resilient crop may be 'genetic engineering' which may allow direct transfer of beneficial gene or gene cassette in the crop of interest for generating desired phenotype. Although this technology serves as an efficient tool in tailoring modifications to produce phenotypes beyond the range available through exploitation of existing genetic variation, but it remains controversial in several countries, including India. A better alternative to this technology has recently been emerged i.e., 'targeted genome editing'. This technology can efficiently manipulate the plant phenotype without integrating any transgene into the plant genome. Of several approaches available for genome editing, CRISPR/cas9 (clustered regularly interspaced short palindromic repeats and CRISPR associated protein 9 nuclease) is the effective genome editing technology used in plant system (Belhaj et al., 2015). Several modifications to the CRISPR/Cas9 system are now available for site-directed modifications in plant genomes such as prime editing etc. (Lin et al., 2020). In this chapter, we briefly discuss the efforts made in the last decades for climate resilient breeding of three major staple crops viz., maize, rice, and wheat. This chapter highlights the genetics of major abiotic stresses tolerance in these crops having focussed on bi-parental QTL mapping and GWAS studies etc. conducted in last decades. This chapter also highlights the different genomic resources available in these crops, high throughput phenotyping methods/platforms and prospects of genomic-assisted breeding, mainly MAS, MARS, GS, transgenic approaches, and genome editing for breeding 'climate resilient' crops or 'climate ready' crops.

3.2 Genomics-Assisted Breeding for Climate Resilient Crops

Genomics offers techniques and tools to address the biggest challenge of increasing food yield, and stability of production under various abiotic stresses through advanced breeding techniques. QTL mapping (based on linkage) and genome wide association studies (based on LD-Linkage Disequilibrium) using different bi-parental populations such as doubled haploid, recombinant inbred lines, near isogenic line populations and, association panels (genotypes preferably collected from different geographical locations) have accelerated the dissection of genetic control of agronomically important traits, potentially allowing MAS (Gupta et al., 2020). Advancement of next-generation sequencing (NGS) methods has enabled the development of large numbers of DNA based markers, such as single nucleotide polymorphisms (SNP), insertion-deletions (InDels), SNP assays and so on etc. in various crops including maize, rice and wheat (Frouin et al., 2018; Gao et al., 2019; Kim et al., 2022; Li et al., 2016a, b; Naveed et al., 2018; Warraich et al., 2020; Wu et al., 2014;

Singh et al., 2015; Unterseer et al., 2014; Xu et al., 2017). NGS technologies have increased the mapping resolution of GWAS which now can be efficiently used to identify the genomic regions associated with traits of interest by performing statistical associations between DNA polymorphisms and trait variations in diverse collection of germplasms/genotypes that are phenotyped for traits of interest. Collectively, these genomic technologies can be used to safeguard the future through improved food security. With rapid advances in genomics tools, the application of genomics to identify and transfer valuable stress tolerance genes from crop relatives to elite crops will increase in pace and assist in meeting the challenge of global food security under changing climatic conditions. The following section summarizes the genetics of tolerance to various abiotic stresses in maize, rice and wheat together with information about breeding methods needed to improve climate resilience in these crops.

3.2.1 Maize

Maize (*Zea mays*) is the major staple crop worldwide. Maize belongs to the family Poaceae and genus *Zea*. Its genome size is approximately 2500 Mb contained in 10 chromosomes (2n = 2x = 20). It accounts for 37.2% of total global cereal production. Global maize production has increased from 313 million thousand tonnes in 1971 to 1162 million thousand tons in 2020 exhibiting an average annual growth of 3.06% (https://knoema.com/atlas/topics/Agriculture/Crops-Production-Quantity-tonnes/Maizeproduction). The USA is the top country by maize production in the world. In 2020, maize production in the USA was 360,252 thousand tonnes that shared around 33% of the total maize production around the world. As aforementioned, it is a staple crop in numerous parts of the word where it is either directly consumed by humans or used for animal feed. Abiotic stresses like drought, heat, nutrient deficiency, and chilling/cold are the major environmental factors that negatively influence maize production. Extreme temperatures, drought and salinity have significant influence on the maize yields. The amount of research studies has increased rapidly related to drought, heat and salinity stresses in maize in recent years (Fig. 3.1).

Genomics Resources Available for Maize

In recent years, genomic resources for functional analyses of the maize genome have been developed very rapidly. The completion of the high-quality maize genome sequence (The Maize Genome Sequencing Project) has significantly accelerated the functional genomics research in maize. Whole-genome sequencing of maize has comprehensively revealed the structural architecture of its genome and provided functional information on genes and SNPs. More than ten maize lines including reference lines B73, W22 and Mo17 have been sequenced and their genome assemblies are now available for public use (www.maizegdb.org). Although, large-scale

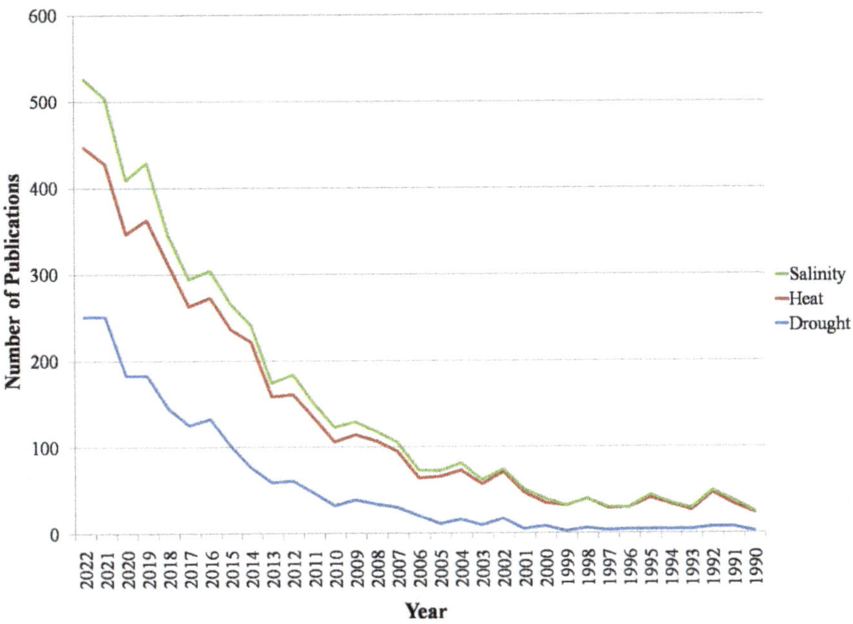

Fig. 3.1 The number of publications per year related to maize abiotic stresses from 01/01/1990 to 15/12/2022. *Source* PubMed

re-sequencing has been done to discover novel allelic variants in maize, but a large part of the information for genetic variation is generally lost by direct mapping of short sequence reads on to the reference genome. Therefore, pan genome datasets have been constructed via deep sequencing and de novo assembly of divergent accessions/genotypes. Pan genome refers to the "union of the gene sets of all the individuals of a clade or a species and it provides a new dimension of genome complexity with the presence/absence variations of genes among these genomes". In a study, a total of 8681 representative transcript assemblies were identified after characterizing the pan-genome of maize, constructed using transcriptome sequencing of 503 maize inbred lines. This study showed that a large portion of variation might lie outside the single reference genome for a particular species (Hirsch et al., 2014). A pan-genome of maize was constructed by performing the reduced representation sequencing of 14,129 maize inbred lines. In this study, a total of 26 million tags were generated and mapped by performing 18 trillion association tests (Lu et al., 2015). In B73 genome, around 40,000 protein coding genes have been identified. By aligning the genome sequences of 15 maize inbreds against the maize reference B73 genome, a total of 6,385,011 SNPs were identified (Xu et al., 2014). These genome wide non-synonymous SNPs can be used in genetic mapping studies and identification of haplotypes. The first-generation haplotype map (of size 3.3 million SNPs and InDels) in maize, called Hapmap1 was developed in 2009 (Gore et al., 2009); followed by second generation haplotype map, HapMap2 (of size 55 million

SNPs) in 2012 (Chia et al., 2012); and third generation haplotype map, Hapmap3 (of size 3.83 million SNPs and InDels) in 2015 (Bukowski et al., 2018). All these datasets are available via the GigaScience repository, GigaDB (https://www.re3data.org/repository/r3d100010478). Among restriction enzyme-based SNP identification, genotype-by-sequencing (GBS) has evolved as a cost-effective high throughput genotyping method. Using GBS, millions of SNPs have been identified and utilized in maize for various purposes (e.g., Gao et al., 2019; Li et al., 2016a, b; Yu et al., 2022).

Using restriction site-associated DNA sequencing (RADseq) technique, thousands of SNPs have been identified and used for different purposes (e.g., Wang et al., 2019a). In addition, thousands of SNPs, obtained from RNA-seq analysis, have also been employed for mapping purposes in maize (Sandhu et al., 2020). In 2014, Illumina developed a high-density Illumina^{R} MaizeSNP50 Beadchip with 56,110 SNPs which was used for characterization of the hundreds of maize inbred lines (Wu et al., 2014). Using Affymetrix^{R} Axiom^{R} platform, a very high-density Maize Genotyping Array composed of 616,201 variants (SNPs and small InDels) was also developed, which is also known as 600 K SNPs genotyping array (MaizeSNP600K) (Unterseer et al., 2014). With 616,201 variants, it was the largest publically available genotyping array in crop species at that time. In 2017, using the same platform, a lesser density array (55 K SNP array) composed of 55,229 SNPs was also developed, which covered variants from both temperate and tropical germplasms (Xu et al., 2017).

These SNP arrays have also been used in genetic characterization of maize inbreds, in GWAS (Guo et al., 2020a; Luo et al., 2019a, b; Xie et al., 2019) and for mapping the traits associated with abiotic stress tolerance in maize through linkage-based QTL mapping approaches (Abdelghany et al., 2019). Pre-selected and validated SNPs may be efficiently exploited for tracking the desirable progenies or plants in MAS, MARS and GS. KASP™ (Kompetitive allele specific PCR) offers customisable genotyping assay to run validated SNPs in applied breeding programs (Kaur et al., 2020). A set of 233 SNPs through KASP assay was used to increase the frequency of favorable alleles for tolerance to drought in maize population through MARS (Abdulmalik et al., 2017). Testcross progenies were selected based on the genomic estimated breeding values (GEBVs) of 1214 SNPs using KASP assay which resulted in rapid genetic gains for drought tolerance in maize (Vivek et al., 2017). Significant information has been obtained from whole genome sequencing of various lines, gene expression profiles, characterization of germplasm and many other useful genomic parameters in maize. Maize functional genomics research required an informatics establishment to make various genomic resources easily accessible and comprehensive to decipher the maize genome. In this view, maize bioinformatics databases based on different bioinformatics resources and tools were developed by the major research groups worldwide. Table 3.1 lists the different bioinformatics tools and database resources with their websites and features (what data they provide) that are helpful for genomic studies maize.

Table 3.1 List of major bioinformatics tools and database resources for genomics studies in maize

S. No.	Name of the database	Website	Features
1	MaizeGDB	https://www.maizegdb.org/	Provides data on gene and genome sequence, stock, phenotype, genotypic and karyotypic variation, and chromosomal mapping dat a
2	MaizeDIG	http://maizedig.maizegdb.org/	Database of images and genomes
3	RGPDB	http://sysbio.unl.edu/RGPDB/	Database of root genes and promoters
4	The TIGR Maize Database	http://maize.jcvi.org/	Provides annotated genomic sequences
5	ZmGDB chromosomal	http://www.plantgdb.org/ZmGDB/	Provides a convenient sequence-centered genome view with a narrow focus on gene structure annotation
6	ProFITS of maize	http://bioinfo.cau.edu.cn/ProFITS	Provides information about protein families involved in the transduction of signalling
7	MaizeSNPDB	http://150.109.59.144:3838/MaizeSNPDB/	For efficient retrieval and analysis of SNPs
8	ZEAMAP	http://www.zeamap.com	Provides information about multiple reference genomes, annotations, comparative genomics, transcriptomes, open chromatin regions, chromatin interactions, high-quality genetic variants, phenotypes, metabolomics, genetic maps, genetic mapping loci etc.
9	Panzea	https://www.panzea.org/	Provides genotypic and phenotypic information for several maize populations
10	MaizeNet	http://www.inetbio.org/maizenet/	Provides a genome-scale co-functional network of maize genes
11	MODEM	http://modem.hzau.edu.cn	Provides multidimensional omics data, including genomic, transcriptomic, metabolic and phenotypic information

High Throughput Phenotyping in Maize

Phenotyping of maize plants for abiotic stress responsive traits under their natural and uncertain growing/environmental conditions remains difficult owing to the high level of phenotypic plasticity of these traits. Whereas phenotyping under controlled conditions can give the detailed information of trait responses to various stresses those cannot be precisely controlled or hard to measure in the field (Rebetzke et al., 2013). Heat waves, for instance, are extremely unpredictable in their timing, and phenotyping for root-based traits, primarily associated with drought tolerance, is much more difficult in the field than in controlled conditions using lysimetric facilities, greenhouses, rhizotrons, and rainout shelters (Basu et al., 2015). As aforementioned, we have tremendous maize genomic resources available with us, but still the impact of this genomics data on maize improvement is far from satisfactory, largely owing to the lack of effective phenotyping. Therefore, the research bottleneck in maize breeding is shifting from genotyping (genomics) to phenotyping (phenomics). This effective and affordable phenotyping can bridge the genotype–phenotype gap for complex traits such as abiotic stress tolerance (Mir et al., 2019).

Various imaging techniques have been used to phenotype the whole maize plants or specific part(s) of them. Thermal imaging for leaf tissues or whole shoot (Araus et al., 2012), visible light imaging for whole plant (Nagel et al., 2012), and 3D imaging for shoot (Klose et al., 2009) have been reported in maize. However, proper caution needs to be taken while decoding the solution for abiotic stress tolerance since the controlled or regulated environment may not actually represent the actual field conditions as well as be less helpful to study the environmental interactions with genotype which are very important for getting insights into the molecular mechanisms underlying abiotic stress tolerance. Instead, dynamic phenotyping in the controlled condition may be developed to represent the actual field conditions where the weather factors are not undynamic but variable all-through the plant life cycle similar to the target production systems.

On the other hand, various remotely controlled unmanned aerial vehicles (UAVs) equipped with appropriate instruments have also been developed to phenotype directly under open field conditions. These UAVs can fly on the field and phenotype the target traits throughout the growing period. Using thermal images, aerial phenotyping provides the NDVI and red, green and blue colour space (RGB) data which can be further used to measure a series of traits such as early vigour, plant density, plant height, canopy cover, leaf size, radiation interception, leaf area index, leaf biomass, leaf senescence and grain yield etc. (e.g., Han et al., 2019; Pradawet et al., 2022; Wang et al., 2019c; Zaman-Allah et al., 2015). Moreover, recently in 2019, one study showed that VIS–NIR–SWIR (visible, near infrared and short-wave infrared range) reflectance spectroscopy can be used as a promising tool for non-destructive, low-cost, and high-throughput analysis of several leaf biochemical and physiological properties (Ge et al., 2019). Using this tool, data on six leaf properties of maize plants including leaf water content, chlorophyll content, specific leaf area, nitrogen, phosphorus, and potassium was recorded under both greenhouse and field conditions, successfully (Ge et al., 2019). This precisely recorded data can be used for accurate identification of the genes/QTLs through QTL mapping and GWAS studies which can be further employed for genomics-assisted breeding for climate resilient maize.

Genetics of Drought Stress Tolerance in Maize

Drought is one of the major challenges to improve the productivity of maize in tropical and subtropical production systems. Rainfed agriculture is the predominant mode of production in subtropical and tropical regions, with a global yield loss of about 25% (Edmeades, 2008). QTL mapping and GWAS are the major approaches to dissect the complex traits like drought tolerance in any crop. Various bi-parental populations such as RILs, F2-derived, near-isogenic lines (NILs), doubled haploid population and others which follow the principle of linkage have been used for coarse and fine mapping of QTLs for traits such as leaf temperature differences, drought tolerance index, shoot fresh weight, shoot dry weight, Normalized Difference Vegetation Index, plant biomass, anthesis-to-silking interval, grain yield, kernel number, 100-kernel fresh weight, plant height, greenness, plant senescence, root capacitance, kernels per row, number of rows per ear, ear length, leaf angle, leaf orientation value, leaf length, leaf width, leaf size and leaf shape, primary branch number and stay green etc. associated with drought tolerance in maize (e.g., Abdelghany et al., 2019; Ajnone-Marsan et al., 1995; Austin and Lee, 1996; Chen et al., 2017; Jiang et al., 2015; Messmer et al., 2009; Zhao et al., 2018b).

The major challenge faced by crop breeders in using QTL mapping results are their dependency upon the population genetic backgrounds and phenotyping environment that restricts their utilization in greater range of environments and populations. To overcome this issue various QTL meta-analysis studies have been conducted in recent decades (Hao et al., 2010; Liu et al., 2019; Zhao et al., 2017, 2018b). These studies revealed the stable and consensus QTLs by merging various QTLs from different experiments irrespective of their population types, genetic backgrounds, evaluated years and locations. Moreover, these analyses further refined the position of QTLs and narrowed down the confidence intervals that may lead to accuracy of MAS. For instances, using QTL meta-analysis, Hao et al., identified 39 consensus QTLs and 48 candidate genes under drought stress. Twenty stable QTLs and 34 candidate genes related to inflorescence development and drought tolerance were identified in maize (Zhao et al., 2017). Later in 2018, twenty-one meta-QTLs and 24 candidate genes controlling leaf architecture traits were also identified (Zhao et al., 2018b). More recently in 2019, 24 meta-QTLs and 47 candidate genes related to grain yield, flowering date and plant height were identified in maize under stress conditions (Liu et al., 2019).

In addition to QTL and QTL meta-analysis, many GWAS studies have also been conducted for identifying marker trait associations (MTAs) in maize under drought conditions (e.g., Guo et al., 2020a; Wang et al., 2019a; Xue et al., 2013). These studies helped in reducing the time taken for population development and provided an opportunity to test more alleles at the same time. Moreover, GWAS is somewhat similar to fine mapping approach since it has the ability to identify genes when genome-wide SNPs associated with the traits are used in genetically diverse genotypes or association panels. For instances, a collection of 210 off-PVP maize inbred lines were genotyped and phenotyped in eight environments in China and a total of 696 traits associated SNPs under the drought conditions were detected in a GWAS analysis, with the phenotypic variation explained by each SNP to the target traits ranged from 10.02 to 25.40% (Wang et al., 2019a). Recently in 2020, using 209 diverse maize accessions, a total of 62 loci for seminal root traits were observed. Further by integrating these GWAS results, co-expression network analysis and the differentially expressed genes, a total of 7 promising candidate genes were also identified (Guo et al., 2020a).

Unfortunately, unequal allele frequencies in the individuals of GWAS panel produce large number of false positives which is considered as a major limitation of GWAS approach. In case of maize, use of NAM (Nested Association Mapping) population was proposed to overcome the limitations of QTL mapping and GWAS or association studies (Yu et al., 2008). It is a population created by systematic crossing of a common line with many founder lines which provides an opportunity to exploit both linkage-disequilibrium and linkage models to map QTLs. This mapping population has been used to identify QTLs associated with drought tolerance in maize (Li et al., 2016a, b). Against this NAM population, a balanced multi-parent cross design i.e., MAGIC (multi-parent advanced generation intercross) population has also been developed in maize with higher resolution, power and elevated minor allele frequency (Kover et al., 2009). Unfortunately, the current statistical models based on these multi-parent populations still have the limitation of including minor alleles in the analysis and therefore new more advanced statistical models are needed to include minor or rare alleles present among the individuals to understand their role in trait phenotype.

Genetics of Heat Stress Tolerance in Maize

Heat stress or high temperature reduces photosynthetic efficiency and thereby increases kernel abortion and reduces starch storage which ultimately results into significant loss in kernel yield (Edreira and Otegui, 2013). Tassels or male flowers are highly sensitive to heat stress (Chen et al., 2012). The effects of heat stress on foliar tissues of the plants are not constant and depending on the intensity of the stress and genotype, the consequences may vary from a temporary reduction in photosynthetic assimilates to extensive tissue and death of the plants (Chen et al., 2012). In addition, heat stress also worsens yield losses caused by drought. Various optimistic models have predicted that heat stress events will become more frequent in the future owing to changes in climate (Coumou and Robinson, 2013). A few QTL mapping studies have been conducted in maize for heat tolerance (Frey et al., 2016; Gao et al., 2019; McNellie et al., 2018; Van Inghelandt et al.. 2019). This lack of genetic mapping studies is understandable, as heat stress is sporadic and of variable severity, which makes heat a challenging stress to study. Heat stress often accompanies drought stress; thus irrigation is mandatory to decrease the confounding effect of drought. Dent genotypes are more tolerant to heat stress compared to Flint genotypes (Frey et al., 2016). Using intra- and interpool Flint and Dent populations, a total of 11 QTLs, six heat-tolerance and 112 heat-responsive candidate genes were identified (Frey et al., 2016). A total of 22 QTLs for two foliar traits, leaf firing and leaf blotching (major components of yield reduction after stress) explaining phenotypic variation from 3.21% to 13.03%, and a single QTL for tassel blasting were identified using two RIL populations (McNellie et al., 2018).

In another study, a total of six QTLs for the heat susceptibility index of the five of the nine studied traits viz., leaf senescence of old leaves, leaf scorching of young leaves, length of the fourth leaf, number of leaves, the plant height, shoot dry weight and the shoot water content, were identified. As, these identified QTLs explained a very small proportion of the phenotypic variance (PVE = 7 to 9%), so, MAS may not be promising for these traits (Van Inghelandt et al., 2019). Recently in 2019, a total of four QTLs for thermo-tolerance of seed-set in maize were identified using a RIL population (Gao et al., 2019). A total of 607 heat responsive genes as well as 39 heat tolerance genes were identified using eight genotypes through GWAS. Recently in 2019, using GWAS, a total of 17 genes associated with 42 SNPs related to thermo-tolerance of seed-set were identified. In addition, near-isogenic lines (NILs) having four out of these 17 genes were developed in a susceptible parent background which showed significant tolerance to heat stress on seed setting stage. Expression profiles of these genes showed that they were induced by heat stress in maize tassels (Gao et al., 2019). More recently in 2022, a total of 66, 27, and 24 SNPs associated with the traits evaluated under combined heat and drought and terminal drought conditions were identified (Osuman et al., 2022). Moreover, the use of genome-wide prediction model to select for heat tolerance was recommended at seedling stages as highest prediction abilities using these models were observed at this stage specifically for the within-population calibrations (Van Inghelandt et al., 2019).

Genetics of Salinity Stress Tolerance in Maize

Salinity stress is another major problem in maize growing regions, and it has been observed that above a specific threshold value (1.7 dS/m) the maize yield reduces as a linear function of salinity until death of plants (Falcon & Naylor, 1998). In general, salinity stress imposes mainly four types of stresses on plants: specific ion toxicities (e.g., Na^+ and Cl^-), osmotic stress, developmental disturbance and ionic imbalance (e.g., Na^+ versus Ca^{2+}; Na^+ versus K^+) (Grattan & Grieve, 1999). Under field conditions, direct selection of superior salinity tolerant genotypes has been challenged by the significant interference of environmental factors. Salinity stress is generally accompanied by alterations in other soil chemical and physical properties such as high pH and sodicity which makes breeding progress more difficult. Owing to the underlying complex genetics of salinity stress tolerance and the troubles in phenotypic selection, the approach of QTL mapping and GWAS has come into the limelight in recent decades as they provide the DNA markers to be used in the improvement of stress tolerance. Several QTLs related to salt tolerance have been identified in different maize populations (Cui et al., 2015; Luo et al., 2017, 2019a; Wang et al., 2012b; Zhang et al., 2019b). For the first time, a total of six QTLs were identified for salt tolerance at seedling stage in maize (Wang et al., 2012b). Followed by this first report, a number of reports have been published where numerous QTLs have been identified for different traits associated with salinity tolerance: 20 additive and 9 epistatic QTLs for traits including germination rate, salt tolerance ranking, shoot dry weight, shoot fresh weight, tissue water content, shoot K^+ concentration, shoot Na^+ concentration, and shoot K^+/Na^+ ratio (Cui et al., 2015); 3 major QTLs for traits including plant height in saline field and ratio of plant height between control and saline fields (Luo et al., 2017); 1 major QTL (*ZmNC1*) which encodes an HKT-type transporter (Zhang et al., 2019b); 1 major QTL (*ZmHKT2*) associated with K^+ homeostasis in saline soils (Cao et al., 2019) and a total of 41 QTLs for salinity tolerance traits at seedling stage such as shoot length, root length, shoot fresh weight, root fresh weight, shoot dry weight, root dry weight and salt tolerance index (Luo et al., 2019a).

Another powerful way to study the genetic mechanisms controlling complex traits is GWAS, although this technique has not been widely used for getting insights into the underlying complex genetics of salinity tolerance in maize. For the first time in 2019, using 445 maize accessions, a total of 57 loci and 49 candidate genes significantly associated with salt tolerance were identified. A total of 44% of the candidate genes were found to be involved in stress responses such as stomata division, ABA signalling, auxin signalling and DNA transcription regulation, indicating that they are main genetic mechanisms of maize salt tolerance (Luo et al., 2019b). In another study, using 150 maize inbred lines, a total of 7 SNP loci and 8 candidate genes, significantly correlated with plant height change rate and fresh weight change rate were identified. Four of these candidate genes were validated by real time quantitative PCR (Xie et al., 2019). A gene i.e., *ZmNC2* was also identified which encodes for a HAK family ion transporter i.e., *ZmHAK4*. This ion transporter provides natural variation for shoot Na^+ exclusion and salt tolerance (Zhang et al., 2019b). Recently in 2020, a total of 57 SNPs and several candidate genes associated with early vigour traits were identified using a total of 399 inbred lines. Out of these, eight candidate genes were further validated using expression analysis (Sandhu et al., 2020). More recently in 2022, a total of 259 highly significant MTAs and several important candidate genes associated with salinity stress tolerance were detected using a panel of 305 diverse maize inbred lines (Zaidi et al., 2022).

Genomics-Assisted Breeding for Developing Climate Resilient Maize

Development of climate-ready varieties with improved resilience to environmental stresses is one important approach to mitigate the yield reduction under stresses because genetic improvement can probably close the yield gaps between stress-affected and optimal conditions. For instance, use of drought tolerant varieties can close around 20–25% of the yield gaps between optimal and drought-affected conditions (Edmeades, 2013). In last decades, much breeding research has been conducted to enhance the performance under stress conditions, with some achievements (Campos et al., 2004). For instance, at CIMMYT, conventional selection for drought in tropical maize populations by concentrating on yield and associated traits has resulted in a gain of approx 100 kg/ha/year (Edmeades, 2013). However, owing to the decreased heritabilities of phenotypes under various stresses, the breeding progress for stress tolerance improvement has been very slow in previous years (Messmer et al., 2009).

Although several QTL mapping experiments have been conducted in maize, little has been published on the successful introgression of QTLs for any stress tolerance. We could find only one study on marker assisted introgression for stress tolerance, where using marker assisted backcrossing programme (MABC), five QTLs associated with various component traits of heat tolerance were introgressed in maize. Compared to control plants, higher mean grain yield of these MABC derived hybrids were observed under severe drought condition (Ribaut & Ragot, 2007). Abiotic stresses are usually the combination of several quantitative traits with a high magnitude of epistatic and environmental interactions. The quantification of phenotypic variation contributed by the QTLs and their possible interactions with other QTLs and environment is a major issue as this is generally confounded with selection of component traits, study design, phenotyping errors, marker coverage, models employed for QTL mapping and so on. Apart from this, although true QTLs are identified, practically, following too many QTLs through MAS remains a discouraging task. Transferring one or two major QTLs may not provide the expected level of phenotypic expression as QTL detected with epistatic interaction may lose the effect in the absence of its counterpart QTL(s). For a successful MAS programme, all above mentioned factors need to be considered while planning QTL mapping experiments.

Marker Assisted Recurrent Selection (MARS), an improvement over MAS, allows concurrent identification and stacking of favourable alleles at a large number of the loci for improving the quantitative traits. It can be either used by directed recombination of the chosen genotypes of a segregating population or by inter-mating the marker genotypes in random. MARS has been successfully employed in maize to improve the drought tolerance (Beyene et al., 2015). In this study, overall gain using MARS across the10 bi-parental tropical maize populations was reported 105 kg/ha/year and 51 kg/ha/year under well watered and water stressed conditions, respectively (Beyene et al., 2015). In another study, the total number of desirable alleles for the drought tolerance increased from 114 in C_0 cycle to 124 in C_3 cycle in maize (Abdulmalik et al., 2017). The frequency of desirable alleles for drought tolerance increased from 0.510 at C_0 cycle to 0.515 at C_2 with average genetic gain of 3% per year by employing MARS in maize (Bankole et al., 2017). Above results demonstrate the potential of MARS for increasing genetic gain under drought stress condition.

Improvement of populations for stress specifically drought tolerance through genomic selection (GS) approach has also been reported in maize. GS simultaneously computes all marker effects to predict total genetic value and by doing so, variation is captured that might not have detected above a defined threshold using conventional statistical approaches. GS works by leveraging whole genome dense marker data so that each QTL/gene is in LD with a marker. Higher prediction accuracy compared to variance explained by the sum of QTL for individual traits has been observed with GS models in maize for drought tolerance (Cerrudo et al., 2018). The average gain from GS per cycle across eight bi-parental populations was 0.086 Mg/ha. And the average grain yield of C_3 cycle–derived hybrids was found to be significantly higher compared to the hybrids derived from C_0 cycle (Beyene et al., 2015). Moreover, C_3 cycle–derived hybrids produced 7.3% higher grain yield compared to those developed via the conventional pedigree breeding method (Beyene et al., 2015). Prediction accuracies of seven genomic selection models (Bayes A, Bayes B, ridge regression, LASSO, random forest, elastic net, reproducing kernel Hilbert space) have been tested for different agronomic traits under drought condition (Shikha et al., 2017). The test crosses of C_2 populations (predicted using GEBVs) showed superiority of 10 to 20% over C_2 (selected based on phenotypic selection) of respective populations and top crosses of C_2 (predicted using GEBVs) showed 4 to 43% superiority of grain yield over that of C2 (selected based on phenotypic selection) of respective populations (Vivek et al., 2017). Recently in 2019, high prediction accuracies were also observed with GS for six different traits associated with drought tolerance (Wang et al., 2019a).

Transcription factors (TFs) may regulate numerous genes that are associated with abiotic stress tolerance. Therefore, a comprehensive study of all TFs related with abiotic stress tolerance regulatory mechanisms in maize may be significantly rewarding. Numerous TFs belonging to different families have been characterized in cereals mainly in maize, rice, and wheat (reviewed in Gahlaut et al., 2016; Kimotho et al., 2019). The detailed information regarding the main transcription factor (TF) families such as MYC/MYB regulons, AP2/EREBP regulons, NAC TFs and regulons, bZIP TFs: AREB/ABF regulon, WRKY TFs and WRKY regulons, Homeodomain-leucine zipper I, heat shock proteins, NF-Y transcription factors, and the interactions of these TFs with the cis-acting elements which are present in the promoter regions of stress responsive genes have been discussed elsewhere (Gahlaut et al., 2016; Kimotho et al., 2019). Genetic engineering of multiple stress regulatory TF genes can be a best approach for the improvement of stress tolerance in plants when compared to concentrating on a single gene alone. Although, identification of commercial genetically engineered or transgenic plants with increased tolerance to various abiotic stress is an expensive, tedious, and lengthy process. Two transgenes, *betA* (encoding choline dehydrogenase) and *TsVP* (encoding V-H+-PPase) were pyramided in maize by cross pollination. Generated transgenic plants exhibited higher H+-PPase activity and glycine-betaine contents compared with the parental lines and contained greater solute accumulation, higher relative water content, and lower cell damage under drought stress condition. These pyramided plants had less growth retardation, shorter silking interval and produced higher yield compared to parental lines (Wei et al., 2011). In addition, over-expression of gene *OsMYB55* in maize reduced the harmful effects of heat and drought stresses resulting in improved plant performance under stressed conditions (Casaretto et al., 2016). More recently in

2019, genetic engineering studies showed that two genes *SAG6* (*GRMZM 2G106056*) and the salt-tolerance-associated gene 4 (*SAG4, GRMZM2G077295*), which encode a double-strand break repair protein MRE11 and protein transport protein, respectively, has significant roles in maize salt tolerance (Luo et al., 2019). Different generalized steps involved in genomics-assisted breeding for climate resilient crop are shown in Fig. 3.2.

Moreover, targeted genome editing using zinc finger nucleases (ZFN) or transcription activator-like effector nuclease (TALEN) or clustered regularly interspaced short palindromic repeats-associated endonucleases (CRISR/Cas) is also becoming versatile tools for crop improvement. Among these three genomes editing technologies, CRISR/Cas has high efficiency, and therefore it is being widely used in genome engineering (Belhaj et al., 2015). For the first time in 2014, targeted genome editing

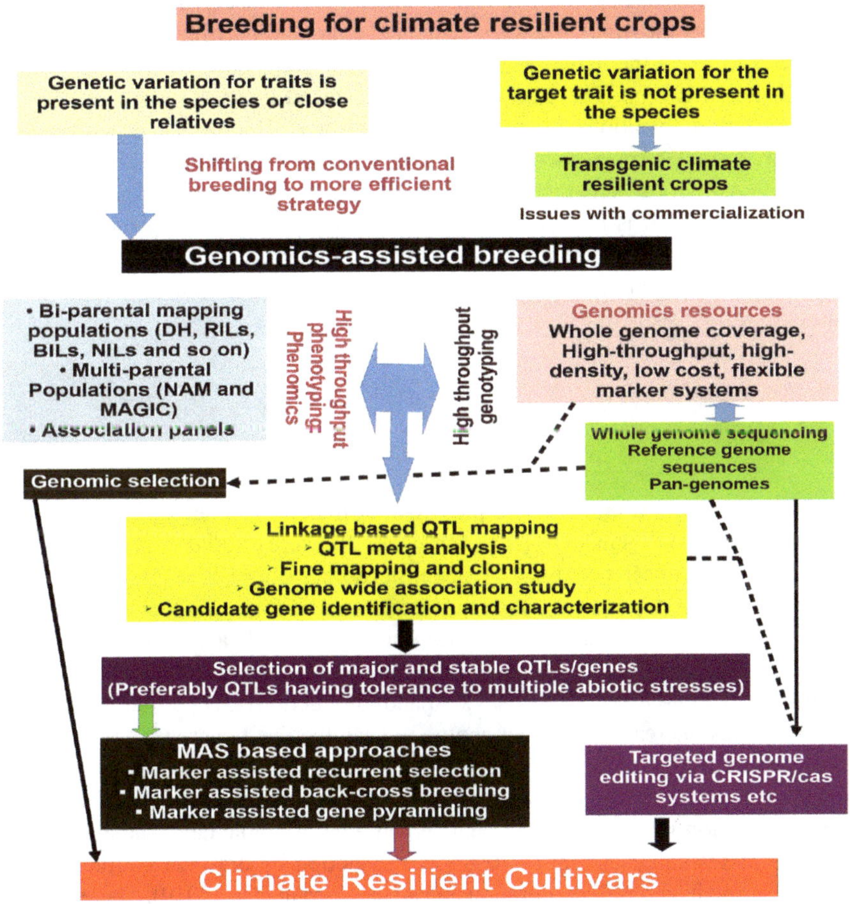

Fig. 3.2 Diagrammatic representation of different steps involved in genomics-assisted breeding for climate resilient crops

by CRISPR/Cas9 was reported in maize (Liang et al., 2014). However, choosing the right candidate genes is crucial for the success of achieving the desired phenotype. These genes can be grouped into two major categories, regulatory and structural genes. Regulatory genes act indirectly by regulating the expression of other genes participating in different cellular processes, whereas structural genes proteins confer tolerance to stresses directly. Moreover, *cis*-regulatory sequences may also be a good choice for achieving tolerance to target stress. Focussing on these structural and/or regulatory along with *cis*-regulatory sequences through CRISPR/cas9 system can be an efficient, robust, and practical approach for developing climate resilient varieties of maize. For instance, CRISPR-based editing of gene such as *ZmHKT2* may provide a practicable strategy for improving salt tolerance in maize. *ZmHKT2* is known to reduce shoot K^+ content by removing K^+ ions from flowing xylem sap (Cao et al., 2019). Creation of novel gene or allele for abiotic stress tolerance via the CRISPR/Cas-mediated editing of promoters is also feasible. For instance, using CRISPR/cas9, novel variants of the *ARGOS 8* were generated either by inserting a novel promoter into the 5′-UTR of the native *ARGOS 8* gene or by replacing the native promoter of the concerned gene. These novel variants produced increased grain yield under flowering stress conditions compared to the wild type. The *ARGOS 8* is a negative regulator of ethylene responses and plants having over-expression of this gene show reduced sensitivity to ethylene and increased grain yield under stress conditions (Shi et al., 2017). It is believed that these targeted genome editing tools will not only significantly contribute towards raising novel plant types having resilience to changing climatic factors but will also help in social acceptance of these products in future.

3.2.2 Rice

Rice (*Oryza sativa* L.), a major staple food for human population, belongs to the family Poaceae and genus *Oryza* (Kaur et al., 2022a, b). It has a genome of size approximately 430 Mb contained in 12 chromosomes (2n = 2x = 24). In 2021, the world's total rice production was estimated around 756 million tonnes. The top 5 rice growing countries (China, India, Indonesia, Bangladesh, and Viet Nam) account for around 72% of the total production (https://knoema.com/atlas/topics/Agriculture/Crops-Production-Quantity-tonnes/Rice-paddyproduction). As per the predictions, rice yield must increase by 1% per annum to keep feeding the growing population.

The abiotic stresses such as cold, drought, heat, submergence, and salt are the major concerns for rice production, and an increasing danger to food security considering population increase, climate change and area of arable land available. These abiotic stresses can cause up to 70% yield reduction by drastically affecting rice survival, growth, grain filling duration and so on (Akram et al., 2019). Therefore, the development of high yielding stress tolerant or climate resilient cultivars is necessary to achieve the target production. As climate change continues to be a leading contributor to stresses in rice production areas, knowing current effective research

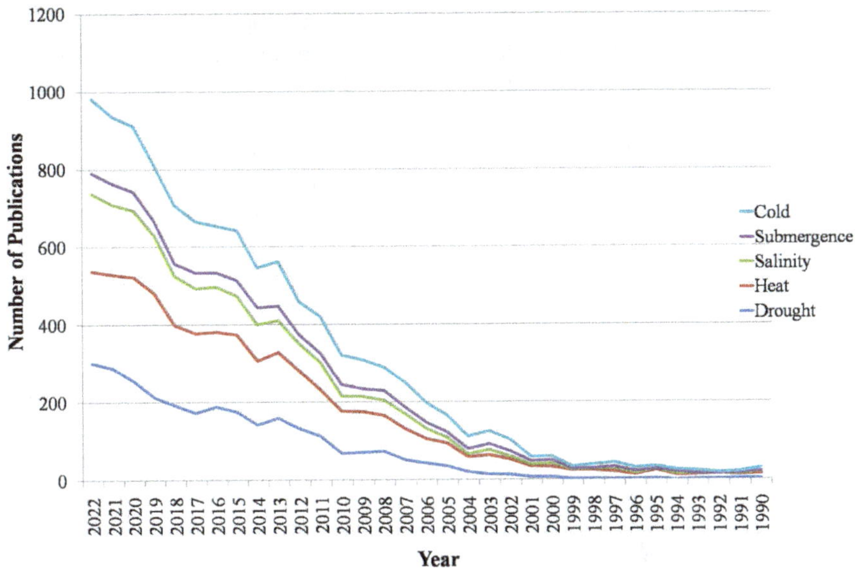

Fig. 3.3 The number of publications per year related to rice abiotic stresses from 01/01/1990 to 15/12/2022. *Source* PubMed (Keywords: rice AND drought; rice AND heat; rice AND salinity; rice AND submergence; rice AND cold) were used to search the number of publications at https://pubmed.ncbi.nlm.nih.gov/)

techniques in rice is necessary. A multidisciplinary approach to introgress the vital QTLs/genes conferring tolerance to these stresses together with selection for high yield under both stressed and optimum conditions would assist to combat with this drastic situation (Muthu et al., 2020; Sandhu et al., 2019). The amount of research studies has increased rapidly related to drought, heat, salinity, submergence, and cold stresses in rice in recent years (Fig. 3.3).

Genomic Resources Available for Rice

Rice has a vast treasure of genomic resources that can hasten the development of useful products including climate resilient rice. Owing to its smaller genome size, availability of gold standards of the reference genome sequence, vast genetic and genomic resources compared to any other crop, rice has emerged as model crop in the twenty-first century (Tyagi & Khurana, 2003). Draft rice genome sequences of two rice subspecies, *O. Sativa* subsp. japonica and indica, using Nipponbare and 93–11 varieties, respectively, were generated using improved de novo sequencing technologies (Goff et al., 2002; Yu et al., 2002). The whole genome sequencing of these varieties has been used for the sequencing of several genotypes because of their high sequence quality. The availability of these high-quality reference genomes has played a significant role in the mining of useful traits by capturing the allelic variants, copy number variations, and presence-absence variants along with the dissection of various quantitative traits (Huang & Han, 2014). The availability of vast information on the rice genomics assisted not only in understanding the 30,000 to 50,000 predicted and annotated rice genes but also in unravelling the genetic interaction regulating the vital agronomical, morphological, and physiological traits under varied environmental conditions. Millions of SNPs and InDels have been identified in rice which have been used for the dissection of complex traits such as abiotic stress tolerance (Frouin et al., 2018; Naveed et al., 2018; Warraich et al., 2020). Moreover, SNP assays/chips have also been developed at various densities, with instances including the 44 K-SNP chip, RICE6K, 50 K SNP chip, 700 K-SNP, C6AIR (Cornell-IR LD Rice Array) and more recently the C7AIR and 580 K SNP arrays (reviewed in Kim et al., 2022). These SNP genotyping systems vary in their informativeness, marker density, cost, data quality, flexibility, and speed. Data generated using these SNP chips can be used for GWAS, DNA fingerprinting, differentiating sub-population groups and genetic diversity analysis.

In 2018, a pan-genome dataset of the *O. sativa–O. rufipogon* species complex was constructed via deep sequencing and de novo assembly of 66 different accessions/genotypes. Around 23 million sequence variants were then identified in the rice genome via inter-genomic comparisons (Zhao et al., 2018a). More recently in 2020, a platinum standard pan-genome resource was constructed to represent the population structure of Asian rice. This generated platinum Standard RefSeqs may be exploited as a template to map re-sequencing data to detect natural variation that exists in the pan genome of Asian rice (Zhou et al., 2020a). An international effort was also made to re-sequence a core collection of 3000 rice accessions collected from 89 countries (3000 Rice Genomes Project, 2014). These 3000 rice accessions were sequenced at the average sequencing depth of 14× and around 18.9 million SNPs were discovered after aligning to the reference genome sequence. Later a Rice Pan-genome Browser (RPAN) tool was also developed to search and visualize this huge data (http://www.rmbreeding.cn/pan3k). This data can serve as a base for large-scale discovery of novel alleles for important traits using various genetic and/or bioinformatics approaches. Other genomic resources available in rice for functional genome analyses has been reviewed elsewhere (Yang et al., 2013b). Table 3.2 lists some representative rice websites that are freely accessible to the rice research community and others.

High Throughput Phenotyping in Rice

The field of HTP is rapidly evolving in rice. There is need to assure that these advances in HTP find applicability in rice breeding programmes and contribute toward improved genetic gain. Present day HTP platforms include several methodologies that employs remote sensing to get non-destructive phenotypic measurements, either in the field or in controlled environments. The most common types of sensors for rice phenotyping include red–green–blue (RGB; Hairmansis et al., 2014), hyperspectral (Bodner et al., 2018), multispectral (Zhang et al., 2018a), thermal (Sagan et al., 2019), fluorescence (Pérez-Bueno et al., 2016), laser-imaging detection and

Table 3.2 List of major bioinformatics tools and database resources for genomics studies in rice

S. No.	Name of the database	Website	Features
1	MSU Rice Genome Annotation Project	https://rice.plantbiology.msu.edu/	Provides sequence and annotation data for the rice genome
2	The Rice Annotation Project	https://rapdb.dna.affrc.go.jp/	Provides an accurate and timely annotation of the rice genome sequence
3	Oryzabase	https://shigen.nig.ac.jp/rice/oryzabase/	Comprehensive of classical rice genetics to recent genomics and from fundamental information to hot topics
4	Rice Expression Database	https://expression.ic4r.org/	An integrated RNA-Seq-derived gene expression database
5	Rice ExpressionProfile Database	https://ricexpro.dna.affrc.go.jp/	Genes expression profiles from microarray analysis
6	RiceVarMap v2.0	https://ricevarmap.ncpgr.cn/v2/	Comprehensive studies of genomic variations and its functional annotations from the sequencing of 4,726 rice accessions
7	TIGR Rice Genome Project BLAST	https://blast.jcvi.org/eukblast/index.cgi?project=osa1	BLAST server for nucleotide and protein alignments
8	OryGenesDB	https://orygenesdb.cirad.fr/data.html	Compare the annotations of Gene loci and models from MSU RGAP and RAP-DB
9	Rice SNP-Seek Database	https://snp-seek.irri.org/	Provides Genotype, Phenotype, and Variety Information for rice
10	Information Commons for Rice	https://ic4r.org/	Provides rice genome sequence, updated rice gene annotations and integrated multiple omics data
11	RPAN: Rice Pan-genome Browser	https://cgm.sjtu.edu.cn/3kricedb/	Provides information regarding the presence/absence variations (PAVs) of genes among 3000 rice genomes
12	RiceGE: Rice Functional Genomic Express Database	https://signal.salk.edu/cgi bin/RiceGE	Provides graphical representation of annotated genes, SNPs, and PAVs on the chromosomes
13	Rice TE Database	https://www.genome.arizona.edu/cgi-bin/rite/index.cgi	Transposable and repeated elements in the genome sequence
14	RiceRelativesG	http://ibi.zju.edu.cn/ricerelativesgd/	Provides large-scale genomic resources from 2 cultivated rice and 11 rice relatives, including 208,321 specific genes and 13,643 genes related to stresses and so on

ranging (LiDAR) (Sun et al., 2018) devices and three-dimensional (3D; Teramoto et al., 2020). Image analysis has great potential to be used for high throughput screening methods in the development of climate resilient rice. For instance, a non-destructive image-based phenotyping protocol has been successfully used to assess salinity tolerance in rice (Hairmansis et al., 2014). Various advanced phenotyping and photonics-based technologies used in rice breeding have been reviewed elsewhere (Yang et al., 2013a). If we talk about advances in root phenotyping, up to year 2015, more than forty image processing software packages for root system measurement were proposed depending on various methods, objectives, and equipment. More recently in 2018, an integrated software named as 'GT-RootS' was developed for automated root system measurement from HTP platform images (Borianne et al., 2018).

Genetics of Drought Stress Tolerance in Rice

Rice is considered as one of the least water-efficient crops because of its requirement of huge amount of water (3000 to 5000 L) to produce very less yield (only 1 kg) (Bouman, 2002). Drought or water deficit stress has been identified as an important problem in upland and rain-fed lowland rice (Fukai & Cooper, 1995). The Genetics of the drought stress tolerance is highly complex in nature and connected with numerous quantitative traits including various biochemical and physiological traits. This stress may come at any growth stage of the rice crop. When it occurs at the vegetative stage of growth, it drastically affects the growth of leaves and elongation of stem. When it comes between the rainfalls, also known as intermittent drought, it affects the development of root system. When it occurs at the end of the crop growing period especially during the flowering stage, also known as terminal drought, it affects the spikelet fertility and grain filling. Large scale systematic studies with different types of mapping populations for identification of QTLs using grain yield as a selection criterion have led to the identification of hundreds of QTLs for grain yield under drought stress and traits associated with drought tolerance in rice (Table 3.3). In addition to QTL analysis, a few meta-analyses have also been conducted to identify MQTLs for yield and other traits under drought stress which represent the hotspot regions for drought stress tolerance in rice (Khahani et al., 2021; Trijatmiko et al., 2014). Moreover, few GWAS studies have also been conducted for dissecting the complex genetic architecture of drought tolerance in rice (Guo et al., 2018; Jiang et al., 2021; Ma et al., 2016). For instance, recently in 2021, a GWAS study was conducted where a total of 111 MTAs were identified for UAV-based leaf-rolling score, plant water content and a new composite trait, drought resistance index by UAV.

Table 3.3 QTLs and MTAs reported for different traits associated with drought tolerance in rice

S. No.	Parents	Population size and type	Number and types of markers	PVE (%)	Chromosome	Number of QTLs	Traits and/or growth stage	References
1	IR64/Azucena	56 DHs	175 RFLPs	16.1	12	1	Drought index	Hemamalini et al. (2000)
				19.8	4	1	Tiller number	
2	CT9993/ IR62266	154 DHs	315 RFLPs, AFLPs, SSR	3.3–17.0	3, 4, 12	5	Root length	Zhang et al. (2001)
				3.6–20.2	1, 2, 4, 6, 9, 10,12	8	Root weight	
				8.5–31.3	1, 2, 3, 4, 6, 7, 8, 9, 12	13	Root thickness	
3	Caiapo/*O. rufipogin*	274 BC_2F_2	125 SSLP and RFLP		11	1	Grain weight	Moncada et al. (2001)
					6	1	Panicle number	
4	IR64/Azucena	90 DHs	260 markers	17.1–52.6	1	5	Plant height	Venuprasad et al. (2002)
5	IR64/Azucena Azucena/Bala	85 DHs		17.1–52.6	1	5	Plant height	Lafitte et al. (2002)
				7.8	3	2	Panicle number	
					3	1	Panicle length	
6	IR64/Azucena	258 BC_2F_2	153 SSR and RFLP	5.9–11.3	1, 2, 3, 5, 12	6	Grain weight	Thomson et al. (2003)
				7.1–21.2	1,2,4	3	Panicle length	
				7.8	3	2	Panicle number	
				7.1–18.6	1, 4, 5, 6, 10	5	Seed setting rate	

(continued)

Table 3.3 (continued)

S. No.	Parents	Population size and type	Number and types of markers	PVE (%)	Chromosome	Number of QTLs	Traits and/or growth stage	References
7	Azucena/Bala CT9993/ IR62266	96 RILs			5	1	Panicle number	Lafitte et al. (2004)
				7.3–15.5	1, 3, 4, 6, 10, 11	9	Grain yield	
				14.8–19.8	4	11	Biomass	
8	IRAT109/ Yuefu	116 DH	RFLP and SSR	2	2	1	Drought index	Li et al. (2005)
9	Zhenshan 97/IRAT109	187 RILs	213 SSR	9.7	2	1	Grain weight	Zou et al. (2005)
				9.17	4	1	Panicle number	
10	Teqing/Lemont	254 ILs	160SSR		11	1	Grain yield	Xu et al. (2005)
11	CT9993/ IR62266	105 DHs	315 AFLPs	14.8–19.8	4	11	Biomass	Kumar et al. (2007)
				7.3–15.5	1, 3, 4, 6, 10, 11	9	Grain yield	
				7.3–15.5	1, 3, 4, 6, 10, 11	9	Grain yield	
12	Way Rarem/ Vandana	436 F_3	126 SSR	18	12	1	Biomass	Bernier et al. (2007)
				37	12	1	Drought index	

(continued)

Table 3.3 (continued)

S. No.	Parents	Population size and type	Number and types of markers	PVE (%)	Chromosome	Number of QTLs	Traits and/or growth stage	References
				33	12	1	Grain yield	
13	Indica/Azucena	RILs	256 RFLP, AFLP & SSR	10–48	1, 4, 8, 9, 11	33	Seedling drought resistance	Zheng et al. (2008)
14	Bala/Azucena	176 RILs	157 RFLP, AFLP & SSR	5–28	1, 2, 3, 5, 7, 9, 10, 11	22	Leaf dimensions, transpiration and leaf rolling	Khowaja and Price (2008)
15	IR20/Nootripathu	250 RILs	101 SSR	5.0–52.2	1, 2, 9, 8	10	Plant height	Gomez et al. (2010)
				4.8–10.0	2, 7, 10	6	Seed setting rate	
				4.9–18.4	2, 4	3	Drought index	
				4.8–32.2	2, 4, 7	6	Canopy temp	
				6.3–36.8	1, 2, 4, 6	8	Biomass	
16	N22/Swarna	292 F_3	140 SSR markers	8.6–20.1	1, 3	2	Biomass	Vikram et al. (2011)
				3.2–16.9	1, 2, 3, 10	6	Grain yield	
				32.6–53.5	1	3	Plant height	
17	N22/IR64	289 F_3	140 SSR	22.6	1	1	Biomass	
18	N22/MTU1010	362 F_3	125 SSR	30.3	1	1	Biomass	
19	CT9993/IR62266	135 DH	145 (RFLP), 17 (SSR), and 153 (AFLP)	14.8–19.8	4	11	Biomass	Sellamuthu et al. (2011)
				22.7	7	1	Drought index	
				7.3–15.5	1, 3, 4, 6, 10, 11	9	Grain yield	

(continued)

Table 3.3 (continued)

S. No.	Parents	Population size and type	Number and types of markers	PVE (%)	Chromosome	Number of QTLs	Traits and/or growth stage	References
				36.6	11	1	Panicle length	
				12.6–30.2	1, 5	3	Panicle number	
				14.7	5	1	Seed setting rate	
				9.5–15.2	1, 11	2	Plant height	
20	IR64/ INRC10192	140 RILs	412 SSR	4.6–20.2	7, 8, 10	3	Biomass	Srividya et al. (2011)
				2.3	2	1	Grain weight	
				5.7	8	1	Grain yield	
				3.9–15.7	6,8	2	Seed setting rate	
				5.4–12.5	1, 5, 7, 8	5	Plant height	
21	Aday Sel/IR64	242 BILs	124 (SSR) markers	4.4–10.2	2, 9	4	Grain yield	Dixit et al. (2012)
	Apo/Swarna			2.2–6.9	2	2	Grain yield	
22	Dhagaddeshi/ Swarna	259 RILs	294 SSR	22.6–22.8	1	2	Grain yield	Ghimire et al. (2012)
				41.8–49.1	1	3	Plant height	
23	Dhagaddeshi/ IR64	251 RILs	269SSR	6.2–9.2	1	2	Grain yield	
				12.8–18.1	1	3	Plant height	
24	IR74371/ Sabitri	294 BILs	682 (SSR)		12	1	Biomass	Mishra et al. (2013)
				3.8–7.5	2, 3, 12	3	Grain yield	
					12	1	Panicle number	

(continued)

Table 3.3 (continued)

S. No.	Parents	Population size and type	Number and types of markers	PVE (%)	Chromosome	Number of QTLs	Traits and/or growth stage	References
				1.1	12	1	Plant height	
25	Guanghui 116/LaGrue	307 RILs	119 (SSR)	6.9–12.3	2,5	2	Grain weight	Zhou et al. (2013)
				6.7–9.7	5, 8	3	Grain yield	
				10.0–18.7	3, 4, 5, 6, 9, 12	7	Seed setting rate	
26	Swarna/ WAB450	202 BILs	134 SSR	14.6–24.8	3, 6	4	Biomass	Saikumar et al. (2014)
				16.9	3	1	Canopy temp	
				7.8–47.9	3,6	5	Flowering time	
				13.0–38.0	3,6,	4	Grain yield	
				18.7–27.0	3	3	Plant height	
26	Kali Aus/ 2*IR64			5.0–9.0	1, 2	5	Grain yield	
	IR55419-04/ 2*TDK1	365 $BC_1F_{3:4}$	600 SSR	4.6–35.6	3, 6	21	Grain yield	Dixit et al. (2014)
27	Apo/3*IR64			4.5	2	1	Plant height	Sandhu et al. (2014)
28	IR64 and Kinandang Patong	26 CSSLs	394 SNPs, 125 SSR	10.2–21.9	6,7	3	Deeper rooting	Uga et al. (2015)

(continued)

Table 3.3 (continued)

S. No.	Parents	Population size and type	Number and types of markers	PVE (%)	Chromosome	Number of QTLs	Traits and/or growth stage	References
29	Teqing/Lemont	224 Teqing background introgression lines	SNP	3.9–8.6	1, 2, 3, 8, 9, 11, 12	13	Spikelet number per panicle; seed fertility; filled grainweight per panicle; plant height; grain yield per plant	Wang et al 2014
		175 ILs in Lemont background	SNP	2.8–8.2	1, 3, 4, 5, 7, 8, 11, 12	12		
30	Zhenshan97B/IRAT109	180 RILs	213 SSR		1, 2, 4, 7, 10	6	Deep roots	Lou et al. (2015)
31	IR64/Cabacu	154 RILS	Illumina 384-plex SNP	8.8–38.6	1, 2, 4, 8, 10	12	Yield and yield traits at the reproductive stage	Trijatmiko et al. (2014)
32	IR77298-5-6–18/2*Sabitri	294BC$_1$-derived	600 (SSR)	2.8–23.4	3	4	Grain yield under severe lowland drought over environments	Yadaw et al. (2013)
33	Kali Aus/2*IR64	300 BC$_1$F$_3$	16 SSR	2–9	2	5	Yield at reproductive stage	Palanog et al. (2014)

(continued)

Table 3.3 (continued)

S. No.	Parents	Population size and type	Number and types of markers	PVE (%)	Chromosome	Number of QTLs	Traits and/or growth stage	References
	Kali Aus/ 2*MTU1010			3–17	2	4		
34	Norungan / TKM9	101BILs	167 SSR		1, 2, 4, 8, 11, 12	23	Photosynthetic rate, transpiration rate, stomatal conductance, RWC, harvest index, plot yield	Ramchander et al. (2016)
35	Jigeng88/4 donors	72 ILs	600 SSR	17.5–65.8 wald	1, 2, 6, 7, 8, 10, 11	13	Yield related traits at reproductive stage	Cui et al. (2018)
36	CR 143-2-2/ Krishnahamsa	190 RILs	201 SSR	60.87	9	1	RWC at reproductive stage	Barik et al. (2018)
37	Cocodrie/N-22	181 RILs	4748 SNP	1.6–35.8	1, 3, 4, 8, 12	14	Root shoot traits at vegetative growth stage	Bhattarai and Subudhi (2018)
38	CR 143-2-2/ Krishnahamsa	190 RILs	401 SSR	45.79–80.18	8, 9, 12	5	Grain yield under reproductive stage drought stress	Barik et al. (2019)
39	CR 143-2-2/ Krishnahamsa	190 RILs	77 SSR	12.37–78.19	1, 3	3	Physiological traits and grain yield reproductive stage drought stress	Barik et al. (2020)

Genetics of Heat Stress Tolerance in Rice

Though, rice was originated from subtropical and tropical regions, heat stress or high temperature (more than 38 °C) can cause drastic yield loss in rice. Heat stress can affect both vegetative and reproductive stages of rice (Katiyar-Agarwal et al., 2003). However, booting, and flowering stages are the most vital stages that may cause complete sterility in rice (Shah et al., 2011). Heat stress may importantly affect the photosynthetic rate, membrane stability, hormone levels, primary and secondary metabolites, respiration and trigger the synthesis of reactive oxygen species (ROS) etc. Hundreds of QTLs for rice heat tolerance have been mapped across the 12 chromosomes using different types of populations (e.g., F2, F2:3 lines, back cross population, near isogenic lines, recombinant inbred lines, chromosomal segmental substitution lines, and doubled haploid population) and different types of molecular marker systems, such as RFLP, SSR, InDels and more recently SNPs and SNP arrays (for instance, Table 3.4).

For the first time in 2003, a QTL mapping study was conducted to map the QTLs for spikelet fertility under heat stress using a DH population derived from R 64/ Azucena cross (Cao et al., 2003). Thereafter, a number of studies using different parents and types of mapping populations have become available for heat tolerance in rice (Buu et al., 2014; Chang-Lan et al., 2005; Hirabayashi et al., 2015; Jagadish et al., 2010; Kilasi et al., 2018; Liu et al., 2017; Shanmugavadivel et al., 2017; Tian et al., 2022; Xiao et al., 2011; Ye et al., 2012, 2015; Zhao et al., 2016; Zhang et al., 2009; Zhu et al., 2017). In some of the studies, phenotyping under heat stress was performed in natural field conditions by delayed sowing (Xiao et al., 2011; Zhao et al., 2016) while others used controlled conditions where high temperature was maintained (Jagadish et al., 2010; Shanmugavadivel et al., 2017). The consistent QTLs for heat tolerance have been identified on chromosome 5 (Kilasi et al., 2018; Shanmugavadivel et al., 2017; Ye et al., 2015) and chromosome 4 (Buu et al., 2014; Chang-Lan et al., 2005; Jagadish et al., 2010; Kilasi et al., 2018; Xiao et al., 2011; Ye et al., 2012; Zhang et al., 2009). A specialized mapping population i.e., CSSLs (chromosomal segmental substitution lines) has also been used for identifying the QTLs for basal dehiscence length under heat stress in rice (Zhao et al., 2016). As different experimental designs, mapping populations, statistical analyses and environments were involved in these studies, somehow these have delivered inconsistent QTL information. In last decades, QTL meta-analysis was proposed to identify the most consistent meta-QTLs across diverse environments and mapping populations (Tian et al., 2022). Recently in 2020, a meta-analysis was performed and a total of 35 most consistent M-QTLs were identified across diverse environments and genetic backgrounds (Raza et al., 2020). Using RNA-seq based transcriptomics and microarray datasets, this study also identified a set of 45 heat-responsive genes, among which 24 were specific to reproductive tissues. Most importantly, all these genes corresponded to different stress associated functions, ranging from abiotic stress sensing to regulating stress responses and included various heat shock genes, TFs, sugar metabolizing, transmembrane transporters, and other abiotic stress related genes. Moreover, various GWAS studies have also been conducted for heat tolerance in rice

Table 3.4 QTLs and MTAs reported for different traits associated with heat tolerance in rice

S. No.	Parents	Mapping population	Number and type of markers	PVE (%)	Chromosome	No. of QTLs	Traits and/or growth stage	References
1	IR 64/Azucena	101 DHs	–	19.2–22.8	1, 11	2	Spikelet fertility	Cao et al. (2003)
2	Nipponbare/ Kasalath	98 BILs	245 RFLP	13.5–17.25	1, 4, 7	3	Heat susceptibility index for grain weight	Chang-Lan et al. (2005)
3	T 2193/T 226	202RIL	181 SSR	29–37.3	3, 9, 12	4	Spikelet fertility	Qingquan et al. (2008)
4	Nagina22/IR64	272 RILs	5 K SNP	6.27–21.29	3, 5, 9, 12	5	Days to 50% flowering	Shanmugavadivel et al (2017)
5	OM5930/N22	310 BC2F2	264 SSRs	10.8–36.2	3, 4, 6, 8, 10, 11	11	Yield and yield-related characters	Buu et al. (2014)
6	EMF20/Nanjing11	146 F2 and 10 F3	154 SSR	4.8–19.7	3, 6, 8	7	Early-morning flowering	Hirabayashi et al. (2015)
7	Bala/Azucena	181 RILs		7.4–14.7	1, 2, 3, 4, 8, 10, 11	18	At anthesis	Jagadish et al. (2010)
8	N22/IR64	158 RILS	4,074 SNP	5.2–20.4	1, 2, 3, 4, 5, 6, 10	15	Root-shot traits	Kilasi et al (2018)
9	Gan-Xiang-Nuo/ Hua-Jing-Xian-74	23 SSSLs	GeneRacer Kit	14.47–51.67	3, 6, 8, 12	4	Flowering stage	Liu et al. (2017)
7	HTS 996/4628	286 RILs	862 SSR	9.3–15.1	4, 6	2	Pollen fertility at flowering stage	Xiao et al. (2011)
8	IR64/N22	52 BC1F1, 158 F2 and 36 BC2F2	384 SNP markers	6.4–29.7	1, 3, 4, 5, 7, 10, 11	21	Spikelet and panicle traits at flowering stage	Ye et al. (2012)
9	IR64/Giza	178 F_2	6 K SNP chips	17.3–21.8	1, 2, 3, 4	4	Spikelet fertility at flowering stage	Ye et al. (2015)
	Milyang23/ Giza178			12.5–20.6	1, 2, 6, 11	4		

(continued)

Table 3.4 (continued)

S. No.	Parents	Mapping population	Number and type of markers	PVE (%)	Chromosome	No. of QTLs	Traits and/or growth stage	References
	IR64//Milyang23/Giza178			11.66–13.9	4, 6, 11	3		
10	996/4628	279 F2	200 SSR	3–17	3, 4	2	Heading stage	Zhang et al. (2009)
11	Sasanishiki/Habataki	37 CSSLs	SSRs and InDels		1, 2, 3, 4, 5, 7, 8, 10, 11	11	At anthesis	Zhao et al. (2016)
12	Sasanishiki/Habataki	39 CSSLs	SSRs and InDels		1, 3, 4, 5, 6, 10, 11	12	At the booting stage	Zhu et al. (2017)

Genome Wide Association Studies

S. No.	Accessions	Number and type of markers	PVE (%)	Chromosome	No. of MTAs	Traits and/or growth stage	References
13	950 worldwide rice varieties	4,109,366 SNPs		2, 3, 6, 7, 8, 9, 10, 11, 12	32	Flowering time and grain-related traits	Huang et al. (2012a, b)
14	167 *indica* landraces and improved varieties	825 SNPs	5–20	All 12	183	Spikelet sterility during anthesis	Lafarge et al. (2017)
15	60 germplasm lines	20 SSRs	1–28	All 12	28	Panicle exsertion and spikelet fertility	Pradhan et al. (2016)
16	48 stable lines	315 SSR	4.26–29.93	1, 3, 4, 5, 8, 11, 12	18	Yield traits at flowering	Prasanth et al. (2016)
17	48 stable lines	49 SSR		1, 2, 3, 4, 6, 8	11	Yield traits at flowering	Prasanth et al. (2017)

(Huang et al., 2012a, b; Lafarge et al., 2017; Pradhan et al., 2016; Prasanth et al., 2016, 2017).

Genetics of Cold/Chilling Stress Tolerance in Rice

Owing to its subtropical and/or tropical origins, rice is sensitive to cold/chilling stress (0–15 °C). Chilling stress negatively affects various physiological and metabolic functions of rice, and thus reduces the yield (Arshad et al., 2017). Moreover, it also affects rice vegetative growth by affecting the photosynthesis, increasing signal substance abundance (ROS and so on), altering membrane rate (Xie et al., 2012); accumulating soluble sugars and compatible osmolytes (proline etc.) (Huang et al., 2012a, b; Zhang et al., 2017b); and generating antioxidants (such as glutathione and ascorbic acid) (Kim and Tai 2011). Chilling stress drastically affects the reproductive stage which results in spikelet sterility, limits gain size, and ultimately leads to reduction of yield (Arshad et al., 2017).

Until now, more than 100 QTLs related to cold tolerance at different growth stages such as germination, seedling stage, vegetative stage, reproductive stage and grain maturity have been identified in different studies using different types of populations such as RILs (Baruah et al., 2009; Jiang et al., 2008; 2017; Kim et al., 2014; Liu et al., 2020; Suh et al., 2010; Yang et al., 2020a), DH (Lou et al., 2007), F_2-F_3 lines (Biswas et al., 2017; Bonnel et al., 2020; Liu et al., 2015; Satoh et al., 2016; Shim et al., 2019), backcross and introgression line (Cheng et al., 2012; Mao et al., 2015; Najeeb et al., 2020; Xiao et al., 2014; Yu et al., 2018). In 2018, a QTL *qCTB10-2* was fine mapped using BC_7F_3 and BC_7F_4 populations, and a total of 17 putative genes were characterized in this QTL region and some of these encode proteins related to the cold stress mechanism in rice (Li et al., 2018a). A semi-dominant QTL, named as *Large Grain Size 1* (*LGS1*) was cloned and its molecular mechanism was also studied. This QTL encodes a TF, *OsGRF4*, which involved in production of larger grains and cold tolerance at the seedling stage (Chen et al., 2019). Two QTL meta-analysis reports have also been reported in rice for cold tolerance at seedling and booting stages, respectively (Kong et al., 2020; Yang et al., 2019b). Some GWAS studies have also been conducted in rice for cold tolerance at several growth stages such as bud burst stage (Zhang et al., 2018b), seedling stage (Lv et al., 2016; Pandit et al., 2017; Schläppi et al., 2017; Song et al., 2018; Wang et al., 2016) (Table 3.5).

Genetics of Submergence Stress Tolerance in Rice

Submergence, another most serious stress that causes large yield loss in rice production mainly in flood affected and rain-fed lowland areas. Modern rice cultivars can be badly damaged by complete submergence owing to oxygen shortage and restricted diffusion of gases which make anoxia in rice plant tissues (Jackson and Ismail, 2015). A submergence tolerant variety named as 'FR13A' was developed via selection from the framer's variety 'Dhallaputia' in 1950 at CRRI (Central Rice Research Institute,

Table 3.5 QTLs and MTAs reported for different traits associated with chilling tolerance in rice

S. No.	Parents	Number and types of markers	Mapping population	PVE (%)	Chromosome	No. of QTLs	Trait and/or stage	References
1	H335/CHA-1	100,307 SNPs	275 RILs	5.96–9.42	9	6	LTG (low temp germinability)	Yang et al. (2020b)
				2.50–3.61		5	Bud stage	
2	TR20/Hwaseong	2240 SSRs	224 F_2	13–39.9	1, 3	2	LTG	Shim et al. (2020)
3	Tolerant Rice1/ Haoannong	704 SNP	ILs 230	16–23.3	1, 5, 7	11	LTG	Najeeb et al. (2020)
4	TR20/Hwaseong	SSRs	224 F_2	20.4–29.9	1	4	LTG	Shim et al. (2019)
5	Changhui 891/02428	70,480 SNPs	124 RILs	6.31–23.33	1, 4, 8, 11	6	LTG	Jiang et al. (2017)
6	Akitakomachi/ Maratteli	97 SSR	120 F_2	5.8–23	1, 3, 11	8	LTG	Satoh et al. (2016)
7	HGKN/Hokuriku-142	232 SSR	162 RILs	5.7–9.3	6, 7, 8, 11	5	LTG	Ranawake et al. (2014)
8	Japonica/ indica cross	102 SSR	162 RILs	5.8–35.6	5, 6, 11	9	Early seedling stage	
9	M-202/IR50	181 SSRs	191 RILs	8.7–41.7	1, 3, 4, 6, 8, 10, 11, 12	15	Vegetative stage	Andaya and Mackill (2003)
10	Lemont/Teqing	RFLPs	269 RILs	5.5–29.8	3, 7, 11	3	Seedling stage	Zhi-Hong et al. (2005)
11	AAV002863/ Zhenshan97B	232 SSRs	193 DHs	6.46–27.42	1,2,8	5	Seedling stage	Lou et al. (2007)
12	Asominori/ IR24	375 RFLPs	71 RILs	9.15–24.51	1, 5,6	3	Seedling stage	Jiang et al. (2008)

(continued)

Table 3.5 (continued)

S. No.	Parents	Number and types of markers	Mapping population	PVE (%)	Chromosome	No. of QTLs	Trait and/or stage	References
13	A58 / W107	263SSR	79 RILs	13.1–27	1, 5, 11, 12	5	Early growth stage	Baruah et al. (2009)
14	Daguandao /IR28	167 SSR	227 RILs	5.5–22.4	3, 8, 11, 12	7	Germination and seedling stages	Wang et al. (2011b)
15	Xiushui 09/ IR2061	142 SSR	240 BC_2F_6	5.69–12.22	1, 7, 11	4	Seedling stage	Cheng et al. (2012)
16	Geumobyeo/ IR66160-121–4-4–2	175 SSR	153 RILs	7.8–8.3	4	2	Seedling stage	Suh et al. (2012)
17	Jinbu/BR29	137 SSR	123 RILs	6.1–16.5	1, 2, 4, 10, 11	6	Seedling	Kim et al. (2014)
18	Danteshwari/ Dagaddeshi	161 SSR	122 RILs	–	1, 3, 6, 9,12	5	At seedling stage	Verma et al. (2014)
19	DX/ Nanjing 11 (NJ)	113 SSR	151 BC_2F_1	9.4–32.1	10	2	Roots at seedling and mature stages	Xiao et al. (2014)
20	Lijiangxintuanheigu/ Sanhuangzhan-2	315 SSR	204 RILs	5–23.1	7, 8, 9, 11, 12	5	At seedling stage	Zhang et al. (2014)
21	(*Oryza sativa* L./*O. rufipogon* Griff.)	600 SSR	229 BILs	5.99–40.07	6, 7, 8, 11, 12	15	At early seedling stage	Luo et al. (2016)
22	Xiang 743/Katy	129 SSR	190 F_2	5.3–14.45	1, 2, 3, 5, 6, 8	7	Early seedling stage	Liu et al. (2015)
	Xiang 743/Dular		181 F2 populations	8.41–38.94	1, 2, 3, 6, 8	5		
23	BR1/Hbj.BVI	620 SSR	151 $F_{2:3}$	7.29–14.96	6, 8, 11, 12	6	Leaf discoloration and % survivability at seedling stage	Biswas et al. (2017)

(continued)

Table 3.5 (continued)

S. No.	Parents	Number and types of markers	Mapping population	PVE (%)	Chromosome	No. of QTLs	Trait and/or stage	References
24	IR24/Asominor	141 SSR	213 RILs	5.662 7–12.787 7	6, 11, 12	3	At seedling stage	Zhang et al. (2018c)
25	Xiang743/Katy	121 SSR	172 RILs	5.05–36.27	1, 2, 3, 8, 10, 12	6	Early-seedling cold tolerance	Liu et al. (2020b)
26	Puita´ INTACL/ H298a/90	159 SSR- and CG-markers	$128F_{2:3}$	14.59–46.51	11, 12	8	At seedling stage	Bonell et al. (2020)
27	DXWR/XieqingzaoB	4740 SLAFs	94 BILs	13.8–22.9	2, 3, 7, 8, 9, 11, 12	15	At seedling stage	Mao et al. (2015)

Genome Wide Association Studies

S.No.	Number and types of markers	AP size and type	PVE (%)	Chromosome	Number of MTAs	Stage	References
28	44 K SNP chip dataset of RDP1	295 rice accessions in RDP1 collected from 82 countries	3.85–8.20	1, 2, 3, 4, 5, 6, 7, 8, 9, 10, 11	67	At seedling stage	Wang et al. (2016)
29	67,511 SNPs	150 accessions of rice landraces in the Ting's rice core collection	26 to 33%,	All twelve	26	At seedling stage	Song et al. (2018)
30	148 SSRs, 3 InDels, SNPs	202 rice cultivars	1.98–13	1, 6, 10, 12	6	LTG	Schläppi et al. (2017)
			3.58–13.34	1, 9, 11	5	Plumule growth rate after cold germination (PGCG)	

(continued)

Table 3.5 (continued)

Genome Wide Association Studies							
S.No.	Number and types of markers	AP size and type	PVE (%)	Chromosome	Number of MTAs	Stage	References
			1.98–12.32	2, 3, 4, 6, 7, 8, 9, 10, 11, 12	24	Low temperature seedling survivability (LTSS)	
			2.96–11.41	3, 6	5	Plumule recovery growth after cold exposure (PGC)	
			4.21–16.55	3, 4, 5, 6, 7, 8, 9, 12	8	Low temperature survival (LTS)	
31	58 SSRs and 2 gene specific markers	66 rice varieties	11.31–28.33	1, 2, 3, 6, 9, 11	19	At seedling stage	Pandit et al. (2017)
32	4,358,600 SNPs	529 rice cultivars		All twelve	132	Natural chilling and cold shock tolerance	Lv et al. (2016)
33	5 K SNP array	249 *indica* rice varieties	2.5–13	All twelve	47	Bud burst stage	Zhang et al. (2018b)

Cuttack, India). Later, this variety was shared with IRRI from which few minor QTLs and one major QTL (*Sub1*) on chromosome 9 for submergence tolerance were identified and mapped by different researchers (Sarkar et al., 2014; Septiningsih et al., 2009; Xu et al., 2006). The identification of QTL/gene, *Sub1* was a major landmark in the history of submergence tolerance breeding. This QTL, *Sub1* comprises of three genes, namely *Sub1A, Sub1B*, and *Sub1C.* In submergence sensitive genotypes, either *Sub1A* gene is lacking or a different allele, *Sub1A-2* is present (Singh et al., 2020a). Gene, *Sub1A* reduces GA (gibbrelic acid) responsiveness and ethylene production, resulting in quiescence of growth under submergence, whereas gene *Sub1C* enhances GA production and its responsiveness causing rapid exhaustion of carbohydrate pool, shoot elongation and bad survival. As submergence is a complex quantitative trait and the QTL/gene, *Sub1* can not completely represent the trait alone (Septiningsih et al., 2012; Singh et al., 2015). Therefore, novel tolerant QTLs that complement *Sub1* needed to be identified and mapped for breeding superior submergence tolerant cultivars. Several secondary QTLs regulating submergence tolerance were then identified and mapped on almost all rice chromosomes (Angaji et al., 2010; Baltazar et al., 2019; Dang et al., 2019; Gonzaga et al., 2016, 2017; Hattori et al., 2007; Hsu and Tung, 2015; Jiang et al., 2006; Manangkil et al., 2013; Septiningsih et al., 2012, 2013; Yang et al., 2019a) (Table 3.6). Till now, no QTL meta-analysis has been conducted for submergence tolerance in rice. Although, various GWAS studies have been conducted in rice where hundreds of SNPs and putative candidate genes associated with submergence tolerance were identified and reported (Dang et al., 2019; Gao et al., 2020; Hsu and Tung, 2015; Nghi et al., 2019; Singh et al., 2020b; Su et al., 2020) (Table 3.6).

Genetics of Salinity Stress Tolerance in Rice

Rice is a most salt sensitive crop among the cereals. In coastal areas, salt stress is a major problem, however, inland is also drastically affected by salinity owing to improper irrigation and drainage system. A reduction in rice yield by 12% per each dS m^{-1} has been observed when soil electrical conductivity (EC) increased far off threshold EC of 33 dS m^{-1} (Geetha et al., 2017). Although, rice is badly affected by salinity from germination to maturity phase, but it is most sensitive to salinity at seedling and reproductive stage which ultimately leads to low grain yield (Chowdhury et al., 2016). Owing to its genetic complexity, salinity tolerance rice breeding has been very slow in last decades. Various QTL mapping studies have been conducted in rice for salinity tolerance at various stages (see Table 3.7). Most of these QTL mapping studies were based on low density markers, hence underlying candidate genes could not be identified (Hongxuan, et al., 1998; Kim et al., 2009; Lang et al., 2008; Lin et al., 2004; Prasad et al., 2000; Takehisa et al., 2004; Thomson et al., 2010; Zang et al., 2008). Recently, high throughput SNP genotyping have facilitated various advantages over previous marker systems and provided dense genomic coverage for the better dissection of the salinity tolerance in rice (Chen et al., 2019; Mazumder et al., 2020; Zhang et al., 2020). Recently in 2020, a MAGIC

Table 3.6 QTLs and MTAs reported for different traits associated with submergence tolerance in rice

S. No.	Parents	Size and type of population	Number and type of Markers	R2 (%)	Chromosome	No. of QTLs	Stage	References
1	USSR5/N22	148 F_2	121 SSR	10.99–15.51	5, 11	2	Anoxia germinability	Jiang et al. (2006)
				4.16–32.77	3, 4, 5, 7, 9, 10, 11	22	LTG	
2	IR72/Madabaru	80 $F_{2:3}$	1,074 SSR and 19 InDel	16.3–52.3	1, 2, 9, 12	4	Seedling stage	Septiningsih et al (2012)
3	Khao Hlan On/IR64	423 BC_2F_2	680 SSR and 1 InDel	1.19–20.59	1, 2, 3, 7, 8, 9	9	During germination	Angaji et al. (2010)
4	IR42/Ma-Zhan Red	175 $F_{2:3}$	118 SSR	7.5–31.7	2, 5, 6, 7	6	During germination	Septiningsih et al. (2013)
5	Nipponbare/Kasalath/ Nipponbare	98 BILs	245 RFLP	1.3–37.9	1, 2, 3, 4, 6, 7, 11	32	Seedling vigor	Manangkil et al. (2013)
6	Nipponbare/IR64	144 RILs	355 SNPs	26.38	1	1	Coleoptile elongation	Hsu and Tung 2015
7	IR42/FR13A	103 RILs	232 SNPs, SSR, STS	4.20–7.50	1, 4, 8, 9, 10	5	Anaerobic germination	Gonzaga et al. (2016)
8	Ciherang-Sub1/IR10F365	115 RILs	6 K SNP	10.2–27.50	2, 3, 8	3	Seedling survival	Gonzaga et al. (2017)
9	Oryza sativa/Oryza rufipogon	94 F_2	SSR	14–36	1, 3, 8, 12	10	Rapid internode elongation	Hattori et al. (2007)
10	YZX/02428	192 RILs	85,742 SNPs	4.28–12.83	Except 11	32	Anaerobic germination	Yang et al. (2019a)
11	IR64/Kharsu	80A $F_{2:3}$	384-plex SNP	8.1–12.6	3, 7	4	Anaerobic germination	Baltazar et al. (2019)

(continued)

Table 3.6 (continued)

S. No.	Parents	Size and type of population	Number and type of Markers	R2 (%)	Chromosome	No. of QTLs	Stage	References
12	Wuyunjing 7 hao/Ludao// Wuyunjing 7 hao	115 BILs	808 SSR	2.56–7.67	Except 10	23	Seedling anoxic tolerance	Dang et al. (2019)

Genome Wide Association Studies

S. No.	AP size and type	Number and type of markers	R^2 (%)	Chrom	No. of MTAs	Stage	References
13	153 diverse accessions	36,901 SNPs		All	88	Anaerobic germination	Hsu and Tung 2015
14	209 natural rice populations	2,123,725 (SNP) markers		1, 2, 3, 5, 6, 7, 8, and 11	26	Anaerobic germination	Su et al. (2020)
15	Natural population 542 accessions	262 markers	2.56–7.67	2, 6, 4, 7, 8, 9, 10, 11, 12	16	Seedling anoxic tolerance	Dang et al. (2019)
16	273 accessions from CREA-Research Centre	GBS-resulted in 246,084 SNPs		1, 2, 3, 5, 6, 10, 11	13	Coleoptile elongation	Nghi et al. (2019)
17	166 cultivated rice accessions	SNPs		1, 2, 3, 5, 6, 7, 11, 12	19	Coleoptile length, shoot length and submergence tolerance index	Gao et al. (2020)
18	179 diverse rice genotypes	36-plex SNP assay			38	Vegetative stage	Singh et al. (2020b)

population was genotyped with a 55KSNP array and a total of 7 QTLs were identified accounting for 7.42–9.38% of the total phenotypic variations. Further, using gene expression analysis, a potential TF gene was revealed (Zhang et al., 2020). Some of the QTLs associated with salinity tolerance have also been fine mapped, for instance, a QTL, *qSL7* was delimited to 252.9 kb region on rice chromosome 7 where a total of 40 putative genes were located, including one gene having major positive role in salt stress tolerance (Jahan et al., 2020). Unfortunately, from a very long time no GWAS study was conducted for salt tolerance until 2015. First GWAS study was performed on 220 rice accessions using 6KSNP chip and a total of 44 SNPs associated with different traits under salinity stress was identified. Since then, several GWAS studies have been published in rice (Muthuramalingam et al., 2022) and hundreds of SNPs and putative candidate genes associated with different traits or expressed at different stages such as root and shoot growth parameters (Frouin et al. (2018); salt tolerance at the germination and seedling stages (Naveed et al., 2018); salinity tolerance at reproductive stage (Warraich et al., 2020); early stage salinity tolerance (Nayyeripasand et al., 2020) have been identified.

Genomics-Assisted Breeding for Developing Climate Resilient Rice

The speed of progress in breeding climate resilient rice was initially slow due to various factors including less understanding of the mechanisms underlying tolerance to various aberrant climatic conditions, complexity of the traits related to the stress, lacking of reliable and repeatable screening techniques, lacking of efficient selection criteria and variation of tolerance with plant growth stages. In recent years, rice genomics related research has been expanded significantly and a number of important genomic resources have become available. These noteworthy capabilities of rice genomics assisted in the rapid understanding of genetics of tolerance to abiotic stresses, improvement of breeding selection strategies, characterization of specific genes and resulted in the successful introgression of numerous agronomical important traits with no or less genetic drag. A large number of genomic resources, genes/QTLs for abiotic stress tolerance in rice have been identified in the last two decades (see Tables 3.3, 3.4, 3.5, 3.6 and 3.7). With these advancements, now rice breeding has become more and more genomics assisted and molecular in nature leading to release of rice cultivars comprising of causal genes/QTLs for tolerances against various abiotic stresses (Singh et al., 2016). For instances, using MABB, the *Sub1* QTL, that confers submergence tolerance was incorporated into several elite cultivars with high accuracy within a short time as compared with traditional breeding techniques.

The first cultivar developed by incorporating *Sub1A* was 'swarna-Sub1', a mega-cultivar in various South-East Asian countries including India (Neeraja et al., 2007). Following the grand success of 'swarna-Sub1', which could withstand around 14–15 days of submergence, other mega varieties such as 'PSB Rc18' (Septiningsih et al., 2015), 'Ciherang' (Septiningsih et al., 2015; Toledo et al., 2015) and 'CO 43' (Rahman et al., 2018) were also improved by introgressing the *Sub1A* genomic

Table 3.7 QTLs reported for different traits associated with salinity tolerance in rice

S. No.	Parents	Size and type of population	Number and type of markers	PVE (%)	Chromosome	No. of QTLs	Stage	Reference
1	Tesanai 2/CB	142 RILs	60 RFLP	11.6	5	1	Seedling survival days	Hongxuan et al. (1998)
2	IR64/Azucena	DHs	RFLP	18.9	6	7	Seedling stage	Prasad et al. (2000)
3	IR4630/IR15324	118 RILs	199 AFLP, 107 RFLP, 84 SSR	6.4–19.6	1, 4, 6, 9	11	Sodium and potassium uptake	Koyama et al. (2001)
4	Nipponbare/ Kasalath	98 BIL	245 RFLP	21–41	1, 2, 3, 7	27	Shoot length; tiller number; shoot fresh weight	Takehisa et al. (2004)
5	NonaBokra/ Koshihikari	133 F_3	161 RFLP	12.4–48.5	1, 4, 6, 7, 9	11	Salt tolerance traits	Lin et al. (2004)
6	IR64/Binam	99 ILs	129 SSR	–	1, 2, 4, 5, 7, 8, 9, 10, 11	35	Seedling stage and tillering stage	Zang et al. (2008)
7	AS996 /IR50404	229 RILs	44 SSRs	–	1, 8	–	Yield related traits	Lang et al. (2008)
8	"Ilpumbyeo"/ "Moroberekan"	117 BC_3F_2	125 SSR	10.5–13.6	1, 6, 7	8	Yield related traits	Kim et al. (2009)
9	IR29/Pokkali	140 RILs	100 SSR		1, 2, 3, 4, 6, 9	17	Salt tolerance traits	Thomson et al. (2010)
10	Jiucaiqing (*japonica*)/IR26	150 RILs	135 SSR	7.8–23.9	1, 4, 6, 11, 12	12	Seedlings survival	Yun et al. (2012)

(continued)

Table 3.7 (continued)

S. No.	Parents	Size and type of population	Number and type of markers	PVE (%)	Chromosome	No. of QTLs	Stage	Reference
11	Teqing/Binam	77 ILs	139 SSRs		2, 3, 4, 5, 6, 8, 9, 10, 11, 12	23	Seedlings survival	Wang et al. (2012b)
12	IR29/Hasawi Saudi	142 RILs	384 SNP	10.6–42.3	1, 2, 6	7	At the Young Seedling Stage	Bimpong et al. (2014)
13	Gharib/Sepidroud	148 F_2	131 SSR, 105 AFLP	0.26–29.71	All 12	41	At Seedling Stage	Ghomi et al. (2013)
14	Cheriviruppu/Pusa Basmati 1	218 F_2	131 SSRs	4–47	1, 7, 8, 10	16	Reproductive-Stage	Hossain et al. (2015)
15	IR36/Pokkali	113 F_2	74 SSR	7.69–72.57	1, 3, 4, 5, 6, 7, 8, 11	17	Na, K and Ca accumulation traits	Khan et al. (2015)
16	IR36/Pokkali	113 F_2	74 SSR	11.52–81.56	2, 3, 7, 8	16	At maturity stage	Khan et al. (2016)
17	IR29/Hasawi	142 RILs	384 SNP	9.3–20.6	1, 2, 4, 6, 8, 9, 12	20	At seedling stage	Bizimana et al. (2017)
18	Cheniere/Nona Bokra	112 ILs	116 SSR	1.2–17.6	All 12	50	Seedling stage	Puram et al. (2018)
19	93–11/PA64s	132 RILs	SNP, SSRs	5.9–23.7	1, 2, 3, 4, 5, 6, 7, 10	38	For shoot length	Jahan et al. (2020)
20	IR29/Pokkali	148 RILs	14,470 SNP	5.3–35.1	1, 3, 4, 6, 7, 8, 12	23	At seedling stage	Chen et al. (2020)
21	Ahlemi Tarom/Neda	96 RILs	40 SSR, 16 ISSR,2 IRAP, iPBS	4–24.8	Except 11	73	At Seedling Stage	Sanchouli et al. (2019)

(continued)

Table 3.7 (continued)

S. No.	Parents	Size and type of population	Number and type of markers	PVE (%)	Chromosome	No. of QTLs	Stage	Reference
22	Kalarata/ Azucena	400 F_2	151 SSRs and InDel	8–25	1, 2, 3, 5, 11	13	At seedling stage	de Ocampo et al. (2020)
23	Kolajoha/Ranjit	68 RILs	3649 SNPs	6–19.5	1, 2, 4, 6, 7, 9, 11, 12	23	At seedling stage	Mazumder et al. (2020)
24	Dianjingyou 1/ SR86	186 F2 BCA	36 SSR	62.6	1	1	Trifoliate stage	Wu et al. (2020)
25	MAGIC population	221 DC1	55 K rice SNP	7.45–9.38	1, 2, 5, 9	7	At the seedling stage	Zhang et al. (2020)
26	WJZ / Nipponbare	181BC_1 F_2	167 SSR	7.07–23.99	3, 6, 8, 10	19	During seed germination	Zeng et al. (2020a)

region through MABB strategy. Although breeding for drought tolerance in rice via introgression of QTLs that regulate drought tolerance and yield under drought is still an open area of research. However, now the scene is slowly changing with fine mapping of the major QTL regions (Vikram et al., 2016) and the identification of candidate genes such as *DEEPER ROOTING 1* (*DRO1*) (Uga et al., 2015). In 2016, the linkage between drought tolerance and plant height was broken by fine mapping of one QTL, *qDTY1.1* region wherein the concerned QTL was mapped to 200–400 kb distal to the *sd1* region (Vikram et al., 2016). These types of efforts have paved the way for incorporation of drought tolerance QTLs in rice cultivars. Some drought-tolerant version of 'IR64', namely 'DRR Dhan 42', 'drought-tolerant MR219' (Shamsudin et al., 2016), 'drought-tolerant Sabitri' (Dixit et al., 2017) were also then developed via MABB programmes. Significant progress has also been made for breeding of heat tolerant rice varieties, cultivar 'Nagina 22, is the most heat-tolerant cultivar tested so far (Pradhan et al., 2016). Various salinity tolerant varieties such as salinity-tolerant version of 'BR28' (Thomson et al., 2007), salinity-tolerant 'Pusa Basmati 1' (Singh et al., 2018), and salinity-tolerant 'Improved White Ponni' (Valarmathi et al., 2019) have been developed through MABB strategy. So far, very little research has been conducted to develop climate resilient varieties/multiple stress tolerance varieties by pyramiding the QTLs that regulate tolerance against multiple stresses such as drought, heat, salinity, submergence and cold via MAS approach. Although, more recently in 2020, four major effect QTLs, *qDTY1.1*, *qDTY2.1, Saltol and Sub1* conferring tolerance against drought, salinity, and submergence stresses, respectively; were introgressed in a popular rice variety 'Improved White Ponni' through a MABB approach. Generated back cross inbred lines showed consistent grain quality and agronomic characters and increased tolerance against drought, salinity and submergence stresses compared with the recipient parent (Muthu et al., 2020).

In recent years, significant progress has also been made for the development of abiotic stress tolerant transgenic rice (Huang et al., 2020; Liang et al., 2019; Wang et al., 2019b; Xu et al., 2020). For instance, drought tolerant transgenic rice was developed by expressing gene, *SlMYB56* in rice that confers tolerance to drought by controlling the ABA signalling and lignin biosynthesis pathways (Xu et al., 2020); heat tolerant transgenic rice was developed by overexpressing a protein disulfide isomerase gene *(MtPDI)* that confers the heat tolerance by increasing the proline content, superoxide dismutase and peroxidase activities; salinity tolerant transgenic rice was developed by over-expressing a ferredoxin-like protein gene *pflp*, that enhances the salt tolerance in plants by reducing the excess Na^+ and ROS accumulation (Huang et al., 2020); and multiple stress tolerant transgenic rice was also developed by expressing the genes such as *Type II Galactinol Synthase 2* (*Type II GolSs*) gene that confers tolerances to cold and drought stresses by increasing the accumulation of raffinose family oligosaccharides (Cui et al., 2020); and *OsARD1* gene that confers tolerances to submergence, drought, and salt stresses by enhancing the ethylene synthesis in rice (Liang et al., 2019). Genes which provide tolerance to multiple stresses can be targeted for the development of climate resilient rice.

Targeted genome editing via CRISPR/Cas technology have also been proved useful in the development of climate resilient rice crop to sustain and enhance its productivity in the changing environmental conditions (Kumar et al., 2020b; Zeng et al., 2020b; Zhang et al., 2019a). In one study, a dstD184–305 mutation was induced by CRISPR-Cas9 method in DST gene in indica rice cultivar 'MTU1010'. This mutant produced broad leaves with decreased stomatal density and showed enhanced leaf water retention. This mutant showed tolerance to drought and salinity stress at seedling stage (Kumar et al., 2020b). By editing a gene, *OsRR22* in rice, salt tolerant mutant lines were also produced (Zhang et al., 2019a). Recently in 2020, a study reported the generation of several rice mutants with excellent cold tolerance and high yield by simultaneously editing three genes, *OsMYB30* (a cold tolerance gene), *GS3* (a grain size gene) and *OsPIN5b* (a panicle length gene) with the CRISPR–Cas9 system (Zeng et al., 2020b). Various CRISPR/Cas mediated schemes for engineering/editing of novel genes have been proposed to generate the plant phenotypes, which can face the adverse climate challenges (see Khan et al., 2022 for an updated review).

3.2.3 *Wheat*

Wheat (*Triticum aestivum* L.) is the largely grown crop in the world which shares about 21% of the global food production. It belongs to the family Poaceae and genus *Triticum* and its genome size is approximately 17 Gb contained in 21 (2n = 6x = 42) chromosomes. The world's total wheat production in 2020 was estimated at 760 million thousand tonnes and the top 5 wheat producing countries are China, India, Russian Federation, the United States of America, and Canada that account for around 64% of the total production (https://knoema.com/atlas/topics/Agriculture/Crops-Production-Quantity-tonnes/Wheatproduction).

Every 12–13 years, the wheat yield regularly has increased to roughly 0.5 tons, which is equal to annual increase of yield about 40 kg/ha. However, this increment likely is not adequate to keep pace with the world population growth and requirement of higher food for rapid urbanization under changing climatic conditions (Tester & Langridge, 2010). Abiotic stresses such as heat, chilling, drought, flooding, salinity, heavy metal toxicity and mineral deficiency are the principal stresses of wheat cultivation. Among these, drought, heat and salinity are directly related to the climatic conditions. Combinedly or individually, these stresses drastically affect the grain quality and reduce the wheat production. Development of stress tolerant wheat cultivars has been a prime objective for breeders. Keeping the most important abiotic stresses viz., drought, heat and salinity into consideration, globally, two most important initiatives have been launched to tackle the issue of improvement of wheat production under these stresses: (a) Heat and Drought Wheat Improvement Consortium (HeDWIC) (http://www.hedwic.org/): this consortium was established by Consultative Group on International Agriculture Research (CGIAR) program on wheat (CRP WHEAT). This is basically a network that enables worldwide coordination of wheat research to adapt to a future with more severe climatic aberrations especially heat and drought

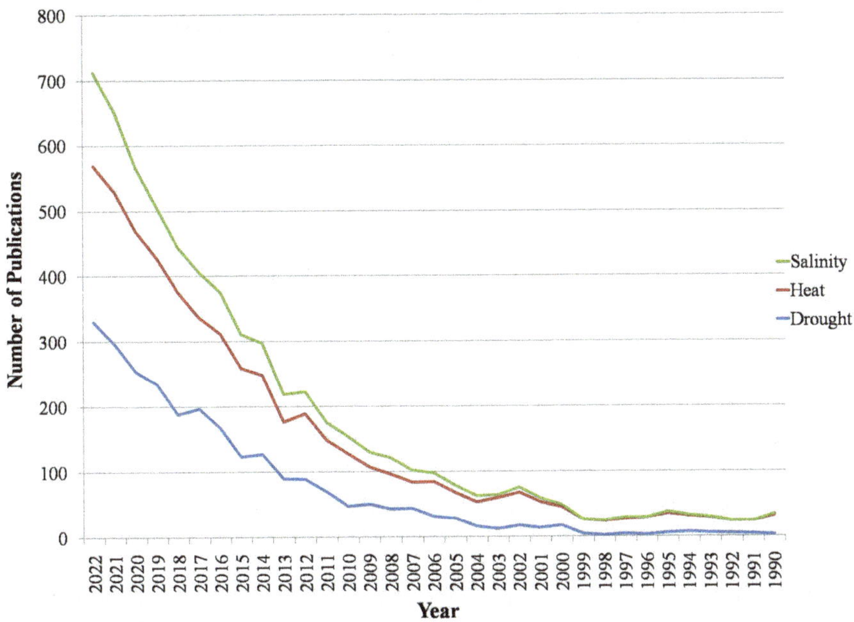

Fig. 3.4 The number of publications per year related to wheat abiotic stresses from 01/01/1990 to 15/12/2022. *Source* PubMed (Keywords: wheat AND drought; wheat AND heat; wheat AND salinity) were used to search the number of publications at https://pubmed.ncbi.nlm.nih.gov/)

and (b) the global Wheat Yield Consortium (WYC) (https://www.cimmyt.org/projects/wheat-yield-consortium/). This consortium conducts research on wheat physiology and genetics to increase the resilience to various abiotic and biotic stresses in wheat and its yield potential, worldwide. In addition to these two major initiatives, the molecular basis of abiotic stress tolerance has received much attention by different groups in various parts of the world, therefore a large number of linkage-based QTL mapping and GWAS studies have also been conducted (http://www.wheatqtldb.net/; Singh et al., 2022). The amount of research studies has increased rapidly related to drought, heat, and salinity stresses in wheat in recent years (Fig. 3.4).

Genomic Resources Available for Wheat

Genomics-assisted breeding programs require elaborated genomic data that can be precisely linked to adaptive traits. Assembly of a reference genome is foremost requirement for modern breeding programs that is used as a basis for comparison between different genotypes or individuals of the species so that allelic variants can be detected, mapped, and ultimately associated with phenotypic variation. Next generation sequencing technologies have improved rapidly, permitting for more accurate and precise reference genome assembly (Harfouche et al., 2019). In last decade, several wheat genome assemblies have been published (e.g., IWGSC, 2014; Zimin et al., 2017). The most detailed assembly, called RefSeqv1.0 was released in 2018. This is a chromosome level assembly which gives access to a total of 107,891 high-confidence genes, including their genomic regions of regulatory sequences (IWGSC, 2018). Recently, a more updated version of the reference sequence of bread wheat (cultivar Chinese Spring), 'IWGSC RefSeq v2.1' have also been released (available at http://wheat-urgi.versailles.inra.fr/Seq-Repository/Assemblies). A recently emerged genomic strategy for assessing large scale genetic variation is Genotyping-by-Sequencing (GBS), one of numerous reduced representation library approaches facilitated by NGS (Elshire et al., 2011). A number of linkage maps developed using thousands of GBS-SNPs are available in GrainGenes database (https://wheat.pw.usda.gov/GG3/node/876#wheat6x). Chinese Spring (CS) is a landrace with large genomic information compared to advanced cultivars, resulting in a very low coverage of markers (SNP and/or InDels) discovered by GBS, hence restricting the application of GBS in modern wheat cultivars. On the other hand, SNP arrays are flexible in respect to the sample type and data point customization that contributes to its high markers coverage and robust and high call rates compared to GBS and other PCR-based markers.

To date, a number of high density SNP genotyping assays/arrays have been developed and utilized for genetic dissection of many traits in common wheat, for instance, the 'Wheat 50 K Triticum TraitBreed array', the 'Wheat Breeders' 35 K Axiom® array', the 'Wheat 55 K SNP array', the 'Axiom®, Wheat 660 K SNP array', the 'Axiom® HD Wheat genotyping (820 K) array', the 'Wheat 15 K SNP array', the 'Illumina Wheat 90 K iSelect SNP genotyping array', and the 'Illumina Wheat 9 K iSelect SNP array' (reviewed in Saini et al., 2022). Recently in 2020, a study was conducted to compare the performance of seven highly used SNP arrays (660 K, 820 K, 90 K, 55 K, 35 K, 15 K, 9 K arrays) in terms of SNP number, density, distribution and so on (Sun et al., 2020). This study showed that the 660 K array has the highest percentage of genome -specific SNPs with precise physical positions and even distribution across the whole genome. This array can substitute other 6 arrays and can be the good choice for targeted genotyping and MAS in wheat (Sun et al., 2020). A few studies also showed that for genomic prediction application, GBS-SNPs are comparable or better than 90 K SNP array-scored SNPs. These contradictory reports showed that desired purposes and objectives must be considered while selecting the best option from various available options of genotyping technologies. Table 3.8 lists the different bioinformatics tools and database resources with their websites and features that are helpful for genomic studies in common wheat. In recent past, to get better insights into the variation in gene presence–absence, a pan-genome data set was also developed by whole genome sequencing of 18 different wheat cultivars (available at http://appliedbioinformatics.com.au/cgi-bin/gb2/gbrowse/WheatPan/). This study predicted the pan-genome of size of 140,500 $\pm$ 102 genes with an average of 128,656 genes in each wheat cultivar. In addition, this study also reported the identification of more than 36 million intervarietal SNPs across the whole genome (Montenegro et al., 2017).

Table 3.8 List of major bioinformatics tools and database resources for genomics studies in common wheat

S. No.	Database	Website	Features
1	KOMUGI DB	https://shigen.nig.ac.jp/wheat/komugi/about/about.jsp	Combines information of wheat genetic resources
2	IWGSC	https://www.wheatgenome.org/About	High quality genome sequence of bread wheat
3	URGI	https://wheat-urgi.versailles.inra.fr/About-us	Maintains databases and tools to study genetic and genomic wheat data
4	PGSB	https://www.helmholtz-muenchen.de/pgsb	Provides data and information resource for individual plant species and offers a platform for integrative and comparative plant genome research also
5	TaSSRDb	http://webtom.cabgrid.res.in/wheatssr/	Integrated online relational database with "three-tier architecture"
6	WHEAT TRAINING	http://www.wheat-training.com/	Provides background information and practical resources to help both budding wheat scientists as well as researchers
7	WheatGenome.info	http://www.wheatgenome.info/	An integrated database and portal for wheat genome information
8	WheatIS	http://www.wheatis.org/index.php	Provide access to the available data resources and bioinformatics tools
9	Wheat MicroRNA Portal	http://wheat.bioinfo.uqam.ca	Allows multiple comparative analyses and reproducible studies on small RNAs and miRNA families in wheat
10	Wheat Expression	http://www.wheat-expression.com/#	Provides information about the expression of genes
11	Wheat Tilling	http://www.wheat-tilling.com/	In silico wheat Target Induced Local Lesions In Genome (TILLING) website
12	WheatQTLdb	http://www.wheatqtldb.net/	Provides information about QTLs and MTAsassociated with different traits

High Throughput Phenotyping in Wheat

Plant architecture and developmental traits including primary and secondary traits under abiotic stresses could accurately be phenotyped in controlled conditions via modern high-throughput and nondestructive platforms/tools. This "phenotyping under controlled conditions" assists in large-scale phenotyping including trait mapping experiments. Since various component traits of abiotic stress tolerance are regulated quantitatively, therefore, via improving the accuracy of phenotyping, we can improve the heritability of these traits. To increase the impact of plant phenotyping and enable cooperation by encouraging communication between stakeholders in academia, government, industry and general public, the International Plant Phenomics Network (IPPN) was established (https://www.plant-phenotyping.org/). This provides the knowledge about modern phenomics platforms and tools which can be used to record the data for various traits associated with abiotic stress tolerance, quality, and others, on thousands of plants in a single day (Rahaman et al., 2015).

The scope of HTP is rapidly progressing in various crops including wheat, current implementations of non-destructive modern platforms and tools include the use of following sophisticated technologies such as "infrared thermography and imagery to scan temperature profiles/transpiration; 3D reconstruction to assess plant growth rate and structure; fluorescent microscopy/spectroscopy to assess photosynthetic rates, magnetic resonance imaging and positron emission tomography to measure growth patterns, root/leaf physiology, water relations, and/or assimilate translocation properties; light detection and ranging (LIDAR) to measure growth rates; nuclear magnetic resonance for monitoring the structure of tissues, mapping water movements, and monitoring sucrose allocation; canopy spectral reflectance for monitoring dynamic complex traits; and digital RGB imaging for recording data on various attributes of roots, shoots, leaves, seeds, and grains" (Gill et al., 2022). Use of these platforms/tools is still in infancy in case of wheat and only few studies are available where either of these platforms has been used for phenotyping of the plants under abiotic stresses (Crain et al., 2017). Two RIL populations were grown for three years under two different treatments heat and drought stress at the CIMMYT, Mexico and these populations were phenotyped at various time intervals throughout the growing season with a portable field phenotyping platform, "Phenocart" that integrates precision GPS, thermal sensors, and spectral reflectance. In this study, both canopy temperature and normalized difference vegetation index (NDVI) were found to be significantly correlated with final grain yield. Moreover, a regular pattern for broad sense heritability and correlation to final grain yield for both canopy temperature and NDVI was observed over the growing season. This study clearly showed that HTP platforms can be used to aid indirect selection via rapid collection of physiological measurements compared with direct selection for final grain yield alone (Crain et al., 2017).

Genetics of Drought Stress Tolerance in Wheat

Insufficient availability of water or drought restricts wheat growth and never permits wheat to show its full yield potential. Among all abiotic stresses, the drought is the most vital factor which reduces the wheat production significantly. Periodic drought is known to affect > 50% of the area under wheat production (Pfeiffer et al., 2005). The development of drought tolerant wheat cultivars can be the most effective way of dealing with drought or water scarcity and therefore this is currently receiving worldwide attention. Drought tolerance is a complex quantitative trait that is governed by several genes/QTLs. Up to now, more than 50 reports on QTL mapping have been published which have reported more than 1200 QTLs spread over all the 21 wheat chromosomes (http://www.wheatqtldb.net/). Maximum number of QTLs have been identified and mapped for as many as 33 agronomic traits (mainly for thousand grain weight and grain yield), followed by around 19 physiological traits (mainly for SPAD/chlorophyll content, water-soluble carbohydrates, and coleoptile length), and few root traits (mainly for root length). Out of the 1200 QTLs, around 70 QTLs are major (contributing ~> 20% PVE), and around 19 QTLs are stable (identified in ≥ 50% environments) (reviewed in Gupta et al., 2020). QTLs detected for root traits show complex interactions with environment (QTL × environment interaction), therefore stable QTLs for these traits are not available. Moreover, > 100 first order Q × Q interactions have also been reported for 10 different drought-responsive physiological and agronomic traits (Gupta et al., 2020), although the phenotypic variance for each pair of epistatic QTLs was mostly low (Khanna Chopra et al., 2019; Yang et al., 2007). However, identified stable QTLs can be targeted for the development of drought tolerant cultivars using MAS (Gupta et al., 2020). More recently in 2020, after 32 field experiments conducted under drought and heat stress conditions in southern Australia, Northwestern Mexico and India, a total of 128 QTLs were identified for 4 different traits: TGW, grain yield, days to heading and grain filling duration. This study further reported the fine mapping of one major grain yield QTL to 2.9 cM region equivalent to 2.2 Mbp genomic region which contained 39 predicted genes (Tura et al., 2020). This QTL can be targeted for wheat breeding.

Two important QTL meta-analyses studies have also been conducted to identify the most consensus and robust meta-QTLs (MQTLs) which can be targeted for future wheat molecular breeding programs (Kumar et al., 2020a; Tanin et al., 2022). In the first study, a meta-analysis was conducted in 2020 using 340 known QTLs from 11 studies (published from 2015 to 2020) which resulted into the identification of 13 drought responsive MQTLs and 228 putative drought responsive candidate genes (Kumar et al., 2020a). The second study was our own study, where we analyzed a total of 3102 QTLs associated with multiple abiotic stress tolerance retrived from 116 mapping studies involving 66 studies for either individual drought or heat stress or combined drought and heat stresses and identified several promising MQTLs for the breeding programs (Tanin et al., 2022). Moreover, up to mid 2020, results of at least 10 studies based on GWAS have also become available, each involving an association panel varying in size from 108 to 382 accessions/genotypes that were screened under drought conditions. Different types of markers such as SSR, GBS-SNPs, DArT and

SNP arrays were utilized in these studies and more than 1150 MTAs were reported for different agronomical and physiological traits (Gupta et al., 2020).

In recent years, several other important GWAS studies for different drought-responsive traits have been conducted in wheat (http://www.wheatqtldb.net/). For instance, using 339 pre-breeding lines, 26 SNPs and several candidate genes associated with different traits such as yield, spike length, number of grains per spikes, TGW and KA were identified (Shokat et al., 2020). A total of 58 SNPs associated with grain yield and 15 with superiority index Pi (measure of stability) were identified in a haplotype based GWAS study conducted using 4302 advanced bread wheat lines. In a more recent study, a total of 460 significant loci were detected for eight yield-related traits such as plant height, spike length, total spikelet number per spike, fertile spikelet number per spike, grain number per spike, spike number per plant, TGW and grain weight per spike under drought stress (Li et al., 2020b). Probable drought-related genes/QTLs, identified through linkage based-QTL mapping and GWAS studies, need to be further characterized, before their use in the development of drought tolerant cultivars. Their functional relevancy to drought stress needs to be shown and finally confirmed with transgenic approaches. Several functionally characterized drought responsive genes are given in Table 3.9.

Genetics of Heat Stress Tolerance in Wheat

Various field and controlled-based studies have predicted that wheat production may decrease by 4.1–6.4% with each global rise of 1 °C in temperature alone, and this scenario can be more harmful in combination with other abiotic stresses (Bennett et al., 2012). As per an earlier estimation, around 58% of the wheat crop experiences heat stress, worldwide (Kosina et al., 2007). Although, heat stress disrupts the duration of all growth stages in wheat, but it extremely affects the reproductive phase of plants. Heat stress alters the plant water status, reduces photosynthetic and metabolic activities, pollen viability and reactive oxygen species (ROS) production (Wollenweber et al., 2003). Duration, frequency and timing of heat stress have an important influence on grain yield (Wardlaw, 2002). Development of heat tolerant wheat cultivars using genes/QTLs for tolerance to heat stress can be a great solution to above all mentioned problems.

Heat tolerance is a complex quantitative trait that is governed by several genes/QTLs. Over the last three decades, many efforts have been made to explain the genetic basis of heat tolerance. For the first time in 1991, heat tolerance QTLs/genes were identified and mapped on wheat chromosomes 3A, 3B, 4A, 4B, and 6A using a langdon chromosome substitution line (Sun & Quick, 1991). Since then, several advancements have been made in molecular marker technologies and related statistical approaches, which have also resulted into the identification of > 440 QTLs related to different heat stress-responsive traits (Kumar et al., 2021). These > 440 QTLs are given in 31 mapping studies which have published by different countries such as Mexico, USA, India, China, and other countries. Maximum number of QTLs

Table 3.9 List of some important functional genes involved in providing tolerance to different abiotic stresses

S. No.	Genes	Features	References
Important functional genes associated with heat stress tolerance			
1	*TamiR159*	*TamiR159* overexpressing plants were more sensitive to heat stress relative to the wild type	Wang et al. (2012a)
2	*TaHsfA6f*	Transgenic plants overexpressing the gene showed improved thermo-tolerance	Xue et al. (2015)
3	*TaMBF1c*	Overexpression of the gene showed higher thermo-tolerance than control plants at both seedling and reproductive stages	Qin et al. (2015)
4	*TaFER-5B*	Transgenic plants exhibited enhanced thermo-tolerance	Zang et al. (2017a)
5	*TaOEP16-2-5B*	Transgenic plants overexpressing the gene exhibited enhanced tolerance to heat stress	Zang et al. (2017b)
6	*TaGASR1*	Overexpressing showed improved tolerance to heat stress and oxidative stress	Zhang et al. (2017a)
7	*TaB2*	Overexpression of TaB2 showed enhanced tolerance to heat stress	Singh and Khurana (2016)
8	*TaHsfC2a*	Overexpression of the gene showed improved thermo-tolerance	Hu et al. (2018)
9	*TaWRKY33*	*TaWRKY33*-transgenic lines showed enhanced tolerance to heat stress	He et al. (2016)
10	*TaNAC2L*	Overexpression of *TaNAC2L* enhanced heat tolerance by activating expression of heat-related genes	Guo et al. (2015)
11	*TaLTP3*	*TaLTP3* -overexpressing plants showed higher thermo-tolerance than control plants at the seedling stage	Wang et al. (2014)
12	*TaHsfA2d*	Transgenic plants overexpressing *TaHsfA2d* exhibited improved thermo-tolerance	Chauhan et al. (2013)

(continued)

Table 3.9 (continued)

S. No.	Genes	Features	References
13	*TaHSF3*	Enhanced tolerance to extreme temperatures	Zhang et al. (2013)
14	*HSP26*	Transgenic plants were more tolerant under continuous high temperature than wild-type plants	Chauhan et al. (2012)
15	*TaPEPKR2*	Enhanced heat and drought tolerance both in wheat and Arabidopsis	Zang et al. (2018)
16	*AtWRKY30*	Enhanced heat tolerance gas-exchange attributes, antioxidant machinery, osmolytes biosynthesis, and stress-related gene expression	El-Esawi et al. (2019)
17	*TaSINA*	Increases in biomass and yield in hot climates	Thomelin et al. (2021)
18	*TaNAC69-1*	Cellular responses to osmotic stress	Baloglu et al. (2012)
19	*TtNAMB-2*	Cellular responses to osmotic stress	Baloglu et al. (2012)
20	*sHSP26*	Seed maturation and response to high temperature stress	Chauhan et al. (2012)
21	*TaRca1*	Involved in photosynthesis	Kumar et al. (2016)
22	*TaHSFA6e*	Gene involved in regulation of heat shock proteins	Kumar et al. (2018)
23	*TaFBA1*	Encodes a homologous F-box protein involved in ROS scavenging	Li et al. (2018b)
24	*RuBisCo activase*	Modulate the activity of RuBisCo and protect the nascent proteins from aggregation	Kumar et al. (2019)
25	*TaHSP23.9*	Works as a protein chaperone	Wang et al. (2020a)
Important functional genes associated with drought stress tolerance			
26	*WRKY2, WRKY19*	Transcription factor: WRKY type TF	Niu et al. (2012)
27	*MYB33*	Transcription factor: R2R3 type MYB TF	Qin et al. (2012)
28	*PIMP1*	Transcription factor: R2R3 type MYB TF	Liu et al. (2011)
29	*NAC (NAM/ATAF /CUC)*	Transcription factor: plant-specific NAC (NAM/ATAF/CUC)	Tang et al. (2012)

(continued)

Table 3.9 (continued)

S. No.	Genes	Features	References
30	*ABC1*	Kinase: protein kinase ABC1 (activity of bc(1) complex)	Wang et al. (2011a)
31	*SnRK2.4*	Kinase: SNF1-type serine/threonine protein kinase	Mao et al. (2010)
32	*SnRK2.7*	Kinase: SNF1-type serine/threonine protein kinase	Zhang et al. (2011)
33	*CP*	Protein degradation: cysteine protease	Zang et al. (2010)
34	*TaSIP*	Salt induced protein with unknown function	Du et al. (2013)
35	*PIP1;1, PIP1;2*	Protective protein: aquaporin	Ayadi et al. (2011)
36	*HVAI*	Protective protein: LEA	Chauhan and Khurana (2011)
37	*ABA08*	ABA catabolism: ABA 8–hydroxylase	Ji et al. (2011)
38	*ATG8*	Autophagy: autophagy related gene 8	Kuzuoglu-Ozturk et al. (2012)
39	*Era1, Sal1*	Enhanced response to ABA, inositol polyphosphate 1-phosphatase	Manmathan et al. (2013)
40	*TaEXPA2*	Exhibited drought tolerant phenotypes, overexpression of TaEXPA2 enhanced the antioxidant capacity	Yang et al. (2020b)
Important functional genes associated with salinity stress tolerance			
41	*TaDi19A*	A salt-responsive gene	Li et al. (2010)
42	*TaOPR1*	Salinity tolerance via enhancement of abscisic acid signaling and ROS scavenging	Dong et al. (2013)
43	*TaCYP81D5*	Salinity tolerance via ROS scavenging	Wang et al. (2020c)
44	*TaAOC1*	Salinity tolerance via jasmonate signalling	Zhao et al. (2014)

have been identified and mapped for as many as 19 agronomic traits (mainly for thousand grain weight, grain number per spike, grain yield, grain weight per spike and plant height) followed by around 14 physiological traits (mainly for canopy temperature, normalized difference vegetative index (NDVI), SPAD/chlorophyll content, grain filling duration, water-soluble carbohydrates, fag leaf temperature depression) (Gupta et al., 2020). Many of these QTLs were either unstable (specific to one environment only) and/or minor; only 18 (5 QTLs for physiological traits and 13 QTLs for agronomic traits) major and stable QTLs ($\geq$ 20% PVE; identified in $\geq$

50% environments) were reported (Gupta et al., 2020). Among physiological traits, canopy temperature has received maximum attention of the wheat breeders as a selection measure for heat tolerance and fortunately major and stable QTLs for canopy temperature have also been reported which can be exploited for breeding programmes (Mason & Singh, 2014).

Recently in 2021, we conducted a meta-analysis of previously available QTLs and identified a total of 85 MQTLs associated with different heat stress-responsive traits (Kumar et al., 2021). These MQTLs and other major and stable QTLs can be targeted by MAS for the development of heat tolerant wheat cutivars.

Further, various epistatic interactions (Q × Q) involving different pairs of QTLs have also been reported (http://www.wheatqtldb.net/). Therefore, these interactions must also be taken into consideration while planning for MAS. GWAS has also been proven to be a good approach for identifying genes underlying heat tolerance. During the last 5 years, at least 15 GWAS were conducted where around 960 MTAs were identified using different association panels having genotypes/accessions ranging from 130 to 2111 (http://www.wheatqtldb.net/). Although many of these MTAs may be false positives because no false positive correction method such as Boneferroni correction was applied in most of these studies. SNPs involved in MTAs were also annotated in a few of these studies and were found to be linked with functional genes for biochemical activities related to abiotic stresses (e.g., Guo et al., 2020b; Pradhan et al., 2020). At least 24 candidate genes have also been identified and characterized for heat stress tolerance that are associated with different phenomena like oxidative stress, photosynthetic light reaction, carbohydrate metabolism, metal binding and so on (Table 3.9).

Genetics for Salinity Stress Tolerance in Wheat

Salinity is another major abiotic stress for wheat which constrains its growth and productivity. It affects the wheat life cycle starting from germination itself. Salt/salinity stress affects more than 800 Mha of global land and led to serious losses to wheat yield in several countries (Wang & Xia, 2018). Development and cultivation of salinity tolerant varieties can be the most promising way to combat the harmful effects of salinity (Ashraf & O'leary, 1996). Conventional breeding programs to improve wheat salt tolerance in wheat could not be much successful owing to the genetic and physiological complexes of the salt tolerance (Zeeshan et al., 2020). However, much can be expected from genomics-assisted breeding in wheat as we already have different genomic resources with us which can be used to better understand the underlying genetics of the trait and prepare the efficient breeding strategies. After the first study in 2004; using QTL mapping approach, around 500 QTLs (excluding those involved in epistatic interactions) for traits associated with salt tolerance have been identified and mapped on all wheat chromosomes so far (Gupta et al., 2020). The QTLs detected for salt tolerance at germination and early growth stages may be different from those detected at the adult plant stage (Yamaguchi & Bulmwald, 2005). Different traits including both root and shoot traits (such as Na^+ exclusion/

content, K^+ content, K^+/Na^+ ratio and so on) have been used in different studies for estimation of salt tolerance. In these studies, the PVE value explained by individual QTL ranged from 8.4 to 38%. Different epistatic QTLs were also identified for traits measured at seedling stage using QTL mapping (Masoudi et al., 2015). However, the contribution of these interactive QTLs to total phenotypic variation was generally low for each trait used in studies. Recently in 2021, using a meta-analysis approach, we defined a genome-wide landscape on the most consistent and stable genomic loci associated with reliable molecular markers and candidate genes for salinity tolerance in wheat (Pal et al., 2021). In wheat, there has been little research for the identification of MTAs using GWAS (reviewed in Pal et al., 2021). For instance, using a set of 150 cultivars, a GWAS study was performed which identified 4 important QTLs on 1BS, 2AL, 2BS and 3AL associated with salt tolerance across the three growth stages and with the ionic traits (leaf K^+ and Na^+ contents). These QTLs explained a good range of phenotypic variation (from 3 to 30.67%). Several candidate genes associated with salt tolerance were also uncovered in this study (Oyiga et al., 2018). Some functionally characterized salinity responsive genes are given in Table 3.9.

Genomics-Assisted Breeding for Developing Climate Resilient Wheat

Despite the great possibilities of the genomics-assisted breeding especially MAS in the genetic improvement of abiotic stress tolerance in wheat, we could not find much documented studies describing the practical application of this MAS for making wheat plants tolerant to abiotic stresses in wheat. Although few promising studies have been published for improving drought tolerance (Gautam et al., 2020; Merchuk-Ovnat et al., 2016; Rai et al., 2018; Todkar et al., 2020), but there is no documented study for heat and salinity tolerance in wheat. There could be many reasons behind the aforementioned facts such as (a) one may not have published the results of MAS, (b) reliability and accuracy of QTL mapping studies may be a major concern, (c) inadequate linkage between QTL and marker, (d) inadequate polymorphism of markers in breeding material, (e) genetic background effects, (f) various interactions such as QTL X environment, (g) high cost of MAS, (h) one major concern may be the “application gap” between research labs and breeding institutes, and the last (i) may be the “knowledge gap” among molecular scientists, plant breeders and other disciplines (Collard & Mackill, 2008).

In addition to these mentioned reasons, there may be other reasons of why even a single application of MAS in heat and salinity tolerance has not been published? For the first time in 2016, three QTLs from wild emmer wheat were successfully introgressed in elite durum and bread wheat cultivars through MAS to enhance drought tolerance. Generated near-isogenic lines (BC_3F_3 and BC_3F_4) showed improved grain yield and biomass under drought stress condition (Merchuk-Ovnat et al., 2016). In 2018, a most important study was published, where transfer of three linked QTLs from cultivar ‘HI500’ in to high-yielding wheat cultivar ‘HD2733’ via marker assisted backcross breeding (MABB) was reported. This study was the first of its kind where five potential drought tolerant varieties were identified for national varietal evaluation programme in the zone (Rai et al., 2018). More recently in 2020, genomic regions associated with drought tolerance traits viz., NDVI, chlorophyll fluorescence, stay green and grain yield were introgressed from 'HI500' in to a high yielding but drought sensitive wheat variety 'GW322' via MABB (Todkar et al., 2020). In another study, a major yield QTL was introgressed into each of the four Indian wheat cultivars (HUW468, HUW234, DBW17 and K307) (Gautam et al., 2020). The generated introgressed lines showed improved grain yield under both drought and normal conditions (Gautam et al., 2020).

Other option to utilize the wealth of genomic data is genome wide selection or genomic selection (GS). GS has been estimated to be twice as effective as MAS in wheat. GS requires precise phenotyping, which has been problematic until the arrival of high throughput phenotyping systems. Phenotypic observations are recorded on the training population (it may be a part of the whole population or a related population) that is then employed by the GS model for predictions. Prediction and selection accuracy can be increased by combining HTP phenotypic data and genomic information in wheat breeding under drought and heat conditions (Crain et al., 2018). Recently in 2020, an integrated R library called BWGS (Bread Wheat Genomic Selection) (https://breedwheat.fr) was developed to allow easy computation of GEBVs for GS, which will allow efficient use of genomic selection in applied breeding programmes (Charmet et al., 2020). In addition, transgenic technology has also been used as a potential tool to develop abiotic stress tolerant plants (Li et al., 2020a; Wang et al., 2020b; Zhou et al., 2020b). A number of genes have been functionally characterized in wheat for tolerance to drought, heat and salinity stress (Table 3.9), these can be used to generate the stress tolerant plants. Moreover, advances in targeted genome editing technologies such as CRISPR/cas9 have provided novel opportunities for wheat improvement. Although it has been used in wheat for targeting various traits of economic importance but, potential of this versatile tool is yet to be explored with regard to abiotic stress tolerance in wheat (Kaur et al., 2022a, b; Wang et al., 2020b, for detailed review).

3.3 Conclusions

Developing climate resilient varieties is crucial to tackle the global climate change in this post genomics era. Delivering and deploying climate resilient crops is difficult as it need to be done in rapid breeding cycles with increased precision. Intensive multi-institutional and/or multidisciplinary efforts are needed to identify and utilize climate resilient germplasm in pipelines of product development. Crop genetics and breeding research are under paradigm shift due to the availability of high-quality reference genome sequences of various genotypes and the broad utilization of high throughput genotyping and phenotyping platforms. Availability of these genomics and phenomics resources in crops provides the potential of accelerating the rate of genetic progress by increasing the selection accuracy and efficiency in breeding programs. The evolving techniques of NGS are making it extremely useful specifically for the accurate identification of the genetic factors that affects the trait via QTL mapping and GWAS analysis tools. In summary, the exponential advancements in plant phenomics, genomics tools and resources, increase in the number of gene discovery/identification studies, and above all the promising targeted genome editing techniques have great potential to deliver future climate resilient varieties, if integrated consciously.

References

3000 Rice Genomes Project. (2014). The 3000 rice genomes project. *GigaScience, 3*(1), 2047–217X.

Abdelghany, M., Liu, X., Hao, L., Gao, C., Kou, S., Su, E., Zhou, Y., Wang, R., Zhang, D., Li, Y., & Song, Y. (2019). QTL analysis for yield-related traits under different water regimes in maize. *Maydica, 64*(2), 10.

Abdulmalik, R. O., Menkir, A., Meseka, S. K., Unachukwu, N., Ado, S. G., Olarewaju, J. D., Aba, D. A., Hearne, S., Crossa, J., & Gedil, M. (2017). Genetic gains in grain yield of a maize population improved through marker assisted recurrent selection under stress and non-stress conditions in West Africa. *Frontiers in Plant Science, 8*, 841.

Ajnone-Marsan, P., Monfredini, G., Ludwig, W. F., Melchinger, A. E., Franceschini, P., Pagnotto, G., & Motto, M. (1995). In an elite cross of maize a major quantitative trait locus controls one-fourth of the genetic variation for grain yield. *Theoretical and Applied Genetics, 90*(3–4), 415–424.

Akram, R., Fahad, S., Masood, N., Rasool, A., Ijaz, M., Ihsan, M. Z., Maqbool, M. M., Ahmad, S., Hussain, S., Ahmed, M., & Kaleem, S. (2019). Plant growth and morphological changes in Rice under abiotic stress. In *Advances in rice research for abiotic stress tolerance* (pp. 69–85). Woodhead Publishing.

Andaya, V. C., & Mackill, D. J. (2003). Mapping of QTLs associated with cold tolerance during the vegetative stage in rice. *Journal of Experimental Botany, 54*(392), 2579–2585.

Angaji, S. A., Septiningsih, E. M., Mackill, D. J., & Ismail, A. M. (2010). QTLs associated with tolerance of flooding during germination in rice (Oryza sativa L.). *Euphytica, 172*(2), 159–168.

Araus, J. L., Serret, M. D., & Edmeades, G. (2012). Phenotyping maize for adaptation to drought. *Frontiers in Physiology, 3*, 305.

Arshad, M. S., Farooq, M., Asch, F., Krishna, J. S., Prasad, P. V., & Siddique, K. H. (2017). Thermal stress impacts reproductive development and grain yield in rice. *Plant Physiology and Biochemistry, 115*, 57–72.

Ashraf, M., & O'leary, J. W. (1996). Responses of some newly developed salt-tolerant genotypes of spring wheat to salt stress: 1. Yield components and ion distribution. *Journal of Agronomy and Crop Science, 176*(2), 91–101.

Austin, D. F., & Lee, M. (1996). Comparative mapping in F 2: 3 and F 6: 7 generations of quantitative trait loci for grain yield and yield components in maize. *Theoretical and Applied Genetics, 92*(7), 817–826.

Ayadi, M., Cavez, D., Miled, N., Chaumont, F., & Masmoudi, K. (2011). Identification and characterization of two plasma membrane aquaporins in durum wheat (Triticum turgidum L. subsp. durum) and their role in abiotic stress tolerance. *Plant Physiology and Biochemistry, 49*(9), 1029–1039.

Baloglu, M. C., Oz, M. T., Oktem, H. A., & Yucel, M. (2012). Expression analysis of TaNAC69-1 and TtNAMB-2, wheat NAC family transcription factor genes under abiotic stress conditions in durum wheat (Triticum turgidum). *Plant Molecular Biology Reporter, 30*(5), 1246–1252.

Baltazar, M. D., Ignacio, J. C. I., Thomson, M. J., Ismail, A. M., Mendioro, M. S., & Septiningsih, E. M. (2019). QTL mapping for tolerance to anaerobic germination in rice from IR64 and the aus landrace Kharsu 80A. *Breeding Science, 69*(2), 227–233.

Bankole, F., Menkir, A., Olaoye, G., Crossa, J., Hearne, S., Unachukwu, N., & Gedil, M. (2017). Genetic gains in yield and yield related traits under drought stress and favorable environments in a maize population improved using marker assisted recurrent selection. *Frontiers in Plant Science, 8*, 808.

Barik, S. R., Pandit, E., Mohanty, S. P., Nayak, D. K., & Pradhan, S. K. (2020). Genetic mapping of physiological traits associated with terminal stage drought tolerance in rice. *BMC Genetics, 21*(1), 1–12.

Barik, S. R., Pandit, E., Pradhan, S. K., Mohanty, S. P., & Mohapatra, T. (2019). Genetic mapping of morpho-physiological traits involved during reproductive stage drought tolerance in rice. *Plos One, 14*(12), e0214979.

Barik, S. R., Pandit, E., Pradhan, S. K., Singh, S., Swain, P., & Mohapatra, T. (2018). QTL mapping for relative water content trait at reproductive stage drought stress in rice. *Indian Journal of Genetics, 78*(4), 401–408.

Baruah, A. R., Ishigo-Oka, N., Adachi, M., Oguma, Y., Tokizono, Y., Onishi, K., & Sano, Y. (2009). Cold tolerance at the early growth stage in wild and cultivated rice. *Euphytica, 165*(3), 459–470.

Basu, P. S., Srivastava, M., Singh, P., Porwal, P., Kant, R., & Singh, J. (2015). High-precision phenotyping under controlled versus natural environments. In *Phenomics in crop plants: Trends, options and limitations* (pp. 27–40). Springer.

Belhaj, K., Chaparro-Garcia, A., Kamoun, S., Patron, N. J., & Nekrasov, V. (2015). Editing plant genomes with CRISPR/Cas9. *Current Opinion in Biotechnology, 32*, 76–84.

Bennett, D., Izanloo, A., Reynolds, M., Kuchel, H., Langridge, P., & Schnurbusch, T. (2012). Genetic dissection of grain yield and physical grain quality in bread wheat (Triticum aestivum L.) under water-limited environments. *Theoretical and Applied Genetics, 125*(2), 255–271.

Bernier, J., Kumar, A., Ramaiah, V., Spaner, D., & Atlin, G. (2007). A large-effect QTL for grain yield under reproductive-stage drought stress in upland rice. *Crop Science, 47*(2), 507–516.

Beyene, Y., Semagn, K., Mugo, S., Tarekegne, A., Babu, R., Meisel, B., Sehabiague, P., Makumbi, D., Magorokosho, C., Oikeh, S., & Gakunga, J. (2015). Genetic gains in grain yield through genomic selection in eight bi-parental maize populations under drought stress. *Crop Science, 55*(1), 154–163.

Bhattarai, U., & Subudhi, P. K. (2018). Identification of drought responsive QTLs during vegetative growth stage of rice using a saturated GBS-based SNP linkage map. *Euphytica, 214*(2), 38.

Bimpong, I. K., Manneh, B., El-Namaky, R., Diaw, F., Amoah, N. K. A., Sanneh, B., Ghislain, K., Sow, A., Singh, R.K., Gregorio, G., & Bizimana, J. B. (2014). Mapping QTLs related to salt tolerance in rice at the young seedling stage using 384-plex single nucleotide polymorphism SNP, marker sets. *Molecular Plant Breeding, 5*.

Biswas, P. S., Khatun, H., Das, N., Sarker, M. M., & Anisuzzaman, M. (2017). Mapping and validation of QTLs for cold tolerance at seedling stage in rice from an indica cultivar Habiganj Boro VI (Hbj. BVI). *3 Biotech, 7*(6), 359.

Bizimana, J. B., Luzi-Kihupi, A., Murori, R. W., & Singh, R. K. (2017). Identification of quantitative trait loci for salinity tolerance in rice (Oryza sativa L.) using IR29/Hasawi mapping population. *Journal of Genetics, 96*(4), 571–582.

Bodner, G., Nakhforoosh, A., Arnold, T., & Leitner, D. (2018). Hyperspectral imaging: a novel approach for plant root phenotyping. *Plant Methods, 14*(1), 84.

Bonell, M. L., Arguissain, G. G., Crepy, M. A., del Carmen Fernández, P., & Heinz, R. A. (2020). Phenotyping and QTL mapping for cold tolerance at the germination and seedling stages in Argentine temperate rice. *Euphytica, 216*(8), 1–15.

Borianne, P., Subsol, G., Fallavier, F., Dardou, A., & Audebert, A. (2018). GT-RootS: An integrated software for automated root system measurement from high-throughput phenotyping platform images. *Computers and Electronics in Agriculture, 150*, 328–342.

Bouman, B. A. (Ed). (2002). *Water-wise rice production* (Vol. 1). International Rice Research Institute.

Bukowski, R., Guo, X., Lu, Y., Zou, C., He, B., Rong, Z., Wang, B., Xu, D., Yang, B., Xie, C., & Fan, L. (2018). Construction of the third-generation Zea mays haplotype map. *Gigascience, 7*(4), gix134.

Buu, B. C., Ha, P. T. T., Tam, B. P., Nhien, T. T., Hieu, N. V., & Phuoc, N. T., et al. (2014). Quantitative trait loci associated with heat tolerance in rice (Oryza sativa L.). *Plant Breeding and Biotechnology, 2*, 14–24.

Cao, L., Zhao, J., Zhan, X., Li, D., He, L., & Cheng, S. (2003). Mapping QTLs for heat tolerance and correlation between heat tolerance and photosynthetic rate in rice. *Chinese Journal of Rice Science, 17*(3), 223–227.

Cao, Y., Liang, X., Yin, P., Zhang, M., & Jiang, C. (2019). A domestication-associated reduction in K+-preferring HKT transporter activity underlies maize shoot K+ accumulation and salt tolerance. *New Phytologist, 222*(1), 301–317.

Casaretto, J. A., El-kereamy, A., Zeng, B., Stiegelmeyer, S. M., Chen, X., Bi, Y. M., & Rothstein, S. J. (2016). Expression of OsMYB55 in maize activates stress-responsive genes and enhances heat and drought tolerance. *BMC Genomics, 17*(1), 1–15.

Cerrudo, D., Cao, S., Yuan, Y., Martinez, C., Suarez, E. A., Babu, R., Zhang, X., & Trachsel, S. (2018). Genomic selection outperforms marker assisted selection for grain yield and physiological traits in a maize doubled haploid population across water treatments. *Frontiers in Plant Science, 9*, 366.

Chang-lan, Z. H. U., Ying-hui, X. I. A. O., Chun-ming, W. A. N. G., Ling, J. I. A. N. G., Hu-Qu, Z., & Jian-min, W. A. N. (2005). Mapping QTL for heat-tolerance at grain filling stage in rice. *Rice Science, 12*(1), 33–38.

Charmet, G., Tran, L. G., Auzanneau, J., Rincent, R., & Bouchet, S. (2020). BWGS: AR package for genomic selection and its application to a wheat breeding programme. *Plos One, 15*(4), e0222733.

Chauhan, H., & Khurana, P. (2011). Use of doubled haploid technology for development of stable drought tolerant bread wheat (Triticum aestivum L.) transgenics. *Plant Biotechnology Journal, 9*(3), 408–417.

Chauhan, H., Khurana, N., Agarwal, P., Khurana, J. P., & Khurana, P. (2013). A seed preferential heat shock transcription factor from wheat provides abiotic stress tolerance and yield enhancement in transgenic Arabidopsis under heat stress environment. *PLoS One, 8*(11), e79577.

Chauhan, H., Khurana, N., Nijhavan, A., Khurana, J. P., & Khurana, P. (2012). The wheat chloroplastic small heat shock protein (sHSP26) is involved in seed maturation and germination and imparts tolerance to heat stress. *Plant, Cell & Environment, 35*(11), 1912–1931.

Chen, J., Xu, W., Velten, J., Xin, Z., & Stout, J. (2012). Characterization of maize inbred lines for drought and heat tolerance. *Journal of Soil and Water Conservation, 67*(5), 354–364.

Chen, L., An, Y., Li, Y. X., Li, C., Shi, Y., Song, Y., Zhang, D., Wang, T., & Li, Y. (2017). Candidate loci for yield-related traits in maize revealed by a combination of metaQTL analysis and regional association mapping. *Frontiers in Plant Science, 8*, 2190.

Chen, T., Zhu, Y., Chen, K., Shen, C., Zhao, X., Shabala, S., Shabala, L., Meinke, H., Venkataraman, G., Chen, Z. H., & Xu, J. (2020). Identification of new QTL for salt tolerance from rice variety Pokkali. *Journal of Agronomy and Crop Science, 206*(2), 202–213.

Chen, X., Jiang, L., Zheng, J., Chen, F., Wang, T., Wang, M., Tao, Y., Wang, H., Hong, Z., Huang, Y., & Huang, R. (2019). A missense mutation in large grain size 1 increases grain size and enhances cold tolerance in rice. *Journal of Experimental Botany, 70*(15), 3851–3866.

Cheng, L. R., Wang, J. M., Uzokwe, V., Meng, L. J., Yun, W. A. N. G., Yong, S. U. N., Zhu, L. H., Xu, J. L., & Li, Z. K. (2012). Genetic analysis of cold tolerance at seedling stage and heat tolerance at anthesis in rice (Oryza sativa L.). *Journal of Integrative Agriculture, 11*(3), 359–367.

Chia, J. M., Song, C., Bradbury, P., Costich, D., de Leon, N., Doebley, J. C., Elshire, R. J., Gaunt, B. S., Geller, L., Glaubitz, J. C., & Gore, M. (2012). Capturing extant variation from a genome in flux: Maize HapMap II. *Nature Genetics, 44*, 803–807.

Chowdhury, A. D., Haritha, G., Sunitha, T., Krishnamurthy, S. L., Divya, B., Padmavathi, G., Ram, T., & Sarla, N. (2016). Haplotyping of rice genotypes using simple sequence repeat markers associated with salt tolerance. *Rice Science, 23*(6), 317–325.

Collard, B. C., & Mackill, D. J. (2008). Marker-assisted selection: an approach for precision plant breeding in the twenty-first century. *Philosophical Transactions of the Royal Society B: Biological Sciences, 363*(1491), 557–572.

Coumou, D., & Robinson, A. (2013). Historic and future increase in the global land area affected by monthly heat extremes. *Environmental Research Letters, 8*(3), 034018.

Crain, J., Mondal, S., Rutkoski, J., Singh, R. P., & Poland, J. (2018). Combining high-throughput phenotyping and genomic information to increase prediction and selection accuracy in wheat breeding. *The Plant Genome, 11*(1), 1–14.

Crain, J., Reynolds, M., & Poland, J. (2017). Utilizing high-throughput phenotypic data for improved phenotypic selection of stress-adaptive traits in wheat. *Crop Science, 57*(2), 648–659.

Cui, D., Wu, D., Somarathna, Y., Xu, C., Li, S., Li, P., Zhang, H., Chen, H., & Zhao, L. (2015). QTL mapping for salt tolerance based on snp markers at the seedling stage in maize (Zea mays L.). *Euphytica, 203*(2), 273–283.

Cui, L. H., Byun, M. Y., Oh, H. G., Kim, S. J., Lee, J., Park, H., Lee, H., & Kim, W. T. (2020). Poaceae type II galactinol synthase 2 from antarctic flowering plant deschampsia antarctica and rice improves cold and drought tolerance by accumulation of raffinose family oligosaccharides in transgenic rice plants. *Plant and Cell Physiology, 61*(1), 88–104.

Cui, Y., Zhang, W., Lin, X., Xu, S., Xu, J., & Li, Z. (2018). Simultaneous improvement and genetic dissection of drought tolerance using selected breeding populations of rice. *Frontiers in Plant Science, 9*, 320.

Dang, X., Li, Y., Zhang, Y., Ji, J., Li, D., Hu, X., Zhu, S., Dong, Z., Liu, E., Wang, H., & Fang, B. (2019). Identification of favorable alleles for rice seedling anoxic tolerance using natural and bi-parental populations. *Euphytica, 215*(8), 140.

de Ocampo, M. P., Thomson, M. J., Mitsuya, S., Yamauchi, A., & Ismail, A.M. (2020). QTL mapping and candidate gene identification in rice using a Kalarata-Azucena population under salt stress.

Dixit, S., Singh, A., Cruz, M. T. S., Maturan, P. T., Amante, M., & Kumar, A. (2014). Multiple major QTL lead to stable yield performance of rice cultivars across varying drought intensities. *BMC Genetics, 15*(1), 16.

Dixit, S., Singh, A., Sandhu, N., Bhandari, A., Vikram, P., & Kumar, A. (2017). Combining drought and submergence tolerance in rice: marker-assisted breeding and QTL combination effects. *Molecular Breeding, 37*(12), 143.

Dixit, S., Swamy, B.M., Vikram, P., Bernier, J., Cruz, M. S., Amante, M., Atri, D., & Kumar, A. (2012). Increased drought tolerance and wider adaptability of qDTY 12.1 conferred by its interaction with qDTY 2.3 and qDTY 3.2. *Molecular Breeding, 30*(4), 1767–1779.

Dong, W., Wang, M., Xu, F., Quan, T., Peng, K., Xiao, L., & Xia, G. (2013). Wheat oxophytodienoate reductase gene TaOPR1 confers salinity tolerance via enhancement of abscisic acid signaling and reactive oxygen species scavenging. *Plant Physiology, 161*(3), 1217–1228.

Du, H. Y., Shen, Y. Z., & Huang, Z. J. (2013). Function of the wheat TaSIP gene in enhancing drought and salt tolerance in transgenic Arabidopsis and rice. *Plant Molecular Biology, 81*(4–5), 417–429.

Edmeades, G.O. (2008). Drought tolerance in maize: An emerging reality. International Service for the Acquisition of Agri-Biotech Applications (ISAAA).

Edmeades, G. O. (2013). Progress in achieving and delivering drought tolerance in maize-an update (p. 130). ISAAA

Edreira, J. R., & Otegui, M. E. (2013). Heat stress in temperate and tropical maize hybrids: A novel approach for assessing sources of kernel loss in field conditions. *Field Crops Research, 142*, 58–67.

El-Esawi, M. A., Al-Ghamdi, A. A., Ali, H. M., & Ahmad, M. (2019). Overexpression of AtWRKY30 transcription factor enhances heat and drought stress tolerance in wheat (Triticum aestivum L.). *Genes, 10*(2), 163.

Elshire, R. J., Glaubitz, J. C., Sun, Q., Poland, J. A., Kawamoto, K., Buckler, E. S., & Mitchell, S. E. (2011). A robust, simple genotyping-by-sequencing (GBS) approach for high diversity species. *PloS one, 6*(5), e19379.

Falcon, W. P., & Naylor, R. L. (1998). The maize transition in Asia: Unlocking the controversy. *American Journal of Agricultural Economics, 80*(5), 960–968.

Frey, F. P., Presterl, T., Lecoq, P., Orlik, A., & Stich, B. (2016). First steps to understand heat tolerance of temperate maize at adult stage: identification of QTL across multiple environments with connected segregating populations. *Theoretical and Applied Genetics, 129*(5), 945–961.

Frouin, J., Languillaume, A., Mas, J., Mieulet, D., Boisnard, A., Labeyrie, A., Bettembourg, M., Bureau, C., Lorenzini, E., Portefaix, M., & Turquay, P. (2018). Tolerance to mild salinity stress in japonica rice: A genome-wide association mapping study highlights calcium signaling and metabolism genes. *PloS One, 13*(1), e0190964.

Fukai, S., & Cooper, M. (1995). Development of drought-resistant cultivars using physiomorphological traits in rice. *Field Crops Research, 40*(2), 67–86.

Gahlaut, V., Jaiswal, V., Kumar, A., & Gupta, P. K. (2016). Transcription factors involved in drought tolerance and their possible role in developing drought tolerant cultivars with emphasis on wheat (Triticum aestivum L.). *Theoretical and Applied Genetics, 129*(11), 2019–2042.

Gao, H., Zhang, C., He, H., Liu, T., Zhang, B., Lin, H., Li, X., Wei, Z., Yuan, Q., Wang, Q., & Yu, C. (2020). Loci and alleles for submergence responses revealed by GWAS and transcriptional analysis in rice. *Molecular Breeding, 40*(8), 1–16.

Gao, J., Wang, S., Zhou, Z., Wang, S., Dong, C., Mu, C., Song, Y., Ma, P., Li, C., Wang, Z., & He, K. (2019). Linkage mapping and genome-wide association reveal candidate genes conferring thermo-tolerance of seed-set in maize. *Journal of Experimental Botany, 70*(18), 4849–4864.

Gautam, T., Saripalli, G., Kumar, A., Gahlaut, V., Gadekar, D. A., Oak, M., Sharma, P. K., Balyan, H. S., & Gupta, P. K. (2020). Introgression of a drought insensitive grain yield QTL for improvement of four Indian bread wheat cultivars using marker assisted breeding without background selection. *Journal of Plant Biochemistry and Biotechnology*, 1–12.

Ge, Y., Atefi, A., Zhang, H., Miao, C., Ramamurthy, R. K., Sigmon, B., Yang, J., & Schnable, J. C. (2019). High-throughput analysis of leaf physiological and chemical traits with VIS–NIR–SWIR spectroscopy: A case study with a maize diversity panel. *Plant Methods, 15*(1), 66.

Geetha, S., Vasuki, A., SeLvam, P. J., Saraswathi, R., Krishnamurthy, S. L., Dhasarathan, M., Thamodharan, G., & Baskar, M. (2017). Development of sodicity tolerant rice varieties through marker assisted backcross breeding. *Electronic Journal of Plant Breeding, 8*(4), 1013–1031.

Ghimire, K. H., Quiatchon, L. A., Vikram, P., Swamy, B. M., Dixit, S., Ahmed, H., Hernandez, J. E., Borromeo, T. H., & Kumar, A. (2012). Identification and mapping of a QTL (qDTY1. 1) with a consistent effect on grain yield under drought. *Field Crops Research, 131*, 88–96.

Ghomi, K., Rabiei, B., Sabouri, H., & Sabouri, A. (2013). Mapping QTLs for traits related to salinity tolerance at seedling stage of rice (Oryza sativa L.): an agrigenomics study of an Iranian rice population. *Omics: A Journal of Integrative Biology, 17*(5), 242–251.

Gill, T., Gill, S. K., Saini, D. K., Chopra, Y., de Koff, J. P., & Sandhu, K. S. (2022). A comprehensive review of high throughput phenotyping and machine learning for plant stress phenotyping. *Phenomics*, 1–28.

Goff, S. A., Ricke, D., Lan, T. H., Presting, G., Wang, R., Dunn, M., Glazebrook, J., Sessions, A., Oeller, P., Varma, H., Hadley, D., Macalma, T., Oliphant, A., & Briggs, S. (2002) A draft sequence of the rice genome (Oryza sativa L. ssp. Japonica). *Science, 296*, 92–100.

Gomez, S. M., Boopathi, N. M., Kumar, S. S., Ramasubramanian, T., Chengsong, Z., Jeyaprakash, P., Senthil, A., & Babu, R. C. (2010). Molecular mapping and location of QTLs for drought-resistance traits in indica rice (Oryza sativa L.) lines adapted to target environments. *Acta Physiologiae Plantarum, 32*(2), 355–364.

Gonzaga, Z. J. C., Carandang, J., Sanchez, D. L., Mackill, D. J., & Septiningsih, E. M. (2016). Mapping additional QTLs from FR13A to increase submergence tolerance in rice beyond SUB1. *Euphytica, 209*(3), 627–636.

Gonzaga, Z. J. C., Carandang, J., Singh, A., Collard, B. C., Thomson, M. J., & Septiningsih, E. M. (2017). Mapping QTLs for submergence tolerance in rice using a population fixed for SUB1A tolerant allele. *Molecular Breeding, 37*(4), 47.

Gore, M. A., Chia, J. M., Elshire, R. J., Sun, Q., Ersoz, E. S., Hurwitz, B. L., Peiffer, J. A., McMullen, M. D., Grills, G. S., Ross-Ibarra, J., & Ware, D. H. (2009). A first-generation haplotype map of maize. *Science, 326*(5956), 1115–1117.

Grattan, S. R., & Grieve, C. M. (1999). Mineral nutrient acquisition and response by plants grown in saline environments. *Handbook of Plant and Crop Stress, 2*, 203–229.

Guo, J., Li, C., Zhang, X., Li, Y., Zhang, D., Shi, Y., Song, Y., Li, Y., Yang, D., & Wang, T. (2020a). Transcriptome and GWAS analyses reveal candidate gene for seminal root length of maize seedlings under drought stress. *Plant Science, 292*, 110380.

Guo, J., Shi, W., Guo, J., Yue, L., Zhuang, L., Zhang, W., Tian, X., Xu, Y., & Zhou, M. (2020b). Genome-wide association studies on heat stress tolerance during grain development in wheat (Triticum aestivum L.).

Guo, Z., Yang, W., Chang, Y., Ma, X., Tu, H., Xiong, F., Jiang, N., Feng, H., Huang, C., Yang, P., & Zhao, H. (2018). Genome-wide association studies of image traits reveal genetic architecture of drought resistance in rice. *Molecular Plant, 11*(6), 789–805.

Guo, W., Zhang, J., Zhang, N., Xin, M., Peng, H., Hu, Z., et al. (2015). The wheat NAC transcription factor TaNAC2L is regulated at the transcriptional and post-translational levels and promotes heat stress tolerance in transgenic Arabidopsis. *PLoS One, 10*(8), e0135667.

Gupta, P. K., Balyan, H. S., Sharma, S., & Kumar, R. (2020). Genetics of yield, abiotic stress tolerance and biofortification in wheat (Triticum aestivum L.). *Theoretical and Applied Genetics*, 1–34.

Hairmansis, A., Berger, B., Tester, M., & Roy, S. J. (2014). Image-based phenotyping for non-destructive screening of different salinity tolerance traits in rice. *Rice, 7*(1), 1–10.

Han, L., Yang, G., Dai, H., Yang, H., Xu, B., Li, H., Long, H., Li, Z., Yang, X., & Zhao, C. (2019). Combining self-organizing maps and biplot analysis to preselect maize phenotypic components based on UAV high-throughput phenotyping platform. *Plant Methods, 15*(1), 1–16.

Hao, Z., Li, X., Liu, X., Xie, C., Li, M., Zhang, D., & Zhang, S. (2010). Meta-analysis of constitutive and adaptive QTL for drought tolerance in maize. *Euphytica, 174*(2), 165–177.

Harfouche, A. L., Jacobson, D. A., Kainer, D., Romero, J. C., Harfouche, A. H., Mugnozza, G. S., Moshelion, M., Tuskan, G. A., Keurentjes, J. J., & Altman, A. (2019). Accelerating climate resilient plant breeding by applying next-generation artificial intelligence. *Trends in biotechnology, 37*(11), 1217–1235.

Hattori, Y., Miura, K., Asano, K., Yamamoto, E., Mori, H., Kitano, H., Matsuoka, M., & Ashikari, M. (2007). A major QTL confers rapid internode elongation in response to water rise in deepwater rice. *Breeding Science, 57*(4), 305–314.

He, G. H., Xu, J. Y., Wang, Y. X., Liu, J. M., Li, P. S., Chen, M., Ma, Y. Z., & Xu, Z. S. (2016). Drought-responsive WRKY transcription factor genes TaWRKY1 and TaWRKY33 from wheat confer drought and/or heat resistance in Arabidopsis. *BMC Plant Biology, 16*(1), 116.

Hemamalini, G. S., Shashidhar, H. E., & Hittalmani, S. (2000). Molecular marker assisted tagging of morphological and physiological traits under two contrasting moisture regimes at peak vegetative stage in rice (Oryza sativa L.). *Euphytica, 112*(1), 69–78.

Hirabayashi, H., Sasaki, K., Kambe, T., Gannaban, R. B., Miras, M. A., Mendioro, M. S., Simon, E. V., Lumanglas, P. D., Fujita, D., Takemoto-Kuno, Y., & Takeuchi, Y. (2015). QEMF3, a novel QTL for the early-morning flowering trait from wild rice, Oryza officinalis, to mitigate heat stress damage at flowering in rice O. sativa. *Journal of Experimental Botany, 66*(5), 1227–1236.

Hirsch, C. N., Foerster, J. M., Johnson, J. M., Sekhon, R. S., Muttoni, G., Vaillancourt, B., Peña-garicano, F., Lindquist, E., Pedraza, M. A., Barry, K., & de Leon, N. (2014). Insights into the maize pan-genome and pan-transcriptome. *The Plant Cell, 26*(1), 121–135.

Hisas, L. (2011). *The food gap: The impacts of climate change on food production: A 2020 perspective* (p. 55). United States of America: FEU-US Universal Ecological Fund.
Hongxuan, L., Yanagihara, S., Jieyun, Z., Senboku, T., Kangle, Z., & Yashima, S. (1998). Identification of QTL for salt tolerance in rice via molecular markers. *Zhongguo Shuidao Kexue, 12*(2), 72–78.
Hossain, H., Rahman, M. A., Alam, M. S., & Singh, R. K. (2015). Mapping of quantitative trait loci associated with reproductive-stage salt tolerance in rice. *Journal of Agronomy and Crop Science, 201*(1), 17–31.
Hsu, S. K., & Tung, C. W. (2015). Genetic mapping of anaerobic germination-associated QTLs controlling coleoptile elongation in rice. *Rice, 8*(1), 1–12.
Hu, X. J., Chen, D., Lynne McIntyre, C., Fernanda Dreccer, M., Zhang, Z. B., Drenth, J., Kalaipandian, S., Chang, H., & Xue, G. P. (2018). Heat shock factor C2a serves as a proactive mechanism for heat protection in developing grains in wheat via an ABA-mediated regulatory pathway. *Plant, Cell & Environment, 41*(1), 79–98.
Huang, H. E., Ho, M. H., Chang, H., Chao, H. Y., & Ger, M. J. (2020). Overexpression of plant ferredoxin-like protein promotes salinity tolerance in rice (Oryza sativa). *Plant Physiology and Biochemistry.*
Huang, J., Sun, S., Xu, D., Lan, H., Sun, H., Wang, Z., Bao, Y., Wang, J., Tang, H., & Zhang, H. (2012a). A TFIIIA-type zinc finger protein confers multiple abiotic stress tolerances in transgenic rice (Oryza sativa L.). *Plant Molecular Biology, 80*(3), 337–350.
Huang, X., & Han, B. (2014). Natural variations and genome-wide association studies in crop plants. *Annual Review of Plant Biology, 65*, 531–551.
Huang, X., Zhao, Y., Li, C., Wang, A., Zhao, Q., Li, W., Guo, Y., Deng, L., Zhu, C., Fan, D., & Lu, Y. (2012b). Genome-wide association study of flowering time and grain yield traits in a worldwide collection of rice germplasm. *Nature Genetics, 44*(1), 32–39.
International Wheat Genome Sequencing Consortium. (2014). A chromosome-based draft sequence of the hexaploid bread wheat (Triticum aestivum) genome. *Science, 345*(6194).
International Wheat Genome Sequencing Consortium. (2018). Shifting the limits in wheat research and breeding using a fully annotated reference genome. *Science, 361.*
Jackson, M. B., & Ismail, A. M. (2015). Introduction to the Special Issue: Electrons, water and rice fields: Plant response and adaptation to flooding and submergence stress. *AoB Plants, 7.*
Jagadish, S. V. K., Cairns, J., Lafitte, R., Wheeler, T. R., Price, A. H., & Craufurd, P. Q. (2010). Genetic analysis of heat tolerance at anthesis in rice. *Crop Science, 50*(5), 1633–1641.
Jahan, N., Zhang, Y., Lv, Y., Song, M., Zhao, C., Hu, H., Cui, Y., Wang, Z., Yang, S., Zhang, A., & Hu, J. (2020). QTL analysis for rice salinity tolerance and fine mapping of a candidate locus qSL7 for shoot length under salt stress. *Plant Growth Regulation, 90*(2), 307–319.
Ji, X., Dong, B., Shiran, B., Talbot, M. J., Edlington, J. E., Hughes, T., White, R. G., Gubler, F., & Dolferus, R. (2011). Control of abscisic acid catabolism and abscisic acid homeostasis is important for reproductive stage stress tolerance in cereals. *Plant Physiology, 156*(2), 647–662.
Jiang, L., Ge, M., Zhao, H., & Zhang, T. (2015). Analysis of heterosis and quantitative trait loci for kernel shape related traits using triple testcross population in maize. *PLoS One, 10*(4), e0124779.
Jiang, L., Liu, S., Hou, M., Tang, J., Chen, L., Zhai, H., & Wan, J. (2006). Analysis of QTLs for seed low temperature germinability and anoxia germinability in rice (Oryza sativa L.). *Field Crops Research, 98*(1), 68–75.
Jiang, L., Xun, M., Wang, J., & Wan, J. (2008). QTL analysis of cold tolerance at seedling stage in rice (Oryza sativa L.) using recombination inbred lines. *Journal of Cereal Science, 48*(1), 173–179.
Jiang, N., Shi, S., Shi, H., Khanzada, H., Wassan, G. M., Zhu, C., Peng, X., Yu, Q., Chen, X., He, X., & Fu, J. (2017). Mapping QTL for seed germinability under low temperature using a new high-density genetic map of rice. *Frontiers in Plant Science, 8*, 1223.
Jiang, Z., Tu, H., Bai, B., Yang, C., Zhao, B., Guo, Z., Liu, Q., Zhao, H., Yang, W., Xiong, L., & Zhang, J. (2021). Combining UAV-RGB high-throughput field phenotyping and genome-wide

association study to reveal genetic variation of rice germplasms in dynamic response to drought stress. *New Phytologist, 232*(1), 440–455.

Katiyar-Agarwal, S., Agarwal, M., & Grover, A. (2003). Heat-tolerant basmati rice engineered by over-expression of hsp101. *Plant Molecular Biology, 51*(5), 677–686.

Kaur, B., Mavi, G. S., Gill, M. S., & Saini, D. K. (2020). Utilization of KASP technology for wheat improvement. *Cereal Research Communications*, 1–13.

Kaur, N., Sharma, S., Hasanuzzaman, M., & Pati, P. K. (2022a). Genome editing: A promising approach for achieving abiotic stress tolerance in plants. *International Journal of Genomics.*

Kaur, R., Kaur, R., Sharma, N., Kumari, N., Khanna, R., & Singh, G. (2022b). Protein profiling in a set of wild rice species and rice cultivars: A stepping stone to protein quality improvement. *Cereal Research Communications*, 1–15.

Khahani, B., Tavakol, E., Shariati, V., & Rossini, L. (2021). Meta-QTL and ortho-MQTL analyses identified genomic regions controlling rice yield, yield-related traits and root architecture under water deficit conditions. *Scientific Reports, 11*(1), 1–18.

Khan, I., Zhang, Y., Akbar, F., & Khan, J. (2022). Abiotic stress tolerance in cereals through genome editing. In *Omics approach to manage abiotic stress in cereals* (pp. 295–319). Springer.

Khan, M. S. K., Saeed, M., & Iqbal, J. (2015). Identification of quantitative trait loci for Na+, K+ and Ca++ accumulation traits in rice grown under saline conditions using F 2 mapping population. *Brazilian Journal of Botany, 38*(3), 555–565.

Khan, M. S. K., Saeed, M., & Iqbal, J. (2016). Quantitative trait locus mapping for salt tolerance at maturity stage in indica rice using replicated F 2 population. *Brazilian Journal of Botany, 39*(2), 641–650.

Khanna-Chopra, R., Singh, K., Shukla, S., Kadam, S., & Singh, N. K. (2019). QTLs for cell membrane stability and flag leaf area under drought stress in a wheat RIL population. *Journal of Plant Biochemistry and Biotechnology*, 1–11.

Khowaja, F. S., & Price, A. H. (2008). QTL mapping rolling, stomatal conductance and dimension traits of excised leaves in the Bala× Azucena recombinant inbred population of rice. *Field Crops Research, 106*(3), 248–257.

Kilasi, N. L., Singh, J., Vallejos, C. E., Ye, C., Jagadish, S. V., Kusolwa, P., & Rathinasabapathi, B. (2018). Heat stress tolerance in rice (Oryza sativa L.): Identification of quantitative trait loci and candidate genes for seedling growth under heat stress. *Frontiers in Plant Science, 9*, 1578.

Kim, D. M., Ju, H. G., Kwon, T. R., Oh, C. S., & Ahn, S. N. (2009). Mapping QTLs for salt tolerance in an introgression line population between japonica cultivars in rice. *Journal of Crop Science and Biotechnology, 12*(3), 121.

Kim, K. W., Nawade, B., Nam, J., Chu, S. H., Ha, J., & Park, Y. J. (2022). Development of an inclusive 580K SNP array and its application for genomic selection and genome-wide association studies in rice. *Frontiers in Plant Science, 13.*

Kim, S. I., & Tai, T. H. (2011). Evaluation of seedling cold tolerance in rice cultivars: A comparison of visual ratings and quantitative indicators of physiological changes. *Euphytica, 178*(3), 437–447.

Kim, S. M., Suh, J. P., Lee, C. K., Lee, J. H., Kim, Y. G., & Jena, K. K. (2014). QTL mapping and development of candidate gene-derived DNA markers associated with seedling cold tolerance in rice (Oryza sativa L.). *Molecular Genetics and Genomics, 289*(3), 333–343.

Kimotho, R. N., Baillo, E. H., & Zhang, Z. (2019). Transcription factors involved in abiotic stress responses in Maize (Zea mays L.) and their roles in enhanced productivity in the post genomics era. *PeerJ, 7*, e7211.

Klose, R., Penlington, J., & Ruckelshausen, A. (2009). Usability study of 3D time-of-flight cameras for automatic plant phenotyping. *Bornimer Agrartechnische Berichte, 69*(93–105), 12.

Kong, W., Zhang, C., Qiang, Y., Zhong, H., Zhao, G., & Li, Y. (2020). Integrated RNA-seq analysis and meta-QTLs mapping provide insights into cold stress response in rice seedling roots. *International Journal of Molecular Sciences, 21*(13), 4615.

Kosina, P., Reynolds, M., Dixon, J., & Joshi, A. (2007). Stakeholder perception of wheat production constraints, capacity building needs, and research partnerships in developing countries. *Euphytica, 157*(3), 475–483.

Kover, P. X., Valdar, W., Trakalo, J., Scarcelli, N., Ehrenreich, I. M., Purugganan, M. D., Durrant, C., & Mott, R. (2009). A multiparent advanced generation inter-cross to fine-map quantitative traits in Arabidopsis thaliana. *PLoS Genetics, 5*(7), e1000551.

Koyama, M. L., Levesley, A., Koebner, R. M., Flowers, T. J., & Yeo, A. R. (2001). Quantitative trait loci for component physiological traits determining salt tolerance in rice. *Plant Physiology, 125*(1), 406–422.

Kumar, A., Saripalli, G., Jan, I., Kumar, K., Sharma, P. K., Balyan, H. S., & Gupta, P. K. (2020a). Meta-QTL analysis and identification of candidate genes for drought tolerance in bread wheat (Triticum aestivum L.). *Physiology and Molecular Biology of Plants*, 1–13.

Kumar, R., Venuprasad, R., & Atlin, G. N. (2007). Genetic analysis of rainfed lowland rice drought tolerance under naturally-occurring stress in eastern India: heritability and QTL effects. *Field Crops Research, 103*(1), 42–52.

Kumar, R. R., Goswami, S., Dubey, K., Singh, K., Singh, J. P., Kumar, A., Rai, G. K., Singh, S. D., Bakshi, S., Singh, B., & Pathak, H. (2019). RuBisCo activase—A catalytic chaperone involved in modulating the RuBisCo activity and heat stress-tolerance in wheat. *Journal of Plant Biochemistry and Biotechnology, 28*(1), 63–75.

Kumar, R. R., Goswami, S., Singh, K., Dubey, K., Rai, G. K., Singh, B., Singh, S., Grover, M., Mishra, D., Kumar, S., & Bakshi, S. (2018). Characterization of novel heat-responsive transcription factor (TaHSFA6e) gene involved in regulation of heat shock proteins (HSPs)—A key member of heat stress-tolerance network of wheat. *Journal of Biotechnology, 279*, 1–12.

Kumar, R. R., Goswami, S., Singh, K., Dubey, K., Singh, S., Sharma, R., Verma, N., Kala, Y. K., Rai, G. K., Grover, M., & Mishra, D. C. (2016). Identification of putative RuBisCo Activase (TaRca1)—the catalytic chaperone regulating carbon assimilatory pathway in wheat (Triticum aestivum) under the heat stress. *Frontiers in Plant Science, 7*, 986.

Kumar, S., Singh, V. P., Saini, D. K., Sharma, H., Saripalli, G., Kumar, S., Balyan, H. S., & Gupta, P. K. (2021). Meta-QTLs, ortho-MQTLs, and candidate genes for thermotolerance in wheat (Triticum aestivum L.). *Molecular Breeding, 41*(11), 1–22.

Kumar, V. S., Verma, R. K., Yadav, S. K., Yadav, P., Watts, A., Rao, M. V., & Chinnusamy, V. (2020b). CRISPR-Cas9 mediated genome editing of drought and salt tolerance (OsDST) gene in indica mega rice cultivar MTU1010. *Physiology and Molecular Biology of Plants*.

Kuzuoglu-Ozturk, D., Yalcinkaya, O. C., Akpinar, B. A., Mitou, G., Korkmaz, G., Gozuacik, D., & Budak, H. (2012). Autophagy-related gene, TdAtg8, in wild emmer wheat plays a role in drought and osmotic stress response. *Planta, 236*(4), 1081–1092.

Lafarge, T., Bueno, C., Frouin, J., Jacquin, L., Courtois, B., & Ahmadi, N. (2017). Genome-wide association analysis for heat tolerance at flowering detected a large set of genes involved in adaptation to thermal and other stresses. *PLoS One, 12*(2), e0171254.

Lafitte, H. R., Courtois, B., & Arraudeau, M. (2002). Genetic improvement of rice in aerobic systems: progress from yield to genes. *Field Crops Research, 75*(2–3), 171–190.

Lafitte, H. R., Price, A. H., & Courtois, B. (2004). Yield response to water deficit in an upland rice mapping population: Associations among traits and genetic markers. *Theoretical and Applied Genetics, 109*(6), 1237–1246.

Lang, N., Buu, B. C., & Ismail, A. (2008). Molecular mapping and marker-assisted selection for salt tolerance in rice (Oryza sativa L.).

Li, C., Sun, B., Li, Y., Liu, C., Wu, X., Zhang, D., Shi, Y., Song, Y., Buckler, E. S., Zhang, Z., & Wang, T. (2016a). Numerous genetic loci identified for drought tolerance in the maize nested association mapping populations. *BMC Genomics, 17*(1), 894.

Li, J., Pan, Y., Guo, H., Zhou, L., Yang, S., Zhang, Z., Yang, J., Zhang, H., Li, J., Zeng, Y., & Li, Z. (2018a). Fine mapping of QTL qCTB10-2 that confers cold tolerance at the booting stage in rice. *Theoretical and Applied Genetics, 131*(1), 157–166.

Li, Q., Wang, W., Wang, W., Zhang, G., Liu, Y., Wang, Y., & Wang, W. (2018b). Wheat F-box protein gene TaFBA1 is involved in plant tolerance to heat stress. *Frontiers in Plant Science, 9*, 521.

Li, S., Chen, N., Li, F., Mei, F., Wang, Z., Cheng, X., Kang, Z., & Mao, H. (2020a). Characterization of wheat homeodomain-leucine zipper family genes and functional analysis of TaHDZ5-6A in drought tolerance in transgenic Arabidopsis. *BMC Plant Biology, 20*(1), 50.

Li, S., Xu, C., Yang, Y., & Xia, G. (2010). Functional analysis of TaDi19A, a salt-responsive gene in wheat. *Plant, Cell & Environment, 33*(1), 117–129.

Li, X., Xu, X., Liu, W., Li, X., Yang, X., Ru, Z., & Li, L. (2020). Dissection of superior alleles for yield-related traits and their distribution in important cultivars of wheat by association mapping. *Frontiers in Plant Science, 11*, 175.

Li, X., Zhou, Z., Ding, J., Wu, Y., Zhou, B., Wang, R., Ma, J., Wang, S., Zhang, X., Xia, Z., & Chen, J. (2016b). Combined linkage and association mapping reveals QTL and candidate genes for plant and ear height in maize. *Frontiers in Plant Science, 7*, 833.

Li, Z., Mu, P., Li, C., Zhang, H., Li, Z., Gao, Y., & Wang, X. (2005). QTL mapping of root traits in a doubled haploid population from a cross between upland and lowland japonica rice in three environments. *Theoretical and Applied Genetics, 110*(7), 1244–1252.

Liang, S., Xiong, W., Yin, C., Xie, X., Jin, Y. J., Zhang, S., Yang, B., Ye, G., Chen, S., & Luan, W. J. (2019). Overexpression of OsARD1 improves submergence, drought, and salt tolerances of seedling through the enhancement of ethylene synthesis in rice. *Frontiers in Plant Science, 10*, 1088.

Liang, Z., Zhang, K., Chen, K., & Gao, C. (2014). Targeted mutagenesis in Zea mays using TALENs and the CRISPR/Cas system. *Journal of Genetics and Genomics, 41*(2), 63–68.

Lin, H. X., Zhu, M. Z., Yano, M., Gao, J. P., Liang, Z. W., Su, W. A., Hu, X. H., Ren, Z. H., & Chao, D. Y. (2004). QTLs for Na+ and K+ uptake of the shoots and roots controlling rice salt tolerance. *Theoretical and Applied Genetics, 108*(2), 253–260.

Lin, Q., Zong, Y., Xue, C., Wang, S., Jin, S., Zhu, Z., Wang, Y., Anzalone, A. V., Raguram, A., Doman, J. L., & Liu, D. R. (2020). Prime genome editing in rice and wheat. *Nature Biotechnology, 38*(5), 582–585.

Liu, H., Zhou, X., Dong, N., Liu, X., Zhang, H., & Zhang, Z. (2011). Expression of a wheat MYB gene in transgenic tobacco enhances resistance to Ralstonia solanacearum, and to drought and salt stresses. *Functional & Integrative Genomics, 11*(3), 431–443.

Liu, Q., Yang, T., Yu, T., Zhang, S., Mao, X., Zhao, J., Wang, X., Dong, J., & Liu, B. (2017). Integrating small RNA sequencing with QTL mapping for identification of miRNAs and their target genes associated with heat tolerance at the flowering stage in rice. *Frontiers in Plant Science, 8*, 43.

Liu, S., Zenda, T., Wang, X., Liu, G., Jin, H., Yang, Y., Dong, A., & Duan, H. (2019). Comprehensive meta-analysis of maize QTLs associated with grain yield, flowering date and plant height under drought conditions. *The Journal of Agricultural Science, 11*, 1–19.

Liu, W., Lu, T., Li, Y., Pan, X., Duan, Y., Min, J., Fu, X., Sheng, X., Xiao, J., Liu, S., & Tan, J. (2015). Mapping of quantitative trait loci for cold tolerance at the early seedling stage in landrace rice Xiang 743. *Euphytica, 201*(3), 401–409.

Liu, W., Pan, X., Xiong, H., Huang, F., Duan, Y., Li, Y., Min, J., Liu, S., Liu, L., Wei, X., & Zhao, W. (2020). QTL mapping and validation of a segment responsible for early-seedling cold tolerance in rice. *Euphytica, 216*(1), 7.

Lou, Q., Chen, L., Mei, H., Wei, H., Feng, F., Wang, P., Xia, H., Li, T., & Luo, L. (2015). Quantitative trait locus mapping of deep rooting by linkage and association analysis in rice. *Journal of Experimental Botany, 66*(15), 4749–4757.

Lou, Q., Chen, L., Sun, Z., Xing, Y., Li, J., Xu, X., Mei, H., & Luo, L. (2007). A major QTL associated with cold tolerance at seedling stage in rice (Oryza sativa L.). *Euphytica, 158*(1–2), 87–94.

Lu, F., Romay, M. C., Glaubitz, J. C., Bradbury, P. J., Elshire, R. J., Wang, T., Li, Y., Li, Y., Semagn, K., Zhang, X., & Hernandez, A. G. (2015). High-resolution genetic mapping of maize pan-genome sequence anchors. *Nature Communications, 6*(1), 1–8.

Luo, M., Zhang, Y., Chen, K., Kong, M., Song, W., Lu, B., Shi, Y., Zhao, Y., & Zhao, J. (2019a). Mapping of quantitative trait loci for seedling salt tolerance in maize. *Molecular Breeding, 39*(5), 64.

Luo, M., Zhao, Y., Zhang, R., Xing, J., Duan, M., Li, J., Wang, N., Wang, W., Zhang, S., Chen, Z., & Zhang, H. (2017). Mapping of a major QTL for salt tolerance of mature field-grown maize plants based on SNP markers. *BMC Plant Biology, 17*(1), 140.

Luo, X., Wang, B., Gao, S., Zhang, F., Terzaghi, W., & Dai, M. (2019b). Genome-wide association study dissects the genetic bases of salt tolerance in maize seedlings. *Journal of Integrative Plant Biology, 61*(6), 658–674.

Luo, X. D., Jun, Z. H. A.O., Dai, L. F., Zhang, F. T., Yi, Z. H. O. U., Yong, W. A. N., & Xie, J. K. (2016). Linkage map construction and QTL mapping for cold tolerance in Oryza rufipogon Griff. at early seedling stage. *Journal of Integrative Agriculture, 15*(12), 2703–2711.

Lv, Y., Guo, Z., Li, X., Ye, H., Li, X., & Xiong, L. (2016). New insights into the genetic basis of natural chilling and cold shock tolerance in rice by genome-wide association analysis. *Plant, Cell & Environment, 39*(3), 556–570.

Ma, X., Feng, F., Wei, H., Mei, H., Xu, K., Chen, S., Li, T., Liang, X., Liu, H., & Luo, L. (2016). Genome-wide association study for plant height and grain yield in rice under contrasting moisture regimes. *Frontiers in Plant Science, 7*, 1801.

Manangkil, O. E., Vu, H. T. T., Mori, N., Yoshida, S., & Nakamura, C. (2013). Mapping of quantitative trait loci controlling seedling vigor in rice (Oryza sativa L.) under submergence. *Euphytica, 192*(1), 63–75.

Manmathan, H., Shaner, D., Snelling, J., Tisserat, N., & Lapitan, N. (2013). Virus-induced gene silencing of Arabidopsis thaliana gene homologues in wheat identifies genes conferring improved drought tolerance. *Journal of Experimental Botany, 64*(5), 1381–1392.

Mao, D., Yu, L., Chen, D., Li, L., Zhu, Y., Xiao, Y., Zhang, D., & Chen, C. (2015). Multiple cold resistance loci confer the high cold tolerance adaptation of Dongxiang wild rice (Oryza rufipogon) to its high-latitude habitat. *Theoretical and Applied Genetics, 128*(7), 1359–1371.

Mao, X., Zhang, H., Tian, S., Chang, X., & Jing, R. (2010). TaSnRK2. 4, an SNF1-type serine/threonine protein kinase of wheat (Triticum aestivum L.), confers enhanced multistress tolerance in Arabidopsis. *Journal of Experimental Botany, 61*(3), 683–696.

Mason, R. E., & Singh, R. P. (2014). Considerations when deploying canopy temperature to select high yielding wheat breeding lines under drought and heat stress. *Agronomy, 4*(2), 191–201.

Masoudi, B., Mardi, M., Hervan, E. M., Bihamta, M. R., Naghavi, M. R., Nakhoda, B., & Amini, A. (2015). QTL mapping of salt tolerance traits with different effects at the seedling stage of bread wheat. *Plant Molecular Biology Reporter, 33*(6), 1790–1803.

Mazumder, A., Rohilla, M., Bisht, D. S., Krishnamurthy, S. L., Barman, M., Sarma, R. N., et al. (2020). Identification and mapping of quantitative trait loci (QTL) and epistatic QTL for salinity tolerance at seedling stage in traditional aromatic short grain rice landrace Kolajoha (Oryza sativa L.) of Assam, India. *Euphytica, 216*, 1–18.

McNellie, J. P., Chen, J., Li, X., & Yu, J. (2018). Genetic mapping of foliar and tassel heat stress tolerance in maize. *Crop Science, 58*(6), 2484–2493.

Merchuk-Ovnat, L., Barak, V., Fahima, T., Ordon, F., Lidzbarsky, G. A., Krugman, T., & Saranga, Y. (2016). Ancestral QTL alleles from wild emmer wheat improve drought resistance and productivity in modern wheat cultivars. *Frontiers in Plant Science, 7*, 452.

Messmer, R., Fracheboud, Y., Bänziger, M., Vargas, M., Stamp, P., & Ribaut, J. M. (2009). Drought stress and tropical maize: QTL-by-environment interactions and stability of QTLs across environments for yield components and secondary traits. *Theoretical and Applied Genetics, 119*(5), 913–930.

Mir, R. R., Reynolds, M., Pinto, F., Khan, M. A., & Bhat, M. A. (2019). High-throughput phenotyping for crop improvement in the genomics era. *Plant Science, 282*, 60–72.

Mishra, K. K., Vikram, P., Yadaw, R. B., Swamy, B. M., Dixit, S., Cruz, M. T. S., Maturan, P., Marker, S., & Kumar, A. (2013). qDTY 12. 1: a locus with a consistent effect on grain yield under drought in rice. *BMC Genetics, 14*(1), 12.

Moncada, P., Martinez, C. P., Borrero, J., Châtel, M., Gauch, H., Jr., Guimaraes, E., Tohmé, J., & McCouch, S. R. (2001). Quantitative trait loci for yield and yield components in an Oryza sativa× Oryza rufipogon BC2F2 population evaluated in an upland environment. *Theoretical and Applied Genetics, 102*(1), 41–52.

Montenegro, J. D., Golicz, A. A., Bayer, P. E., Hurgobin, B., Lee, H., Chan, C. K. K., Visendi, P., Lai, K., Doležel, J., Batley, J., & Edwards, D. (2017). The pangenome of hexaploid bread wheat. *The Plant Journal, 90*(5), 1007–1013.

Muthu, V., Abbai, R., Nallathambi, J., Rahman, H., Ramasamy, S., Kambale, R., Thulasinathan, T., Ayyenar, B., & Muthurajan, R. (2020). Pyramiding QTLs controlling tolerance against drought, salinity, and submergence in rice through marker assisted breeding. *PloS One, 15*(1), e0227421.

Muthuramalingam, P., Jeyasri, R., Rakkammal, K., Satish, L., Shamili, S., Karthikeyan, A., Valliammai, A., Priya, A., Selvaraj, A., Gowri, P., & Wu, Q. S. (2022). Multi-omics and integrative approach towards understanding salinity tolerance in rice: A review. *Biology, 11*(7), 1022.

Nagel, K. A., Putz, A., Gilmer, F., Heinz, K., Fischbach, A., Pfeifer, J., Faget, M., Blossfeld, S., Ernst, M., Dimaki, C., & Kastenholz, B. (2012). GROWSCREEN-Rhizo is a novel phenotyping robot enabling simultaneous measurements of root and shoot growth for plants grown in soil-filled rhizotrons. *Functional Plant Biology, 39*(11), 891–904.

Najeeb, S., Ali, J., Mahender, A., Pang, Y. L., Zilhas, J., Murugaiyan, V., Vemireddy, L. R., & Li, Z. (2020). Identification of main-effect quantitative trait loci (QTLs) for low-temperature stress tolerance germination-and early seedling vigor-related traits in rice (Oryza sativa L.). *Molecular Breeding, 40*(1), 10.

Naveed, S. A., Zhang, F., Zhang, J., Zheng, T. Q., Meng, L. J., Pang, Y. L., Xu, J. L., & Li, Z. K. (2018). Identification of QTN and candidate genes for salinity tolerance at the germination and seedling stages in rice by genome-wide association analyses. *Scientific Reports, 8*(1), 1–11.

Nayyeripasand, L., Garoosi, G. A., & Ahmadikhah, A. (2020). Genome-wide association study (gwas) to identify salt-tolerance qtls carrying novel candidate genes in rice during early vegetative stage.

Neeraja, C. N., Maghirang-Rodriguez, R., Pamplona, A., Heuer, S., Collard, B. C., Septiningsih, E. M., Vergara, G., Sanchez, D., Xu, K., Ismail, A. M., & Mackill, D. J. (2007). A marker-assisted backcross approach for developing submergence-tolerant rice cultivars. *Theoretical and Applied Genetics, 115*(6), 767–776.

Nghi, K. N., Tondelli, A., Valè, G., Tagliani, A., Marè, C., Perata, P., & Pucciariello, C. (2019). Dissection of coleoptile elongation in japonica rice under submergence through integrated genome-wide association mapping and transcriptional analyses. *Plant, Cell & Environment, 42*(6), 1832–1846.

Niu, C. F., Wei, W. E. I., Zhou, Q. Y., Tian, A. G., Hao, Y. J., Zhang, W. K., Ma, B., Lin, Q., Zhang, Z. B., Zhang, J. S., & Chen, S. Y. (2012). Wheat WRKY genes TaWRKY2 and TaWRKY19 regulate abiotic stress tolerance in transgenic Arabidopsis plants. *Plant, Cell & Environment, 35*(6), 1156–1170.

Osuman, A. S., Badu-Apraku, B., Karikari, B., Ifie, B. E., Tongoona, P., & Danquah, E. Y. (2022). Genome-wide association study reveals genetic architecture and candidate genes for yield and related traits under terminal drought, combined heat and drought in tropical maize germplasm. *Genes, 13*(2), 349.

Oyiga, B. C., Sharma, R. C., Baum, M., Ogbonnaya, F. C., Léon, J., & Ballvora, A. (2018). Allelic variations and differential expressions detected at quantitative trait loci for salt stress tolerance in wheat. *Plant, Cell & Environment, 41*(5), 919–935.

Pal, N., Saini, D. K., & Kumar, S. (2021). Meta-QTLs, ortho-MQTLs and candidate genes for the traits contributing to salinity stress tolerance in common wheat (Triticum aestivum L.). *Physiology and Molecular Biology of Plants, 27*(12), 2767–2786.

Palanog, A. D., Swamy, B. M., Shamsudin, N. A. A., Dixit, S., Hernandez, J. E., Boromeo, T. H., Cruz, P. C. S., & Kumar, A. (2014). Grain yield QTLs with consistent-effect under reproductive-stage drought stress in rice. *Field Crops Research, 161*, 46–54.

Pandit, E., Tasleem, S., Barik, S. R., Mohanty, D. P., Nayak, D. K., Mohanty, S. P., Das, S., & Pradhan, S. K. (2017). Genome-wide association mapping reveals multiple QTLs governing tolerance response for seedling stage chilling stress in indica rice. *Frontiers in Plant Science, 8*, 552.

Pérez-Bueno, M. L., Pineda, M., Cabeza, F. M., & Barón, M. (2016). Multicolor fluorescence imaging as a candidate for disease detection in plant phenotyping. *Frontiers in Plant Science, 7*, 1790.

Pfeiffer, W. H., Trethowan, R. M., Ginkel, M. V., Ortiz-Monasterio, I., & Rajaram, S. (2005). Breeding for abiotic stress tolerance in wheat. Abiotic stresses plant resistance through breeding and molecular approaches (No. CIS-4737. CIMMYT.).

Pradawet, C., Khongdee, N., Pansak, W., Spreer, W., Hilger, T., & Cadisch, G. (2022). Thermal imaging for assessment of maize water stress and yield prediction under drought conditions. *Journal of Agronomy and Crop Science*.

Pradhan, S., Babar, M. A., Bai, G., Khan, J., Shahi, D., Avci, M., Guo, J., McBreen, J., Asseng, S., Gezan, S., & Baik, B. K. (2020). Genetic dissection of heat-responsive physiological traits to improve adaptation and increase yield potential in soft winter wheat. *BMC Genomics, 21*, 1–15.

Pradhan, S. K., Barik, S. R., Sahoo, A., Mohapatra, S., Nayak, D. K., Mahender, A., Meher, J., Anandan, A., & Pandit, E. (2016). Population structure, genetic diversity and molecular marker-trait association analysis for high temperature stress tolerance in rice. *PLoS One, 11*(8), e0160027.

Prasad, S. R., Bagali, P. G., Hittalmani, S., & Shashidhar, H. E. (2000). Molecular mapping of quantitative trait loci associated with seedling tolerance to salt stress in rice (Oryza sativa L.). *Current Science*, 162–164.

Prasanth, V. V., Babu, M. S., Basava, R. K., Tripura Venkata, V. T., Mangrauthia, S. K., Voleti, S. R., & Neelamraju, S. (2017). Trait and marker associations in Oryza nivara and O. rufipogon derived rice lines under two different heat stress conditions. *Frontiers in Plant Science, 8*, 1819.

Prasanth, V. V., Basava, K. R., Babu, M. S., Tripura Venkata, V. T., Devi, S. R., Mangrauthia, S. K., Voleti, S. R., & Sarla, N. (2016). Field level evaluation of rice introgression lines for heat tolerance and validation of markers linked to spikelet fertility. *Physiology and Molecular Biology of Plants, 22*(2), 179–192.

Puram, V. R. R., Ontoy, J., & Subudhi, P. K. (2018). Identification of QTLs for salt tolerance traits and prebreeding lines with enhanced salt tolerance in an introgression line population of rice. *Plant Molecular Biology Reporter, 36*(5), 695–709.

Qin, D., Wang, F., Geng, X., Zhang, L., Yao, Y., Ni, Z., Peng, H., & Sun, Q. (2015). Overexpression of heat stress-responsive TaMBF1c, a wheat (Triticum aestivum L.) Multiprotein Bridging Factor, confers heat tolerance in both yeast and rice. *Plant Molecular Biology, 87*(1–2), 31–45.

Qin, Y., Wang, M., Tian, Y., He, W., Han, L., & Xia, G. (2012). Over-expression of TaMYB33 encoding a novel wheat MYB transcription factor increases salt and drought tolerance in Arabidopsis. *Molecular Biology Reports, 39*(6), 7183–7192.

Qingquan, C., Sibin, Y., & Chunhai, L. (2008). Identification of QTLs for heat tolerance at flowering stage in rice. *Scientia Agricultura Sinica*.

Rahaman, M., Chen, D., Gillani, Z., Klukas, C., & Chen, M. (2015). Advanced phenotyping and phenotype data analysis for the study of plant growth and development. *Frontiers in Plant Science, 6*, 619.

Rahman, H., Dakshinamurthi, V., Ramasamy, S., Manickam, S., Kaliyaperumal, A. K., Raha, S., et al. (2018). Introgression of submergence tolerance into CO 43, a popular rice variety of India, through marker-assisted backcross breeding. *Czech Journal of Genetics and Plant Breeding, 54*(3), 101–108.

Rai, N., Bellundagi, A., Kumar, P. K., Kalasapura Thimmappa, R., Rani, S., Sinha, N., krishna, H., Jain, N., Singh, G. P., Singh, P. K., & Chand, S. (2018). Marker-assisted backcross breeding for improvement of drought tolerance in bread wheat (Triticum aestivum L. em Thell). *Plant Breeding, 137*(4), 514–526.

Ramchander, S., Raveendran, M., & Robin, S. (2016). Mapping QTLs for physiological traits associated with drought tolerance in rice (Oryza sativa L.). *Journal of Investigative Genomics, 3*(3), 00052.

Ranawake, A. L., Manangkil, O. E., Yoshida, S., Ishii, T., Mori, N., & Nakamura, C. (2014). Mapping QTLs for cold tolerance at germination and the early seedling stage in rice (Oryza sativa L.). *Biotechnology & Biotechnological Equipment, 28*(6), 989–998.

Raza, Q., Riaz, A., Bashir, K., & Sabar, M. (2020). Reproductive tissues-specific meta-QTLs and candidate genes for development of heat-tolerant rice cultivars. *Plant Molecular Biology, 104*(1), 97–112.

Rebetzke, G. J., Chenu, K., Biddulph, B., Moeller, C., Deery, D. M., Rattey, A. R., Bennett, D., Barrett-Lennard, E. G., & Mayer, J. E. (2013). A multisite managed environment facility for targeted trait and germplasm phenotyping. *Functional Plant Biology, 40*(1), 1–13.

Ribaut, J. M., & Ragot, M. (2007). Marker-assisted selection to improve drought adaptation in maize: the backcross approach, perspectives, limitations, and alternatives. *Journal of Experimental Botany, 58*(2), 351–360.

Sagan, V., Maimaitijiang, M., Sidike, P., Eblimit, K., Peterson, K. T., Hartling, S., Esposito, F., Khanal, K., Newcomb, M., Pauli, D., & Ward, R. (2019). UAV-based high resolution thermal imaging for vegetation monitoring, and plant phenotyping using ICI 8640 P, FLIR Vue Pro R 640, and thermomap cameras. *Remote Sensing, 11*(3), 330.

Saikumar, S., Gouda, P. K., Saiharini, A., Varma, C. M. K., Vineesha, O., Padmavathi, G., & Shenoy, V. V. (2014). Major QTL for enhancing rice grain yield under lowland reproductive drought stress identified using an O. sativa/O. glaberrima introgression line. *Field Crops Research, 163*, 119–131.

Saini, D. K., Chopra, Y., Singh, J., Sandhu, K. S., Kumar, A., Bazzer, S., & Srivastava, P. (2022). Comprehensive evaluation of mapping complex traits in wheat using genome-wide association studies. *Molecular Breeding, 42*(1), 1–52.

Sanchouli, S., Neghab, M., Sabouri, H., & Zare Mehrjerdi, M. (2019). Genetic structure of salinity tolerance in rice at seedling stage. *Journal of Genetic Resources, 5*(1), 22–30.

Sandhu, D., Pudussery, M. V., Kumar, R., Pallete, A., Markley, P., Bridges, W. C., & Sekhon, R. S. (2020). Characterization of natural genetic variation identifies multiple genes involved in salt tolerance in maize. *Functional & Integrative Genomics, 20*(2), 261–275.

Sandhu, N., Dixit, S., Swamy, B. P. M., Raman, A., Kumar, S., Singh, S. P., Yadaw, R. B., Singh, O. N., Reddy, J. N., Anandan, A., & Yadav, S. (2019). Marker assisted breeding to develop multiple stress tolerant varieties for flood and drought prone areas. *Rice, 12*(1), 8.

Sandhu, N., Singh, A., Dixit, S., Cruz, M. T. S., Maturan, P. C., Jain, R. K., & Kumar, A. (2014). Identification and mapping of stable QTL with main and epistasis effect on rice grain yield under upland drought stress. *BMC Genetics, 15*(1), 63.

Sarkar, R. K., Das, K. K., Panda, D., Reddy, J. N., Patnaik, S. S. C., Patra, B. C., & Singh, D. P. (2014). Submergence tolerance in rice: Biophysical constraints, physiological basis and identification of donors (p. 36). Central Rice Research Institute.

Satoh, T., Tezuka, K., Kawamoto, T., Matsumoto, S., Satoh-Nagasawa, N., Ueda, K., Sakurai, K., Watanabe, A., Takahashi, H., & Akagi, H. (2016). Identification of QTLs controlling low-temperature germination of the East European rice (Oryza sativa L.) variety Marateli. *Euphytica, 207*(2), 245–254.

Schläppi, M. R., Jackson, A. K., Eizenga, G. C., Wang, A., Chu, C., Shi, Y., Shimoyama, N., & Boykin, D. L. (2017). Assessment of five chilling tolerance traits and GWAS mapping in rice using the USDA mini-core collection. *Frontiers in Plant Science, 8*, 957.

Sellamuthu, R., Liu, G. F., Ranganathan, C. B., & Serraj, R. (2011). Genetic analysis and validation of quantitative trait loci associated with reproductive-growth traits and grain yield under drought stress in a doubled haploid line population of rice (Oryza sativa L.). *Field Crops Research, 124*(1), 46–58.

Septiningsih, E. M., Collard, B. C., Heuer, S., Bailey-Serres, J., Ismail, A. M., & Mackill, D. J. (2013). Applying genomics tools for breeding submergence tolerance in rice. *Translational genomics for crop breeding: abiotic stress, yield and quality, 2*, 9–30.

Septiningsih, E. M., Hidayatun, N., Sanchez, D. L., Nugraha, Y., Carandang, J., Pamplona, A. M., Collard, B. C., Ismail, A. M., & Mackill, D. J. (2015). Accelerating the development of new submergence tolerant rice varieties: The case of Ciherang-Sub1 and PSB Rc18-Sub1. *Euphytica, 202*(2), 259–268.

Septiningsih, E. M., Pamplona, A. M., Sanchez, D. L., Neeraja, C. N., Vergara, G. V., Heuer, S., Ismail, A. M., & Mackill, D. J. (2009). Development of submergence-tolerant rice cultivars: the Sub1 locus and beyond. *Annals of Botany, 103*(2), 151–160.

Septiningsih, E. M., Sanchez, D. L., Singh, N., Sendon, P. M., Pamplona, A. M., Heuer, S., & Mackill, D. J. (2012). Identifying novel QTLs for submergence tolerance in rice cultivars IR72 and Madabaru. *Theoretical and Applied Genetics, 124*(5), 867–874.

Shah, F., Huang, J., Cui, K., Nie, L., Shah, T., Chen, C., & Wang, K. (2011). Impact of high-temperature stress on rice plant and its traits related to tolerance. *The Journal of Agricultural Science, 149*(5), 545–556.

Shamsudin, N. A. A., Swamy, B. M., Ratnam, W., Cruz, M. T. S., Sandhu, N., Raman, A. K., & Kumar, A. (2016). Pyramiding of drought yield QTLs into a high quality Malaysian rice cultivar MRQ74 improves yield under reproductive stage drought. *Rice, 9*(1), 21.

Shanmugavadivel, P. S., Sv, A. M., Prakash, C., Ramkumar, M. K., Tiwari, R., Mohapatra, T., & Singh, N. K. (2017). High resolution mapping of QTLs for heat tolerance in rice using a 5K SNP array. *Rice, 10*(1), 28.

Shi, J., Gao, H., Wang, H., Lafitte, H. R., Archibald, R. L., Yang, M., Hakimi, S. M., Mo, H., & Habben, J. E. (2017). ARGOS 8 variants generated by CRISPR-Cas9 improve maize grain yield under field drought stress conditions. *Plant Biotechnology Journal, 15*(2), 207–216.

Shikha, M., Kanika, A., Rao, A. R., Mallikarjuna, M. G., Gupta, H. S., & Nepolean, T. (2017). Genomic selection for drought tolerance using genome-wide SNPs in maize. *Frontiers in Plant Science, 8*, 550.

Shim, K. C., Kim, S., Le, A. Q., Lee, H. S., Adeva, C., Jeon, Y. A., Luong, N. H., Kim, W. J., Akhtamov, M., & Ahn, S. N. (2019). Fine mapping of a low-temperature Germinability QTL qLTG1 using introgression lines derived from Oryza rufipogon. *Plant Breeding and Biotechnology, 7*(2), 141–150.

Shim, K. C., Kim, S. H., Lee, H. S., Adeva, C., Jeon, Y. A., Luong, N. H., Kim, W. J., Akhtamov, M., Park, Y. J., & Ahn, S. N. (2020). Characterization of a new qLTG3–1 allele for low-temperature germinability in rice from the wild species oryza rufipogon. *Rice, 13*(1), 10.

Shokat, S., Sehgal, D., Liu, F., & Singh, S. (2020). GWAS analysis of wheat pre-breeding germplasm for terminal drought stress using next generation sequencing technology.

Singh, A., & Khurana, P. (2016). Molecular and functional characterization of a wheat B2 protein imparting adverse temperature tolerance and influencing plant growth. *Frontiers in Plant Science, 7*, 642.

Singh, A., Singh, Y., Mahato, A. K., Jayaswal, P. K., Singh, S., Singh, R., Yadav, N., Singh, A. K., Singh, P. K., Singh, R., & Kumar, R. (2020a). Allelic sequence variation in the Sub1A, Sub1B and Sub1C genes among diverse rice cultivars and its association with submergence tolerance. *Scientific Reports, 10*(1), 1–18.

Singh, A. M. M. R. D., Krishnamurthy, B. S., Tilak, M. B. R. N. S., & Sharma, R. (2020b). Identification and mapping of quantitative trait loci (QTL) and epistatic QTL for salinity tolerance at seedling stage in traditional aromatic short grain rice landrace Kolajoha (Oryza sativa L.) of Assam, India. *Euphytica, 216*, 75.

Singh, K., Saini, D. K., Saripalli, G., Batra, R., Gautam, T., Singh, R., Pal, S., Kumar M., Jan, I., Singh, S., Kumar, A., Sharma, H., Chaudhary, J., Kumar, K., Kumar, S., Singh, V. K., Singh, V. P., Kumar D., Sharma, S., Kumar, S., Kumar, R., Sharma, S., Gaurav, S. S., Sharma, P. K., Balyan, H. S., & Gupta, P. K. (2022). WheatQTLdb V2.0: A supplement to the database for wheat QTL. *Molecular Breeding, 42*, 56.

Singh, N., Jayaswal, P. K., Panda, K., Mandal, P., Kumar, V., Singh, B., Mishra, S., Singh, Y., Singh, R., Rai, V., & Gupta, A. (2015). Single-copy gene based 50 K SNP chip for genetic studies and molecular breeding in rice. *Scientific Reports, 5*, 11600.

Singh, R., Singh, Y., Xalaxo, S., Verulkar, S., Yadav, N., Singh, S., Singh, N., Prasad, K. S. N., Kondayya, K., Rao, P. R., & Rani, M. G. (2016). From QTL to variety-harnessing the benefits of QTLs for drought, flood and salt tolerance in mega rice varieties of India through a multi-institutional network. *Plant Science, 242*, 278–287.

Singh, V. K., Singh, B. D., Kumar, A., Maurya, S., Krishnan, S. G., Vinod, K. K., Singh, M. P., Ellur, R. K., Bhowmick, P. K., & Singh, A. K. (2018). Marker-assisted introgression of Saltol QTL enhances seedling stage salt tolerance in the rice variety "Pusa Basmati 1". *International Journal of Genomics*, 2018.

Song, J., Li, J., Sun, J., Hu, T., Wu, A., Liu, S., Wang, W., Ma, D., & Zhao, M. (2018). Genome-wide association mapping for cold tolerance in a core collection of rice (Oryza sativa L.) landraces by using high-density single nucleotide polymorphism markers from specific-locus amplified fragment sequencing. *Frontiers in Plant Science, 9*, 875.

Srividya, A., Ramanarao, P. V., Sridhar, S., Jayaprada, M., Anuradha, G., Srilakshmi, B., Reddy, H. K., Hariprasad, A. S., Siddiq, E. A., & Vemireddy, L. R. (2011). Molecular mapping of QTLs for drought related traits at seedling stage under PEG induced stress conditions in rice. *American Journal of Plant Sciences, 2*(02), 190.

Su, L., Yang, J., Li, D., Peng, Z., Xia, A., Yang, M., Luo, L., Huang, C., Wang, J., Wang, H., & Chen, Z. (2020). Dynamic genome-wide association analysis and identification of candidate genes involved in anaerobic germination tolerance in rice

Suh, J. P., Jeung, J. U., Lee, J. I., Choi, Y. H., Yea, J. D., Virk, P. S., Mackill, D. J., & Jena, K. K. (2010). Identification and analysis of QTLs controlling cold tolerance at the reproductive stage and validation of effective QTLs in cold-tolerant genotypes of rice (Oryza sativa L.). *Theoretical and Applied Genetics, 120*(5), 985–995.

Suh, J. P., Lee, C. K., Lee, J. H., Kim, J. J., Kim, S. M., Cho, Y. C., Park, S. H., Shin, J. C., Kim, Y. G., & Jena, K. K. (2012). Identification of quantitative trait loci for seedling cold tolerance using RILs derived from a cross between japonica and tropical japonica rice cultivars. *Euphytica, 184*(1), 101–108.

Sun, C., Dong, Z., Zhao, L., Ren, Y., Zhang, N., & Chen, F. (2020). The Wheat 660K SNP array demonstrates great potential for marker-assisted selection in polyploid wheat. *Plant Biotechnology Journal, 18*(6), 1354–1360.

Sun, J., Yang, L., Wang, J., Liu, H., Zheng, H., Xie, D., Zhang, M., Feng, M., Jia, Y., Zhao, H., & Zou, D. (2018). Identification of a cold-tolerant locus in rice (Oryza sativa L.) using bulked segregant analysis with a next-generation sequencing strategy. *Rice, 11*(1), 24.

Sun, Q. X., & Quick, J. S. (1991). Chromosomal locations of genes for heat tolerance in tetraploid wheat. *Cereal Research Communications*, 431–437.

Takehisa, H., Shimodate, T., Fukuta, Y., Ueda, T., Yano, M., Yamaya, T., Kameya, T., & Sato, T. (2004). Identification of quantitative trait loci for plant growth of rice in paddy field flooded with salt water. *Field Crops Research, 89*(1), 85–95.

Tang, Y., Liu, M., Gao, S., Zhang, Z., Zhao, X., Zhao, C., Zhang, F., & Chen, X. (2012). Molecular characterization of novel TaNAC genes in wheat and overexpression of TaNAC2a confers drought tolerance in tobacco. *Physiologia Plantarum, 144*(3), 210–224.

Tanin, M. J., Saini, D. K., Sandhu, K. S., Pal, N., Gudi, S., Chaudhary, J., et al. (2022). Consensus genomic regions associated with multiple abiotic stress tolerance in wheat and implications for wheat breeding. *Scientific Reports, 12*(1), 13680.

Teramoto, S., Takayasu, S., Kitomi, Y., Arai-Sanoh, Y., Tanabata, T., & Uga, Y. (2020). High-throughput three-dimensional visualization of root system architecture of rice using X-ray computed tomography. *Plant Methods, 16*, 1–14.

Tester, M., & Langridge, P. (2010). Breeding technologies to increase crop production in a changing world. *Science, 327*(5967), 818–822.

Thomelin, P., Bonneau, J., Brien, C., Suchecki, R., Baumann, U., Kalambettu, P., Langridge, P., Tricker, P., & Fleury, D. (2021). The wheat Seven in absentia gene is associated with increases in biomass and yield in hot climates. *Journal of Experimental Botany, 72*(10), 3774–3791.

Thomson, M. J., de Ocampo, M., Egdane, J., Rahman, M. A., Sajise, A. G., Adorada, D. L., Tumimbang-Raiz, E., Blumwald, E., Seraj, Z. I., Singh, R. K., & Gregorio, G. B. (2010). Characterizing the Saltol quantitative trait locus for salinity tolerance in rice. *Rice, 3*(2–3), 148–160.

Thomson, M. J., Ocampo, D. M., Egdane, J., Katimbang, M., Singh, R. K., Gregorio, G., & Ismail, M. (2007, January). QTL mapping and marker-assisted backcrossing for improved salinity tolerance in rice. In *Plant and animal genomes XV conference, San Diego, CA* (pp. 13–17).

Thomson, M. J., Tai, T. H., McClung, A. M., Lai, X. H., Hinga, M. E., Lobos, K. B., Xu, Y., Martinez, C. P., & McCouch, S. R. (2003). Mapping quantitative trait loci for yield, yield components and morphological traits in an advanced backcross population between Oryza rufipogon and the Oryza sativa cultivar Jefferson. *Theoretical and Applied Genetics, 107*(3), 479–493.

Tian, T., Chen, L., Ai, Y., & He, H. (2022). Selection of candidate genes conferring blast resistance and heat tolerance in rice through integration of Meta-QTLs and RNA-Seq. *Genes, 13*(2), 224.

Todkar, L., Harikrishna, G. P., Jain, N., Singh, P. K., & Prabhu, K. V. (2020). Introgression of drought tolerance QTLs through marker assisted backcross breeding in wheat (Triticum aestivum L.). *Indian Journal of Genetics and Plant Breeding, 80*(2), 209–212.

Toledo, A. M. U., Ignacio, J. C. I., Casal, C., Gonzaga, Z. J., Mendioro, M. S., & Septiningsih, E. M. (2015). Development of improved Ciherang-Sub1 having tolerance to anaerobic germination conditions.

Trijatmiko, K. R., Prasetiyono, J., Thomson, M. J., Cruz, C. M. V., Moeljopawiro, S., & Pereira, A. (2014). Meta-analysis of quantitative trait loci for grain yield and component traits under reproductive-stage drought stress in an upland rice population. *Molecular Breeding, 34*(2), 283–295.

Tura, H., Edwards, J., Gahlaut, V., Garcia, M., Sznajder, B., Baumann, U., Shahinnia, F., Reynolds, M., Langridge, P., Balyan, H. S., & Gupta, P. K. (2020). QTL analysis and fine mapping of a QTL for yield-related traits in wheat grown in dry and hot environments. *Theoretical and Applied Genetics, 133*(1), 239–257.

Tyagi, A. K., & Khurana, J. P. (2003). Plant molecular biology and biotechnology research in the post-recombinant DNA era. In *Biotechnology in India I* (pp. 91–121). Springer.

Uga, Y., Kitomi, Y., Yamamoto, E., Kanno, N., Kawai, S., Mizubayashi, T., & Fukuoka, S. (2015). A QTL for root growth angle on rice chromosome 7 is involved in the genetic pathway of DEEPER ROOTING 1. *Rice, 8*(1), 1–8.

Unterseer, S., Bauer, E., Haberer, G., Seidel, M., Knaak, C., Ouzunova, M., Meitinger, T., Strom, T. M., Fries, R., Pausch, H., & Bertani, C. (2014). A powerful tool for genome analysis in maize: Development and evaluation of the high density 600 k SNP genotyping array. *BMC Genomics, 15*(1), 823.

Valarmathi, M., Sasikala, R., Rahman, H., Jagadeeshselvam, N., Kambale, R., & Raveendran, M. (2019). Development of salinity tolerant version of a popular rice variety improved white ponni through marker assisted back cross breeding. *Plant Physiology Reports, 24*(2), 262–271.

Van Inghelandt, D., Frey, F. P., Ries, D., & Stich, B. (2019). QTL mapping and genome-wide prediction of heat tolerance in multiple connected populations of temperate maize. *Scientific Reports, 9*(1), 1–16.

Venuprasad, R., Shashidhar, H. E., Hittalmani, S., & Hemamalini, G. S. (2002). Tagging quantitative trait loci associated with grain yield and root morphological traits in rice (Oryza sativa L.) under contrasting moisture regimes. *Euphytica, 128*(3), 293–300.

Verma, S. K., Xalxo, M. S., Saxena, R. R., & Verulkar, S. B. (2014). Identification of QTLs for cold tolerance at seedling stage in rice (Oryza sativa L.). *Indian Journal of Genetics, 74*(1), 86–89.

Vikram, P., Swamy, B. M., Dixit, S., Ahmed, H. U., Cruz, M. T. S., Singh, A. K., & Kumar, A. (2011). qDTY 1.1, a major QTL for rice grain yield under reproductive-stage drought stress with a consistent effect in multiple elite genetic backgrounds. *BMC Genetics, 12*(1), 89.

Vikram, P., Swamy, B. M., Dixit, S., Trinidad, J., Sta Cruz, M. T., Maturan, P. C., Amante, M., & Kumar, A. (2016). Linkages and interactions analysis of major effect drought grain yield QTLs in rice. *PLoS One, 11*(3), e0151532.

Vivek, B. S., Krishna, G. K., Vengadessan, V., Babu, R., Zaidi, P. H., Kha, L. Q., Mandal, S. S., Grudloyma, P., Takalkar, S., Krothapalli, K., & Singh, I. S. (2017). Use of genomic estimated breeding values results in rapid genetic gains for drought tolerance in maize. *The Plant Genome, 10*(1), 1–8.

Wang, C., Jing, R., Mao, X., Chang, X., & Li, A. (2011a). TaABC1, a member of the activity of bc 1 complex protein kinase family from common wheat, confers enhanced tolerance to abiotic stresses in Arabidopsis. *Journal of Experimental Botany, 62*(3), 1299–1311.

Wang, D., Liu, J., Li, C., Kang, H., Wang, Y., Tan, X., Liu, M., Deng, Y., Wang, Z., Liu, Y., & Zhang, D. (2016). Genome-wide association mapping of cold tolerance genes at the seedling stage in rice. *Rice, 9*(1), 61.

Wang, F., Zang, X. S., Kabir, M. R., Liu, K. L., Liu, Z. S., Ni, Z. F., Yao, Y. Y., Hu, Z. R., Sun, Q. X., & Peng, H. R. (2014). A wheat lipid transfer protein 3 could enhance the basal thermo-tolerance and oxidative stress resistance of Arabidopsis. *Gene, 550*(1), 18–26.

Wang, J., Gao, X., Dong, J., Tian, X., Wang, J., Palta, J. A., Xu, S., Fang, Y., & Wang, Z. (2020a). Over-expression of the heat-responsive wheat gene TaHSP23. 9 in transgenic Arabidopsis conferred tolerance to heat and salt stress. *Frontiers in Plant Science, 11*, 243.

Wang, K., Gong, Q., & Ye, X. (2020b). Recent developments and applications of genetic transformation and genome editing technologies in wheat. *Theoretical and Applied Genetics*, 1–20.@@@

Wang, M., & Xia, G. (2018). The landscape of molecular mechanisms for salt tolerance in wheat. *The Crop Journal, 6*(1), 42–47.

Wang, M., Yuan, J., Qin, L., Shi, W., Xia, G., & Liu, S. (2020c). Ta CYP 81D5, one member in a wheat cytochrome P450 gene cluster, confers salinity tolerance via reactive oxygen species scavenging. *Plant Biotechnology Journal, 18*(3), 791–804.

Wang, N., Liu, B., Liang, X., Zhou, Y., Song, J., Yang, J., Yong, H., Weng, J., Zhang, D., Li, M., & Nair, S. (2019a). Genome-wide association study and genomic prediction analyses of drought stress tolerance in China in a collection of off-PVP maize inbred lines. *Molecular Breeding, 39*(8), 113.

Wang, X., Chen, J., Liu, C., Luo, J., Yan, X., Ai, A., Cai, Y., Xie, H., Ding, X., & Peng, X. (2019b). Over-expression of a protein disulfide isomerase gene from Methanothermobacter thermautotrophicus, enhances heat stress tolerance in rice. *Gene, 684*, 124–130.

Wang, X., Zhang, R., Song, W., Han, L., Liu, X., Sun, X., Luo, M., Chen, K., Zhang, Y., Yang, H., & Yang, G. (2019c). Dynamic plant height QTL revealed in maize through remote sensing phenotyping using a high-throughput unmanned aerial vehicle (UAV). *Scientific Reports, 9*(1), 1–10.

Wang, Y., Sun, F., Cao, H., Peng, H., Ni, Z., Sun, Q., & Yao, Y. (2012a). TamiR159 directed wheat TaGAMYB cleavage and its involvement in anther development and heat response. *PloS one, 7*(11), e48445.

Wang, Z., Cheng, J., Chen, Z., Huang, J., Bao, Y., Wang, J., & Zhang, H. (2012b). Identification of QTLs with main, epistatic and QTL× environment interaction effects for salt tolerance in rice seedlings under different salinity conditions. *Theoretical and Applied Genetics, 125*(4), 807–815.

Wang, Z., Wang, F., Zhou, R., Wang, J., & Zhang, H. (2011b). Identification of quantitative trait loci for cold tolerance during the germination and seedling stages in rice (Oryza sativa L.). *Euphytica, 181*(3), 405.

Wardlaw, I. F. (2002). Interaction between drought and chronic high temperature during kernel filling in wheat in a controlled environment. *Annals of Botany, 90*(4), 469–476.

Warraich, A. S., Krishnamurthy, S. L., Sooch, B. S., Vinaykumar, N. M., Dushyanthkumar, B. M., Bose, J., & Sharma, P. C. (2020). Rice GWAS reveals key genomic regions essential for salinity tolerance at reproductive stage. *Acta Physiologiae Plantarum, 42*(8), 1–15.

Wei, A., He, C., Li, B., Li, N., & Zhang, J. (2011). The pyramid of transgenes TsVP and BetA effectively enhances the drought tolerance of maize plants. *Plant Biotechnology Journal, 9*(2), 216–229.

Wollenweber, B., Porter, J. R., & Schellberg, J. (2003). Lack of interaction between extreme high-temperature events at vegetative and reproductive growth stages in wheat. *Journal of Agronomy and Crop Science, 189*(3), 142–150.

Wu, X., Li, Y., Shi, Y., Song, Y., Wang, T., Huang, Y., et al. (2014). Fine genetic characterization of elite maize germplasm using high-throughput SNP genotyping. *Theoretical and Applied Genetics, 127*, 621–631.

Wu, F., Yang, J., Yu, D., & Xu, P. (2020). Identification and validation a major QTL from "Sea Rice 86" seedlings conferred salt tolerance. *Agronomy, 10*(3), 410.

Xiao, N., Huang, W. N., Zhang, X. X., Gao, Y., Li, A. H., Dai, Y., Yu, L., Liu, G. Q., Pan, C. H., Li, Y. H., & Dai, Z. Y. (2014). Fine mapping of qRC10-2, a quantitative trait locus for cold tolerance of rice roots at seedling and mature stages. *PLoS One, 9*(5), e96046.

Xiao, Y. H., Yi, P. A. N., Luo, L. H., Deng, H. B., Zhang, G. L., Tang, W. B., & Chen, L. Y. (2011). Quantitative trait loci associated with pollen fertility under high temperature stress at flowering stage in rice (Oryza sativa). *Rice Science, 18*(3), 204–209.

Xie, G., Kato, H., & Imai, R. (2012). Biochemical identification of the OsMKK6–OsMPK3 signalling pathway for chilling stress tolerance in rice. *Biochemical Journal, 443*(1), 95–102.

Xie, Y., Feng, Y., Chen, Q., Zhao, F., Zhou, S., Ding, Y., Song, X., Li, P., & Wang, B. (2019). Genome-wide association analysis of salt tolerance QTLs with SNP markers in maize (Zea mays L.). *Genes & Genomics, 41*(10), 1135–1145.

Xu, C., Ren, Y., Jian, Y., Guo, Z., Zhang, Y., Xie, C., Fu, J., Wang, H., Wang, G., Xu, Y., & Li, P. (2017). Development of a maize 55 K SNP array with improved genome coverage for molecular breeding. *Molecular Breeding, 37*(3), 20.

Xu, J., Yuan, Y., Xu, Y., Zhang, G., Guo, X., Wu, F., Wang, Q., Rong, T., Pan, G., Cao, M., & Tang, Q. (2014). Identification of candidate genes for drought tolerance by whole-genome resequencing in maize. *BMC Plant Biology, 14*(1), 83.

Xu, J. L., Lafitte, H. R., Gao, Y. M., Fu, B. Y., Torres, R., & Li, Z. K. (2005). QTLs for drought escape and tolerance identified in a set of random introgression lines of rice. *Theoretical and Applied Genetics, 111*(8), 1642–1650.

Xu, K., Xu, X., Fukao, T., Canlas, P., Maghirang-Rodriguez, R., Heuer, S., Ismail, A. M., Bailey-Serres, J., Ronald, P. C., & Mackill, D. J. (2006). Sub1A is an ethylene-response-factor-like gene that confers submergence tolerance to rice. *Nature, 442*(7103), 705–708.

Xu, W., Tang, W., Wang, C., Ge, L., Sun, J., Qi, X., He, Z., Zhou, Y., Chen, J., Xu, Z., & Ma, Y. Z. (2020). SiMYB56 confers drought stress tolerance in transgenic rice by regulating lignin biosynthesis and ABA signaling pathway. *Frontiers in Plant Science, 11*, 785.

Xue, G. P., Drenth, J., & McIntyre, C. L. (2015). TaHsfA6f is a transcriptional activator that regulates a suite of heat stress protection genes in wheat (Triticum aestivum L.) including previously unknown Hsf targets. *Journal of Experimental Botany, 66*(3), 1025–1039.

Xue, Y., Warburton, M. L., Sawkins, M., Zhang, X., Setter, T., Xu, Y., Grudloyma, P., Gethi, J., Ribaut, J. M., Li, W., & Zhang, X. (2013). Genome-wide association analysis for nine agronomic traits in maize under well-watered and water-stressed conditions. *Theoretical and Applied Genetics, 126*(10), 2587–2596.

Yadaw, R. B., Dixit, S., Raman, A., Mishra, K. K., Vikram, P., Swamy, B. M., Cruz, M. T. S., Maturan, P. T., Pandey, M., & Kumar, A. (2013). A QTL for high grain yield under lowland drought in the background of popular rice variety Sabitri from Nepal. *Field Crops Research, 144*, 281–287.

Yamaguchi, T., & Blumwald, E. (2005). Developing salt-tolerant crop plants: Challenges and opportunities. *Trends in Plant Science, 10*(12), 615–620.

Yang, D. L., Jing, R. L., Chang, X. P., & Li, W. (2007). Identifcation of quantitative trait loci and environmental interactions for accumulation and remobilization of water-soluble carbohydrates in wheat (Triticum aestivum L.) stems. *Genetics, 176*, 571–584.

Yang, J., Li, D., Liu, H., Liu, Y., Huang, M., Wang, H., Chen, Z., & Guo, T. (2020a). Identification of QTLs involved in cold tolerance during the germination and bud stages of rice (Oryza sativa L.) via a high-density genetic map. *Breeding Science*, 19127.

Yang, J., Sun, K., Li, D., Luo, L., Liu, Y., Huang, M., Yang, G., Liu, H., Wang, H., Chen, Z., & Guo, T. (2019a). Identification of stable QTLs and candidate genes involved in anaerobic germination tolerance in rice via high-density genetic mapping and RNA-Seq. *BMC Genomics, 20*(1), 1–15.

Yang, J., Zhang, G., An, J., Li, Q., Chen, Y., Zhao, X., Wu, J., Wang, Y., Hao, Q., Wang, W., & Wang, W. (2020b). Expansin gene TaEXPA2 positively regulates drought tolerance in transgenic wheat (Triticum aestivum L.). *Plant Science,* 110596.

Yang, L. M., Liu, H. L., Zhao, H. W., Wang, J. G., Sun, J., Zheng, H. L., Lei, L., & Zou, D. T. (2019b). Mapping quantitative trait loci and meta-analysis for cold tolerance in rice at booting stage. *Euphytica, 215*(5), 89.

Yang, W., Duan, L., Chen, G., Xiong, L., & Liu, Q. (2013a). Plant phenomics and high-throughput phenotyping: Accelerating rice functional genomics using multidisciplinary technologies. *Current Opinion in Plant Biology, 16*(2), 180–187.

Yang, Y., Li, Y., & Wu, C. (2013b). Genomic resources for functional analyses of the rice genome. *Current Opinion in Plant Biology, 16*(2), 157–163.

Ye, C., Argayoso, M. A., Redoña, E. D., Sierra, S. N., Laza, M. A., Dilla, C. J., Mo, Y., Thomson, M. J., Chin, J., Delaviña, C. B., & Diaz, G. Q. (2012). Mapping QTL for heat tolerance at flowering stage in rice using SNP markers. *Plant Breeding, 131*(1), 33–41.

Ye, C., Tenorio, F. A., Argayoso, M. A., Laza, M. A., Koh, H. J., Redoña, E. D., Jagadish, K. S., & Gregorio, G. B. (2015). Identifying and confirming quantitative trait loci associated with heat tolerance at flowering stage in different rice populations. *BMC Genetics, 16*(1), 41.

Yu, G., Cui, Y., Jiao, Y., Zhou, K., Wang, X., Yang, W., Xu, Y., Yang, K., Zhang, X., Li, P. & Yang, Z. (2022). Comparison of sequencing-based and array-based genotyping platforms for genomic prediction of maize hybrid performance. *The Crop Journal.*

Yu, J., Holland, J. B., McMullen, M. D., & Buckler, E. S. (2008). Genetic design and statistical power of nested association mapping in maize. *Genetics, 178*(1), 539–551.

Yu, J., Hu, S., Wang, J., Wong, G. K. S., Li, S., Liu, B., Deng, Y., Dai, L., Zhou, Y., Zhang, X., & Cao, M. (2002). A draft sequence of the rice genome (Oryza sativa L. ssp. indica). *Science, 296*(5565), 79–92.

Yu, S., Li, M., Xiao, Y., Huang, D., & Chen, D. (2018). Mapping QTLs for cold tolerance at seedling stage using an Oryza sativa× O. rufipogon backcross inbred line population. *Czech Journal of Genetics and Plant Breeding, 54*(2), 59–64.

Yun, W., Jinping, Z., Yong, S., Jauhar, A., Jianlong, X., & Zhikang, L. (2012). Identification of genetic overlaps for salt and drought tolerance using simple sequence repeat markers on an advanced backcross population in rice. *Crop Science, 52*(4), 1583–1592.

Zaidi, P. H., Shahid, M., Seetharam, K., & Vinayan, M. T. (2022). Genomic regions associated with salinity stress tolerance in tropical maize (Zea Mays L.). *Frontiers in Plant Science, 13.*

Zaman-Allah, M., Vergara, O., Araus, J. L., Tarekegne, A., Magorokosho, C., Zarco-Tejada, P. J., Hornero, A., Albà, A. H., Das, B., Craufurd, P., & Olsen, M. (2015). Unmanned aerial platform-based multi-spectral imaging for field phenotyping of maize. *Plant Methods, 11*(1), 1–10.

Zang, J., Sun, Y., Wang, Y., Yang, J., Li, F., Zhou, Y., Zhu, L., Jessica, R., Mohammadhosein, F., Xu, J., & Li, Z. (2008). Dissection of genetic overlap of salt tolerance QTLs at the seedling and tillering stages using backcross introgression lines in rice. *Science in China Series C: Life Sciences, 51*(7), 583–591.

Zang, Q. W., Wang, C. X., Li, X. Y., Guo, Z. A., Jing, R. L., Zhao, J., & Chang, X. P. (2010). Isolation and characterization of a gene encoding a polyethylene glycol-induced cysteine protease in common wheat. *Journal of Biosciences, 35*(3), 379–388.

Zang, X., Geng, X., He, K., Wang, F., Tian, X., Xin, M., Yao, Y., Hu, Z., Ni, Z., Sun, Q., & Peng, H. (2018). Overexpression of the wheat (Triticum aestivum L.) TaPEPKR2 gene enhances heat and dehydration tolerance in both wheat and Arabidopsis. *Frontiers in Plant Science, 9,* 1710.

Zang, X., Geng, X., Liu, K., Wang, F., Liu, Z., Zhang, L., Zhao, Y., Tian, X., Hu, Z., Yao, Y., & Ni, Z. (2017a). Ectopic expression of TaOEP16-2-5B, a wheat plastid outer envelope protein gene, enhances heat and drought stress tolerance in transgenic Arabidopsis plants. *Plant Science, 258*, 1–11.

Zang, X., Geng, X., Wang, F., Liu, Z., Zhang, L., Zhao, Y., Tian, X., Ni, Z., Yao, Y., Xin, M., & Hu, Z. (2017b). Overexpression of wheat ferritin gene TaFER-5B enhances tolerance to heat stress and other abiotic stresses associated with the ROS scavenging. *BMC Plant Biology, 17*(1), 1–13.

Zeeshan, M., Lu, M., Naz, S., Sehar, S., Cao, F., & Wu, F. (2020). Resemblance and difference of seedling metabolic and transporter gene expression in high tolerance wheat and barley cultivars in response to salinity stress. *Plants, 9*(4), 519.

Zeng, P., Zhu, P., Qian, L., Qian, X., Mi, Y., Lin, Z., Dong, S., Aronsson, H., Zhang, H., & Cheng, J. (2020a). Identification and fine mapping of a novel qGR6. 2 locus controlling rice salt tolerance during seed germination.

Zeng, Y., Wen, J., Zhao, W., Wang, Q., & Huang, W. (2020b). Rational improvement of rice yield and cold tolerance by editing the three genes OsPIN5b, GS3, and OsMYB30 with the CRISPR–Cas9 system. *Frontiers in Plant Science, 10*, 1663.

Zhang, A., Liu, Y., Wang, F., Li, T., Chen, Z., Kong, D., Bi, J., Zhang, F., Luo, X., Wang, J., & Tang, J. (2019). Enhanced rice salinity tolerance via CRISPR/Cas9-targeted mutagenesis of the OsRR22 gene. *Molecular Breeding, 39*(3), 47.

Zhang, D., Zhou, X., Zhang, J., Lan, Y., Xu, C., & Liang, D. (2018a). Detection of rice sheath blight using an unmanned aerial system with high-resolution color and multispectral imaging. *PloS One, 13*(5), e0187470.

Zhang, G. L., Chen, L. Y., Xiao, G. Y., Xiao, Y. H., Chen, X. B., & Zhang, S. T. (2009). Bulked segregant analysis to detect QTL related to heat tolerance in rice (Oryza sativa L.) using SSR markers. *Agricultural Sciences in China, 8*(4), 482–487.

Zhang, H., Mao, X., Jing, R., Chang, X., & Xie, H. (2011). Characterization of a common wheat (Triticum aestivum L.) TaSnRK2. 7 gene involved in abiotic stress responses. *Journal of Experimental Botany, 62*(3), 975–988.

Zhang, J., Zheng, H. G., Aarti, A., Pantuwan, G., Nguyen, T. T., Tripathy, J. N., Sarial, A. K., Robin, S., Babu, R. C., Nguyen, B. D., & Sarkarung, S. (2001). Locating genomic regions associated with components of drought resistance in rice: comparative mapping within and across species. *Theoretical and Applied Genetics, 103*(1), 19–29.

Zhang, L., Geng, X., Zhang, H., Zhou, C., Zhao, A., Wang, F., Zhao, Y., Tian, X., Hu, Z., Xin, M., & Yao, Y. (2017a). Isolation and characterization of heat-responsive gene TaGASR1 from wheat (Triticum aestivum L.). *Journal of Plant Biology, 60*(1), 57–65.

Zhang, M., Liang, X., Wang, L., Cao, Y., Song, W., Shi, J., Lai, J., & Jiang, C. (2019b). A HAK family Na+ transporter confers natural variation of salt tolerance in maize. *Nature Plants, 5*(12), 1297–1308.

Zhang, M., Ye, J., Xu, Q., Feng, Y., Yuan, X., Yu, H., Wang, Y., & Yang, Y. (2018b). Genome-wide association study of cold tolerance of Chinese indica rice varieties at the bud burst stage. *Plant Cell Reports, 37*(3), 529–539.

Zhang, S., Xu, Z. S., Li, P., Yang, L., Wei, Y., Chen, M., Li, L., Zhang, G., & Ma, Y. (2013). Overexpression of TaHSF3 in transgenic Arabidopsis enhances tolerance to extreme temperatures. *Plant Molecular Biology Reporter, 31*(3), 688–697.

Zhang, S., Zheng, J., Liu, B., Peng, S., Leung, H., Zhao, J., Wang, X., Yang, T., & Huang, Z. (2014). Identification of QTLs for cold tolerance at seedling stage in rice (Oryza sativa L.) using two distinct methods of cold treatment. *Euphytica, 195*(1), 95–104.

Zhang, X., Peng, Q., Zhang, D., Wu, J., Wang, J., Huang, P., & Zhu, S. (2018c). QTL analysis of cold-tolerance at seedling stage in rice. *Agricultural Biotechnology, 7*(5), 80–82.

Zhang, Y., Ponce, K.S., Meng, L., Chakraborty, P., Zhao, Q., Guo, L., Gao, Z., Leng, Y., & Ye, G. (2020). QTL identification for salt tolerance related traits at the seedling stage in indica rice using a multi-parent advanced generation intercross (MAGIC) population. *Plant Growth Regulation*, 1–9.

Zhang, Z., Li, J., Li, F., Liu, H., Yang, W., Chong, K., & Xu, Y. (2017b). OsMAPK3 phosphorylates OsbHLH002/OsICE1 and inhibits its ubiquitination to activate OsTPP1 and enhances rice chilling tolerance. *Developmental cell, 43*(6), 731–743.

Zhao, L., Lei, J., Huang, Y., Zhu, S., Chen, H., Huang, R., Peng, Z., Tu, Q., Shen, X., & Yan, S. (2016). Mapping quantitative trait loci for heat tolerance at anthesis in rice using chromosomal segment substitution lines. *Breeding Science*, 15084.

Zhao, Q., Feng, Q., Lu, H., Li, Y., Wang, A., Tian, Q., Zhan, Q., Lu, Y., Zhang, L., Huang, T., & Wang, Y. (2018). Pan-genome analysis highlights the extent of genomic variation in cultivated and wild rice. *Nature Genetics, 50*(2), 278–284.

Zhao, X., Peng, Y., Zhang, J., Fang, P., & Wu, B. (2017). Mapping QTLs and meta-QTLs for two inflorescence architecture traits in multiple maize populations under different watering environments. *Molecular Breeding, 37*(7), 91.

Zhao, X., Peng, Y., Zhang, J., Fang, P., & Wu, B. (2018). Identification of QTLs and Meta-QTLs for seven agronomic traits in multiple maize populations under well-watered and water-stressed conditions. *Crop Science, 58*(2), 507–520.

Zhao, Y., Dong, W., Zhang, N., Ai, X., Wang, M., Huang, Z., Xiao, L., & Xia, G. (2014). A wheat allene oxide cyclase gene enhances salinity tolerance via jasmonate signaling. *Plant Physiology, 164*(2), 1068–1076.

Zheng, B. S., Yang, L., Mao, C. Z., Huang, Y. J., & Wu, P. (2008). Mapping QTLs for morphological traits under two water supply conditions at the young seedling stage in rice. *Plant Science, 175*(6), 767–776.

Zhi-Hong, Z., Li, S., Wei, L., Wei, C., & Ying-Guo, Z. (2005). A major QTL conferring cold tolerance at the early seedling stage using recombinant inbred lines of rice (Oryza sativa L.). *Plant Science, 168*(2), 527–534.

Zhou, S., Zhu, M., Wang, F., Huang, J., & Wang, G. (2013). Mapping of QTLs for yield and its components in a rice recombinant inbred line population. *Pakistan Journal of Botany, 45*(1), 183–189.

Zhou, Y., Chebotarov, D., Kudrna, D., Llaca, V., Lee, S., Rajasekar, S., Mohammed, N., Al-Bader, N., Sobel-Sorenson, C., Parakkal, P., & Arbelaez, L. J. (2020a). A platinum standard pan-genome resource that represents the population structure of Asian rice. *Scientific Data, 7*(1), 1–11.

Zhou, Y., Chen, M., Guo, J., Wang, Y., Min, D., Jiang, Q., Ji, H., Huang, C., Wei, W., Xu, H., & Chen, X. (2020b). Overexpression of soybean DREB1 enhances drought stress tolerance of transgenic wheat in the field. *Journal of Experimental Botany, 71*(6), 1842–1857.

Zhu, S., Huang, R., Wai, H. P., Xiong, H., Shen, X., He, H., & Yan, S. (2017). Mapping quantitative trait loci for heat tolerance at the booting stage using chromosomal segment substitution lines in rice. *Physiology and Molecular Biology of Plants, 23*(4), 817–825.

Zimin, A. V., Puiu, D., Hall, R., Kingan, S., Clavijo, B. J., & Salzberg, S. L. (2017). The first near-complete assembly of the hexaploid bread wheat genome, Triticum aestivum. *Gigascience, 6*(11), gix097.

Zou, G. H., Mei, H. W., Liu, H. Y., Liu, G. L., Hu, S. P., Yu, X. Q., Li, M. S., Wu, J. H., & Luo, L. J. (2005). Grain yield responses to moisture regimes in a rice population: Association among traits and genetic markers. *Theoretical and Applied Genetics, 112*(1), 106–113.

Chapter 4
Perceptions on Disease and Pest Status of Major Cultivated Crops in Indian Himalayas Under Changing Climate

K. K. Mishra, A. R. N. S. Subbanna, H. Rajashekara, Amit U. Paschapur, B. Jeevan, Ashish K. Singh, and Chandan Maharana

4.1 Introduction

The Indian Himalaya is one of the most important mountain ecosystems covering the largest snow and ice outside the Polar Regions. Often termed as the "*Third Pole or Water tower of Asia*", it is origin of three major rivers Indus, Ganges and Brahmaputra of Asia. Major fraction of the world's population resides in the great gangetic plains and depends on availability of these resources for water (Kulkarni et al., 2007). These mountains are imparting primary role in sustaining livelihood of over 300 million people directly or indirectly (Schild, 2008). Himalayan mountains have been declared as one of the 34 biological hotspots in the world and harbour maximum biological wealth (Anonymous, 2011). The hill agriculture is one of the most vulnerable ecosystems to the climate change and it mainly depends on animal husbandry, marginal agriculture and horticulture. Presently, Himalayan ecosystem is mainly facing challenges due to increasing dryness, mean temperature during winter, changes in rainfall and unexpected frosts and storms (Dash & Hunt, 2007; Renton, 2009), consequently affecting the entire biodiversity range, especially horticultural and agricultural crops (Kala, 2013; Renton, 2009). Being richest biodiversity, the Himalaya is one of the most vulnerable mountain ecosystems to climate change (Bawa et al., 2010; Xu et al., 2009) and very meagre information is available on systematic analysis of climate change and its effect on the Himalayan ecosystems, biodiversity and local people's livelihoods (Shrestha et al., 2012). The annual average temperature of India has remained consistently above normal temperature from 1993 to 2011 (except 1997). In India, annual average temperature as a whole was found

K. K. Mishra (✉) · A. R. N. S. Subbanna · H. Rajashekara · A. U. Paschapur · B. Jeevan · A. K. Singh · C. Maharana
Crop Protection Division, ICAR-Vivekananda Parvatiya Krishi Anusandhan Sansthan, Almora, Uttarakhand 263 601, India
e-mail: Krishna.Mishra@icar.gov.in

S. Sheraz Mahdi et al. (eds.), *Adapting to Climate Change in Agriculture-Theories and Practices*, https://doi.org/10.1007/978-3-031-28142-6_4

to be 0.55 °C more of normal temperature. According to the IMD future projection, monsoon rainfall is expected to increase by 15–30% by the end of the twenty-first century. The mean annual temperature is predicted to rise by 3–6 °C, but the northern India will be warmer than the other part of the country.

Agriculture shares approximately 13.7% of gross domestic product (GDP) of India. It was predicted that by the year 2020, food grain requirement will be almost 30–50% more in comparison to the demand during 2000 (Paroda & Kumar, 2000). In addition to increased competition for land, water and labour, Indian agriculture is also facing one of the most important challenges because of changing climatic conditions. The earth's temperature has increased by 0.74 °C between 1906 and 2005 due to increase in anthropogenic emission of greenhouse gases (report of Intergovernmental Panel on Climate Change, IPCC of the United Nations). It is presumed that by the end of twenty-first century, temperature increase is likely to be 1.8–4.0 °C, leading to frequent hot extremes, floods, droughts, cyclones and gradual recession of glaciers, and finally in greater instability in food production. The crop production losses in India by 2100 could be 10–40% in spite of beneficial effects of higher CO_2 concentration on crop growth. It is also predicted that by the end of the twenty-first century, rainfall will increase by 15–40% and the mean annual temperature by 3–6 °C (NATCOM, 2004). Several studies established a decreasing trend of agricultural crop production with changing climate (Aggarwal, 2003; Aggarwal & Sinha, 1993; Mall & Aggarwal, 2002; Rao & Sinha, 1994; Saseendran et al., 2000). Aggarwal (2008) reported that with rise of 1 °C temperature throughout the growing period with current land use, there will be projected loss of 4–5 million tonnes in wheat production. Samra and Singh (2002) reported a loss of more than 10% in food production by severe drought during 2002. The northwest Himalayan regions consisting of Uttarakhand, Himachal Pradesh, Jammu and Kashmir and Leh contribute significantly to ecology and economy of Indian hill agriculture. Singh et al. (2010) reported that the mountains are rich repository of biodiversity and water. But, they are also the most fragile environments on earth. They are sources of ecosystem goods and services on which regional and global communities depend on. Apart from causing direct impacts on crop productivity; climate change is threatening food production in hill agriculture through increasing pest population and their damage potential by expanding distribution, enhancing survivability and allowing adaptable insect pests to temperate climate (Bebber et al., 2013). In comparison to insect-pest species found in tropical and temperate regions, the change in climate has significant role on species found in temperate regions. The increase in temperature and decrease in amount of precipitation is leading to altitudinal shift and intrusion of insect pests into high altitude agricultural lands (Bhutiyani et al., 2007; Kumar et al., 2008a). According to Stoeckli et al. (2012) the climate change is also associated with several changes in insect pest physiology and population dynamics like;

(a) Changes in diversity, distribution and abundance of insect-pests
(b) Changes in geographical distribution of insect-pests
(c) Increased number of overwintering insects
(d) Rapid population growth and increase in number of generations

(e) Invasion on new alternative host plants
(f) Changes in host plant resistance to insect pests
(g) Increased risk of invasive pest species
(h) Emergence and dissemination of vector borne diseases.

In the present article we attempted to give elaborate information regarding recent disease and pest outbreaks (minor pest assuming the status of major pest due to several biotic and abiotic factors) in the Uttarakhand, Himalayan region due to changing climatic conditions.

4.2 Observed Climate Change in Northwest Himalayas

The average temperature during 1955–2007 of Almora (29° 35′ N, 79° 35′ E, 1640 m amsl), Uttarakhand, India showed an increasing trend. The average increase in temperature i.e. 17.55 °C (1955–2007) was recorded to be 0.46 °C. This observation showed rise of average temperature in Uttarakhand state. However, annual rainfall of three places of district Almora, i.e. Almora, Manora Peak and Hawalbagh showed a decreasing trend (UCCC, 2011). The occurrence of drought incidences has shown the increasing trend in the recent past. During year 1964–2000, the total drought incidences were 16, and 5 were severe. But within 8 years duration (2001–2009), a total 7 drought years occurred and out of which 3 were severe (Panday et al., 2009). Maximum and minimum temperatures were observed increasing and decreasing, respectively (Panday et al., 2003). During 1935–1971, 1971–2004 and 2004–2005 time duration, the Gangotri glaciers of Uttarakhand showed a retreating trend of 26.5 m/year, 17.1 m/year and 12.5 m/year, respectively (Bali et al., 2010; Kumar et al., 2008b). The increased anthropogenic pressure in the mountain areas has changed the existing land use pattern and caused hazardous events like landslides, which led to significant loss of lives, resources and property (Upreti, 2010). Due to climate change, a change in the biotic (pest and diseases) and abiotic stress have been recorded and expected particularly in major crops grown in NW Himalayas.

4.3 Impact of Climate Change on Insect Pests

Insect pests are dependent on temperature and expected to respond to climate change by shifting their geographic distributions to take advantage of new niches that become available. Besides, alterations in concentration of different gases viz., CO_2, O_3 etc. indirectly affects the feeding behaviours of insects that ultimately affect the productivity. The prediction studies and subsequently modeling of these variations will help in identifying the insect species likely to become pests in the near future. However, due to realized climate changes in recent past, there are evidences of changes in pest spectrum, species diversity and richness, minor pests becoming major etc. which are detailed in the following sections crop wise.

4.3.1 White Grubs

White grubs are the important polyphagous pests of India. However, due to high prevalence of rainfed cultivation under Himalayan region, India their damage is frequently noticed in the region, irrespective to the crop with a damage potential of around 30%. Under conducive environmental conditions, the pest has a potential to cause complete crop failure. Due to prevailing varied ecological conditions and cropping patterns, a huge diversity of species including approximately 78 species of white grubs have been reported in Uttarakhand Hills alone (Sushil et al., 2004). Different species are reported to be endemic to crops like potato incurring huge economic losses (Chandel et al., 2015). In particular, to Uttarakhand Himalayan region, VL light trap-1, developed by ICAR-VPKAS, Almora is a technology used for monitoring and mass trapping of adult beetles during the beetle emergence period (June to August). A ten-year light trap catches of the white grub beetles since 2010 (Table 4.1), clearly evidenced a change in the spectrum of species where *Anomala dimidiata*, a major species with a catch composition of 58.4% become 11.2% by 2020. This position was taken over by other species of *Anomala* viz., *A. tristis* and *Anomala* spp.

Table 4.1 Species composition of different white grub beetles trapped in VL Light trap 1

Species	Percent species composition										
	2010	2011	2012	2013	2014	2015	2016	2017	2018	2019	2020
Anomala dimidiata	58.4	67.2	63.1	79.9	78.8	85.6	44.2	16.4	6.6	13.1	11.9
Anomala tristis	1.4	1.5	2.7	1.4	1.3	0.4	1.3	1.9	1.9	5.3	5.2
Anomala sp.	3.2	3.6	2.2	1.3	3.1	1.8	8.8	12.1	7.7	13.8	17.9
Mimela fulgidividata	0.8	1.1	1.2	1.0	0.9	1.3	1.6	2.3	2.1	1.6	0.8
Heteronychus lioderus	0.3	0.6	1.8	1.7	0.4	0.4	1.0	1.5	1.3	1.8	0.2
Holotrichia seticollis	1.3	1.4	1.2	0.2	0.4	0.3	0.3	0.2	0.2	0.2	1.1
Lepidiota stigma	4.9	0.0	2.5	0.0	1.0	0.0	1.5	0.2	2.6	0.2	2.2
Xylotrupes Gideon	4.7	1.9	2.7	1.0	1.1	0.9	1.5	2.3	2.6	4.0	2.8

The emergence and abundance of white grub beetles in a given season is a function of rain fall and temperature. Studies in the region reported that high temperatures during beetle emergence time (May–June) and low soil temperatures during peak winters (December, January and February) favoured the white grub beetle emergence (Sushil et al., 2004). Whereas, rains during April, October and November were detrimental to beetle emergence (Stanley et al., 2009) which breaks the diapause resulting in death due to low temperatures.

4.3.2 Grasshoppers (Oxya spp., Heiroglyphus spp. and Chrotugonus spp.)

Grasshoppers have recently gained the status of major pest in paddy, millets soybean and vegetable crops grown in Himalayan conditions. They are highly polyphagus pests infesting both cultivated and uncultivated plants. Both nymphs and adults defoliate the plants leaving only the midribs and stalks. Under Uttarakhand hill region, they are usually seen during the peak vegetative stages of the crop during *kharif* seasons. In general, the damage of the pest is unnoticed due to their small numbers. However, during *kharif*, 2016 and 2017 very high incidence of grasshoppers was observed in different villages of Nainital district during 2nd week of August to September (Fig. 4.1). The population levels were comparable with that of locusts accounting to more than 10 adults or nymphs per square meter area. This high incidence of grasshoppers is attributed to the pre-monsoon showers received during the years. This unnatural rainfall resulted in profuse growth of vegetation which favoured the growth and multiplication of the pest.

Fig. 4.1 Incidence of grasshoppers in Janoli village of Nainital district during *kharif* 2016

Fig. 4.2 Field view of damage caused by Brown Plant Hopper

4.3.3 *Brown Plant Hoppers in Rice*

Brown plant hoppers (BPH) is a major sucking pest of rice in plain region causing extensive damage to rice cultivation. Both nymphs and adults congregate at the base of the rice plant just above the water level and suck the sap. The affected plants dry up in circular patches leading to typical symptom called 'hopper burn' (Fig. 4.2). Till 2008, the pest was not observed in Himalayan conditions and made its first appearance in *kharif*, 2009 in low hills of Uttarakhand Himalayas, but not severe. In the following year i.e., *kharif*, 2010 the pest reached mid hills of Uttarakhand Himalayas and appeared in severe form. The damage levels ranged from 30 to 50%. This is clear evidence of pest migration/expansion in respect of temperature alleviations more precisely increase in maximum temperature coupled with low rain fall observed during the years (Stanley et al., 2009). In addition, delay in sowing time due to late monsoons during the season also aggravated the pest damage. A similar migration of BPH from hotter parts to cooler areas due to global warming was reported in Japan (Heong et al., 1995). Studies also reported elevated temperatures positively affect the fecundity, multiplication, growth parameters of BPH (Pandi et al., 2018) although the pest is evident in further years no noticeable damage was observed.

4.3.4 *Fall Army Worm (Spodoptera frugiperda)*

Fall army worm is a most notorious pest of recent times invading Indian subcontinent. Although pest was observed initially during 2018 in southern parts of India (Telangana), the pest has made its appearance in Himalayan states by *kharif*, 2019 and became major threat to maize production. By *kharif*, 2020 the pest reached most of the maize growing areas including high hill region causing damage up to 80%. The caterpillars feed on central whirl of the plants during night time and cause dead

Fig. 4.3 Symptoms of FAW on maize **a** field view **b** closer view and **c** larvae

hearts, series of pin holes on newly furled leaves and severe defoliation of the young and old leaves (Fig. 4.3). The caterpillars also bore into cobs and feed on seeds thus causing severe yield losses. Monitoring and early management of the pest is necessary for containing its damage.

4.3.5 South American Pin Worm in Tomato (Tuta absoluta)

The tomato pin worm, *Tuta absoluta* is one of the serious pests of tomato causing extensive damage in many countries. The larva mine between the epidermis of leaves and make irregular blotches. It also feeds on stem, buds, calyx and fruits. In fruits, they make pin holes and mine inside. It is reported in brinjal, potato, capsicum, etc. Adults are small brown moths (5–7 mm) with silvery and black spots and reportedly lay 250 eggs. The pest was first reported in India in 2014. In Uttarakhand Himalayas, the pest was first observed during May 2018 in village Bhagartola, Bageshwar district in tomato crops grown under polyhouses (Fig. 4.4). The polyhouse conditions provide ideal microclimate for the multiplication and development of the pest and thus appeared in sever form only in polyhouse conditions, whereas, pest is negligible in open field conditions which might be due to prevailing low temperature and poor survival during winters. However, modelling of pest distribution in respect of climatological data clearly depicts severe infestation levels in Indian Himalayan region (Fand et al., 2020) which can be attributed to expected increasing temperatures in the region in association with climate change.

Fig. 4.4 Symptoms of *Tuta absoluta* on tomato **a** leaves **b** fruits and **c** larvae

4.4 Recent Invasions of Insect Pests Observed in Uttarakhand Himalayan Conditions

4.4.1 *Shoot Fly in Millets and Wheat*

Shoot flies are the minor and sporadic pests infesting different millets (Finger millet, Proso millet, Kodo Millet and Barnyard Millet) of the region. In Uttarakhand region, more than six species of shoot flies (*Atherigona* sp.) are known to infect wheat and barley crops in seedling stage. The maggots bore into the central shoot, make a horizontal cut at the base and feed on decaying tissues. Feeding by the maggots cause dead hearts and these dead hearts can be easily pulled out from the whorl and emit foul smell. The damage of the pest is restricted up to 30 days of the crop. In general, the damage by this pest becomes unnoticed due to very low population and microscopic size. However, in the recent past (*Rabi* 2018 and 2019) the per cent infestation of shoot flies in wheat ranged from 20–30% and 10–20% in case of barley. The fish meal bait trap installed @ one trap per 200 m^2could trap more than 58 flies per trap with in a span of 3 days in finger millet, thus indicating its severity and possible pest outbreaks in the near future.

4.4.2 *Maize Cob Worm (Helicoverpa armigera) in Wheat*

The infestation of *Helicoverpa* in wheat and barley was observed in recent years (Patel, 2016). The early-stage caterpillars feed on leaves by scraping the chlorophyll content, while, the late stages found feeding on ear head. The infestation is severe during milky and dough stage of the crop, where the infestation is seen as individual chaffy grains with circular holes in wheat panicle. Although the per cent infestation of the pest is between 2 and 5% in the last two years but the chance of *Helicoverpa* reaching pest status is not far because of changing environmental conditions.

4.4.3 Mustard Aphid (*Lipaphis erysimi*)

Aphids in mustard are key and persistent pests. Both the nymphs and adults suck the sap from tender shoot and capsules (Fig. 4.5). Due to feeding of aphids seeds shrink and yield is drastically reduced. Up to 100% yield loss can be observed in case of severe infestations. Although the infestation is observed during late in the season, in years with high temperatures, the pest had its appearance early in the season causing severe damage. This early infestation also escapes predation by coccinellids incurring huge losses.

4.4.4 Soybean Sucking Bug (*Chauliops* spp.)

The leaf bug is pest of hilly regions and not found in plains. The nymphs and adults suck the sap from under surface of the leaf and cause white spots on leaves there by reducing photosynthetic potential. In severe cases, the leaves turn brown and dry up. Severe infestation during early crop growth stage can lead to complete crop failures. The insects prefer shady areas near the bunds. The population of the pest drastically reduces during rainy days and at the time of harvest (Fig. 4.5). The bug is also known to infect rajma, red gram and horse gram.

4.4.5 Wheat Aphids (*Microsiphum miscanthis* and *Sitobion avenae*)

The infestation of aphids begins from second fortnight of January. The winged females give birth to wingless nymphs that suck the sap from underside of the leaf and as the infestation continues, the aphids also infect ear heads and cause shrinking of seeds and thus reduction in germination percentage of the seeds. The aphid infestation in the *Rabi*–2018 varied between 15 and 20%, while in *rabi*-2019 the incidence rose to 30% due to extended winters (cold weather continuing till the end of March month).

4.4.6 Greenhouse Whitefly (*Trialeurodes vaporariorum*)

The whitefly infects wide variety of crops including pulses and vegetables. Recently, due to increased cultivation of crops in greenhouses, the pest population has shoot up rapidly. Whiteflies are hard to control pests that infect tomato, capsicum, chilli, French bean, cauliflower and brinjal in Uttarakhand, Himalayan region. The nymphs and adults suck the sap from under surface of the leaf and excrete honey dew, on

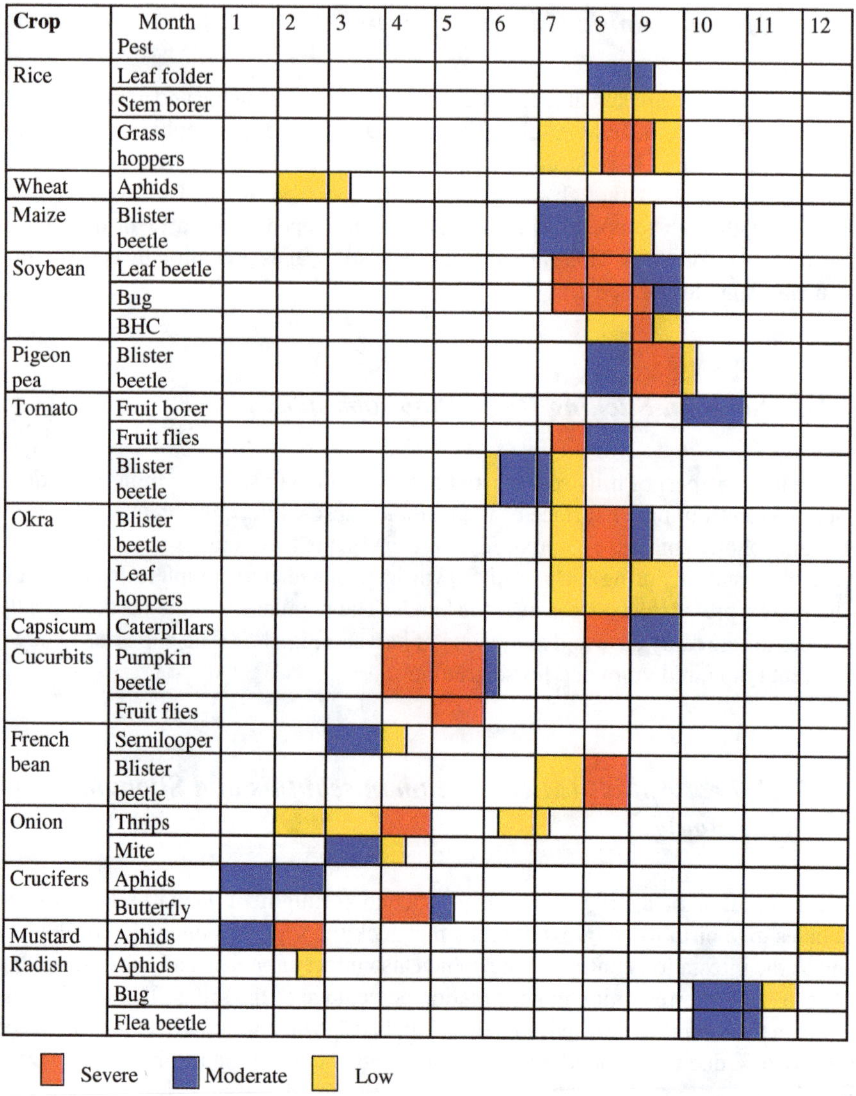

Fig. 4.5 Progress of pests in different cereal and vegetable crops

which sooty mould develops and hinders the photosynthetic activity of plants. They also transmit large number of viral diseases in plants, thus causing severe yield losses. The greenhouse whitefly has also developed resistant to several novel insecticides and the only management option left with the farmers is to regularly fumigate the greenhouses to reduce the pest load.

4.4.7 *Black Cut Worm or Army Worm (Agrotis segetum)*

It is sporadic pest infecting all the vegetable and field crops in nursery. The adults lay up to 300–400 eggs on plant debris and soil. The caterpillars cut the young seedlings at collar region at night time and defoliate the plants. During day time they hide in soil and are difficult to find. The damage is severe in uplands, rain fed areas and sandy soils. The damage in nursery may sometimes reach up to 100%. Seedling root treatment and broadcasting food poison bait in the evening hours could manage the pest. BSKE powder spread is also an effective management strategy.

4.4.8 *Mites in Chilli and Capsicum (Tetranychus spp. and Polyphagotarsonemus spp.)*

The infection of mites is severe in greenhouse conditions. The nymphs, larvae and adults of mites suck the sap from under surface of leaf and cause downward curling and crinkling of the leaves. The infected leaves show dark green colouration on upper surface and bronzing on lower surface with elongated petiole. Early crop infestation can cause crop mortality and 100% crop loss.

4.5 Diseases of Major Crops

4.5.1 *Wheat Diseases*

Outbreak of yellow rust in the northern part of the India has been recorded during *rabi* 2010–2011 and ranged from 5 to 80S. The disease started in second fortnight of February in most of areas and reached up to 80S during end of March month (Fig. 4.6a and 4.10). In hilly areas, the severity was upto 80S with maximum prevalence in locally collected wheat varieties. However, the recommended varieties like VL *Gehun* 804, VL *Gehun* 829, and VL *Gehun* 907 were free from disease (Mishra & Rajashekara, 2019). In Kangra, Hamirpur, Bilaspur and Mandi districts of Himachal Pradesh, stripe rust was observed ranging from 10 to 80S. The varieties viz. HPW 184, HPW 42, HPW 155 and VL *Gehun* 907 were found resistant and free from disease. In Jammu and Kashmir, prevalence of stripe rust ranged from 5 to 80S. Varieties like PBW 343, PBW 550, PBW 502 and DBW 17 showed 30–60% disease prevalence, however, Raj 3,077 and Raj 3,765 were found resistant and hardly affected by yellow rust (10S) (Kant et al., 2011). A total of 897 samples of three rusts from fourteen Indian states, and Nepal were pathotyped at ICAR-Indian Institute of Wheat and Barley Research, Regional Station, Flowerdale, Shimla, Himachal Pradesh during 2019–20. Pathotyping revealed that the frequency of *Puccinia striiformis tritici* (Pst; Yellow rust) pathotype 238S119 was maximum (44.06%) followed by pathotype

Fig. 4.6 Symptoms of **a** yellow rust **b** black rust

46S119 (33.2%) and 110S119 (18.98%) of the analyzed samples. Pathotypes 77-9, 77-5 and 77-1 were the most predominant in *triticina* (Pt; leaf rut) population and were identified in 50.3, 28.2 and 7.1% of the samples, respectively. Pathotype 11 of *P. graminis tritici* (Pgt; black rust) was the most frequent and was observed in 88.2% of the samples analyzed (Fig. 4.6b) followed by 40A (4.7%) and 40-2 (3.9%) (Prasad et al., 2020).

4.5.2 Rice Diseases

With changing climate pattern, the occurrence of diseases has changed drastically in hilly region. The initial establishment and development of severity depends mainly on micro-climate prevailing during given cropping season (Rajashekara et al., 2019). Rice blast (Fig. 4.7) and brown leaf spot diseases are endemic in nature in hilly rice growing areas and false smut (Fig. 4.7), leaf scald and sheath blight diseases are fluctuating every year depending on the prevailing environmental conditions during the growing season (Fig. 4.10). High humidity is reported to induce fungal epidemics. Blast and sheath rot incidence in rice increases with increased winter temperature. Minimum temperature (15–20 °C) and average temperature between 22 and 25 °C, along with high relative humidity of 90% and maximum number of rainy days are conducive for blast development in rice at Uttar Pradesh hills (Bhatt, 1992). Brown leaf spot has positive correlation with September rainfall and August maximum temperature. A temperature of 25–30 °C and a RH of 86–100% are favourable to brown spot incidence in rice. Due to changed climatic situations, crop will be susceptible or climate will be favouring brown spot pathogen for another

Fig. 4.7 Different rice disease symptoms **a** leaf blast **b** neck blast **c** node blast **d** false smut

10 days during rice growing period. The false smut disease favoured by cloudy weather, high relative humidity (> 95%), low temperature (25 to 30 °C), water stress and rainy days at the time of flowering (Raji et al., 2016; Sanghera et al., 2012) and late sowing and application of higher nitrogen doses favours the development of disease (Ahonsi et al., 2000). Sheath blight of rice infection is favoured by warm temperatures (~28–32 °C), high humidity (~95%) and high levels of nitrogen fertiliser (Savary et al., 1995; Singh et al., 2019).

4.5.3 Barley Diseases

Cold arid trans-Himalayan Leh region is highly vulnerable to climate change, and serious environmental threats often result in reduction in crop productivity due to agricultural pest and plant pathogens. Yellow rust, powdery mildew, leaf spot blotch/ blight, covered smut, loose smut, foot/root rot and cereal cyst nematode causing molya disease were observed during field survey (Vaish et al., 2011) and the most destructive diseases in barley were found to be yellow rust, molya and foot/root rot. They also reported that the enhanced temperature would lower the occurrence of yellow rust, whilst the occurrence of leaf spot blotch/blight might become severe.

4.6 Recent Changes in Disease Scenario in Uttarakhand Himalayas

- During the month of February in *rabi* 2018–2019, severe incidence of rust disease in *Parthenium* at ICAR-VPKAS Experimental Farm, Hawalbagh (29° 56′ N, 79° 40′ E, and 1250 m MSL), Almora, Uttarakhand was recorded (Anonymous, 2019). The typical rust symptoms include brown color pustules on leaf and stem (Fig. 4.8). The entire foliage showed burnt appearance and complete drying of

(a) (b) (c)

Fig. 4.8 Symptoms of rust on *Parthenium* **a** leaves **b** stem and **c** uredospores

plants. Frequent monitoring of pathogen was done for teliospore development but it could not be found during the season. This season only uredia could be observed on the infected plants. This report will envisage in development of effective bio-control method for management of noxious weed.

- During February 2019, severe incidence of garlic rust was observed at garlic fields of Mukteshwar, District Nainital, Uttarakhand, India (unpublished data). The symptoms included orange colored pustules on the leaf and leaf sheath. The samples were collected and subjected for microscopic studies. Only uredinial pustules which contain numerous uredospores which were orange in colour, spherical to ellipsoidal, echinulate and are crowded in mass were observed (Fig. 4.9). Frequent monitoring of infected plants was done for teliospore development, however, it was not found in entire season.

- Occurrences of zonate leaf spot (Hooda et al., 2009) and sclerotium rot of maize caused by *Gloeocercospora sorghii* and *Sclerotium rolfsii,* respectively were also recorded first time at Experimental farm, Hawalbagh (29° 56′ N, 79° 40′ E, 1250 m MSL), Almora, Uttarakhand, India.

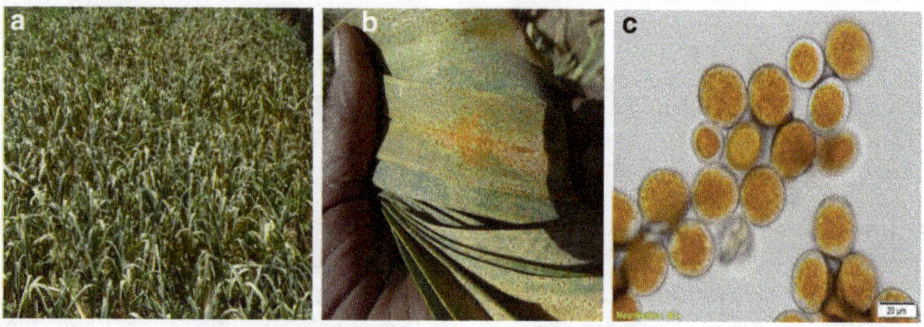

Fig. 4.9 Symptoms of rust on Garlic **a** field view **b** closer view and **c** uredospores

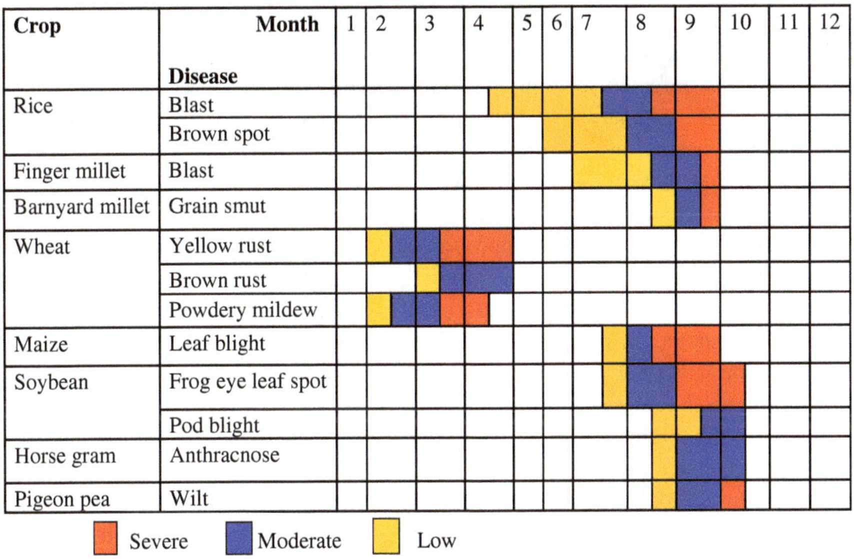

Fig. 4.10 Progress of disease in different cereal crops

***Sclerotium rolfsii* of Maize**

4.7 Impact of Climate Change on Plant Parasitic Nematodes

The impact of global climate change and their mechanism of influence are on soil biota particularly nematodes are little understood. Nematodes are the most important biotic constraints affecting crop yields. Global warming due to increased greenhouse gases may have direct impact on nematode's development rate, survivability and interaction with host. Studies have shown the potential of increase in spread of nematodes to new area besides the plant parasites, the free living, predators, and insect parasite have shown their potential in influencing crop productivity by regulating the nutrient recycling and insect pest suppression. Climate changes directly may affect the dispersal behaviour in agro-ecosystem through infective propagules including eggs, juveniles, infected seedling etc. and indirectly through insects (*Monochamus* transmitting—*Bursaphelenchus xylophilus*) and passive dispersal by flood water, storms etc. Changes may be more significant on the individual species related to their biological behaviour such as development, fecundity life cycle and survivability. Other than this, the diversity and functional status in trophic structure are predicted to be modified. The climate change predicted in the current scenario is likely to have strong influence on nematode's population in Antarctic region by modifying abundance, species diversity, host pattern and new species introduction (Nielsen et al., 2011). The climate change in future will expand the area of cultivation suitable for different crop so the nematode prevalent in those crops will be benefitted by presence of host and so the more yield loss could be predicted. With this the most affected crop will be the Soybean infested by soybean cyst nematode (*Heterodera glycines*) in Canada (Marseille et al., 2019).

The mean global temperature expected to be changed in coming time and it plays a most important role in nematodes development and geographical distribution. The temperature elevated helps in increase of embryonic development rate in the nematodes up to upper threshold limit of 37 °C. So, if the temperature increases, there will be faster rate of multiplication with in less time (Weaver et al., 2010). PPNs have optimum temperature preferendum for its biological activities such as host finding, feeding and reproduction and survival and even a slight change in temperature may influence its lifecycle (increase or decrease in duration) significantly (Evans & Perry, 2009; Papadopoulou & Triantaphyllou, 1982; Tzortzakakis & Trudgill, 2005). Therefore, it may presume that the increase in global temperature may decrease the lifecycle duration of nematode and may increase the number of generation rate of nematodes in a season. High temperature in Pine forest growing regions favors the pine wood nematodes (*Bursaphelenchus mucronatus*) and its vector beetle infestation causing 'Pine wilt disease' (Rebetez & Dobbertin, 2004).

Climate change might play role in interaction host and pathogen in which nematodes may interfere positively with the host by breaking down the existing resistance. For instance the Mi gene a well known nematode resistance gene lost under influence of elevated temperature (Jablonska et al., 2007). The virus vector nematodes (*Xiphinema, Longidous*) have also shown the potential to spread due to change in temperature (Neilson & Boag, 1996). The increase in temperature influences the risk of nematodes on crop host in diverse climate. Risk analysis of coffee nematode (*Meloidogyne incognita*) races 1, 2 and 4 suggests increase in future with the current temperature changing scenarios and the number of generations will also increase in the coming time (Ghini et al., 2008). Climate change may play a very prominent role in geographical distribution of nematodes. As the mercury rises in higher altitude the flow of pest and diseases may be introduced in the higher altitudes (Rosenzweig & Liverman, 1992). The prediction indicted that severity and distribution of *Heterodera glycines*, *Pratylcnhus* spp. and *B. Xylophilus* and *Globodera rostochiensis* increases due to climate change (Boland et al., 2004; Carter et al., 1996; Niblack, 2005). In the recent past the Rice root knot nematode (*M. graminicola*) has been spreading fast from upland region to all rice growing areas of India (Prasad & Somasekhar, 2009; Somasekhar & Prasad, 2009). Apart from these the ongoing survey in the north western Himalayan region of India has shown the incidence of nematodes in polyhouse grown crops and rice. Increase in CO_2 concentration enhances the root proliferation and it can be anticipated that the increase in root mass may influence the plant parasitic nematodes (Rogers et al., 1994; Yeates et al., 1997, 1999, 2003). Abundance of Nematodes like *Tylenchus* and *Longidorus*, *Meloidogyne* have been found to be increased in response to the CO_2 but on genera like *Trichodorus, Paratylenchus,* and Hoplolaimidae have shown no response to CO_2 concentration (Yeates et al., 2003). Shift of climate from one region to other such as tropicalization of temperate region may increase the chances of favourable climatic condition in proliferation of tropical nematode to temperate region and vice versa. Research conducted in the polar region indicates that due to global warming the ice cap is on way of melting which may lead to increase in sea level. In future the pattern of change in climate among polar countries like Canada, Siberia and Greenland is predicted. Therefore, the change in nematode frequency associated with different crop in these regions can be predicted. In a study conducted at subarctic soils reveals that the elevated soil temperature enhanced the population density of nematode to almost twice of the control and the nematode community was found to be affected strongly due to climate change (Ruess et al., 1999).

PPNs are managed by integration of different management practices but we cannot rely on the current practices in the dynamic climate change scenario. Although there is little data available on nematode population change with change in climate, therefore, it is matter of interest to investigate and in addition nematodes are most dominant group inhabiting below ground and their vast diversity offers a potential to asses climate change impact in diverse climatic regime. Therefore, understanding of climate change impact on nematode pest abundance and infestation in different crop is essential to revise the adopted management practices and develop a novel strategy to manage the menace as the climatic condition shifts.

4.8 Conclusion

Crop pests infestation in major cultivated crops in hills indicate the altitudinal shift to temperate regions, due to changing climatic conditions. The pests have gained the alarming status in recent years due to decrease in amount and frequency of rainfall, higher average temperature, and increased area of crops which provide favourable climatic condition for insect-pests to propagate in large numbers. Moreover, the introduction of exotic pests like tomato pin worm, fall armyworm are known to cause severe threats to hill agriculture in years to come and need proper monitoring and forecasting technologies to check their spread before they establish in a new favourable climate. The farmers in Uttarakhand, Himalayas rely mainly on natural control strategies and depends on conventional pesticides, which pave path for faster resistance development by pests. So, use of novel and safer pesticides with different modes of action can help farmers in managing the pests successfully in eco-friendly way. It is clear that with changing climatic conditions, different components of agriculture got affected, therefore, there is an urgent need to study the extent of variability in North West Himalayas with climate change along with scientific interventions to combat the yield losses.

References

Aggarwal, P. K. (2003). Impact of climate change on Indian agriculture. *Journal of Plant Biology, 30*, 189–198.

Aggarwal, P. K. (2008). Global climate change and Indian agriculture: Impacts, adaptation and mitigation. *Indian Journal of Agricultural Sciences, 78*, 911–919.

Aggarwal, P. K., & Sinha, S. K. (1993). Effect of probable increase in carbon dioxide and temperature on productivity of wheat in India. *Journal of Agricultural Meteorology, 48*, 811–814.

Ahonsi, M. O., Adeoti, A. A., Erinle, I. D., Alegbejo, T. A., Singh, B. N., & Sy, A. A. (2000). Effect of variety and sowing date on false smut incidence in upland rice in Edo state, Nigeria. *International Rice Research Notes, 25*, 14.

Anonymous. (2011). R: A language and environment for statistical computing. R Foundation for Statistical Computing, Vienna, Austria.

Anonymous. (2019). First report of rust disease in *Parthenium* weed from Uttarakhand hills, India. *ICAR-VPKAS Newsletter*. *23*(1), 4.

Bali, R., Agarwal, K. K., Ali, S. N., & Srivastava, P. (2010). Is the recessional pattern of Himalayan glaciers suggestive of anthropogenically induced global warming? *Arabian Journal of Geosciences, 4*(7–8), 1087–1093.

Bawa, K. S., Koh, L. P., Lee, T. M., Liu, J., & Ramakrishnan, P. (2010). China, India, and the environment. *Science, 327*, 1457–1459.

Bebber, D. P., Ramotowski, M. A., & Gurr, S. J. (2013). Crop pests and pathogens move polewards in a warming world. *Nature Climate Change, 3*(11), 985.

Bhatt, J. C. (1992). Epidemiology of foliar blast of rice in Uttar Pradesh hills. *Oryza, 29*, 350–354.

Bhutiyani, M. R., Kale, V. S., & Pawar, N. J. (2007). Long-term trends in maximum, minimum and mean annual air temperatures across the Northwestern Himalaya during the twentieth century. *Climatic Change, 85*(1–2), 159–177.

Boland, G. J., Melzer, M. S., Hopkin, A., Higgins, V., & Nassuth, A. (2004). Climate change and plant diseases in Ontario. *Canadian Journal of Plant Pathology, 26*, 335–350.

Carter, T. R., Saarikko, R. A., & Niemi, K. J. (1996). Assessing the risks and uncertainties of regional crop potential under a changing climate in Finland. *Agricultural and Food Science, 5*, 329–350.

Chandel, R. S., Pathania, M., Verma, K. S., Bhatacharyya, B., Vashisth, S., & Kumar, V. (2015). The ecology and control of potato whitegrubs of India. *Potato Research, 58*(2), 147–164.

Dash, S. K., & Hunt, J. C. R. (2007). Variability of climate change in India. *Current Science, 93*, 782–788.

Evans, A. F., & Perry, R. N. (2009). Survival mechanisms. In R. N. Perry, M. Moens, & J. L. Starr (Eds.), *Root knot nematodes* (pp. 201–219). CABI.

Fand, B. B., Shashank, P. R., Suroshe, S. S., Chandrashekar, K., Meshram, N. M., & Timmanna, H. N. (2020). Invasion risk of the South American tomato pinworm Tutaabsoluta (Meyrick) (Lepidoptera: Gelechiidae) in India: Predictions based on MaxEnt ecological niche modelling. *International Journal of Tropical Insect Science*, 1–11.

Ghini, R., Hamada, E., Junior, M. J. P., Marengo, J. A., & Goncalves, R. R. V. (2008). Risk analysis of climate change on coffee nematodes and leaf miner in Brazil. *Pesquisa Agropecuária Brasileira, 43*, 187–194.

Heong, K. L., Song, Y. H., Pimamarn, S., Zhang, R., & Bae, S. D. (1995). Global warming and rice arthropod communities. In S. Peng (Ed.), *Climate change and rice* (pp. 326–335). Springer-Verlag.

Hooda, K. S., Pant, S. K., Mahajan, V., Khati, P., & Bhatt, J. C. (2009). A first record of zonate leaf spot in maize from Uttarakhand, India. *Journal of Mycology and Plant Pathology, 39*, 541.

Jablonska, B., Ammiraju, J. S. S., Bhattarai, K. K., Mantelin, S., Ilarduya, O. M., Roberts, P. A., & Kaloshian, I. (2007). The Mi-9 gene from *Solanum arcanum* conferring heat-stable resistance to root-knot nematodes is a homolog of Mi1. *Plant Physiology, 143*, 1044–1054.

Kala, C. P. (2013). Climate change and challenges of biodiversity conservation. In C. P. Kala, & C. S. Silori (Eds.), *Biodiversity, communities and climate change* (pp. 259–269). The Energy and Resources Institute.

Kant, L., Jain, S. K., Mahanta, D., Agrawal, P. K., & Bhatt, J. C. (2011). Do winter x spring wheat derivatives have the capability to tolerate the climate change conditions? In *3rd international group meeting on wheat productivity enhancement under changing climate* (283 p.), *February 9–12*.

Kulkarni, A. V., Bahuguna, I. M., Rathore, B. P., Singh, S. K., Randhawa, S. S., Sood, R. K., & Dhar, S. (2007). Glacial retreat in Himalaya using Indian Remote Sensing satellite data. *Current Science, 69*, 74.

Kumar, K., Dumka, R. K., Miral, M. S., Satyal, G. S., & Pant, M. (2008a). Estimation of retreat rate of Gangotri glacier using rapid static and kinematic GPS survey. *Current Science, 94*, 258–262.

Kumar, K., Joshi, S., & Joshi, V. (2008b). Climate variability, vulnerability, and coping mechanism in Alaknanda catchment, Central Himalaya, India. *Ambio, 37*, 286–291.

Mall, R. K., & Aggarwal, P. K. (2002). Climate change and rice yields in diverse agro-environments of India: Evaluation of impact assessment models. *Climate Change, 52*, 315–330.

Marseille, A. G., Bourgeois, G., Brodeur, J., & Mimee, B. (2019). Simulating the impacts of climate change on soybean cyst nematode and the distribution of soybean. *Agricultural and Forest Meteorology, 264*, 178–187.

Mishra, K. K., & Rajashekara, H. (2019). Important diseases of wheat and their management. In J. Stanley, K. K. Mishra, A. R. N. S. Subbanna, H. Rajashekara, & A. Pattanayak (Eds.), *Integrated pest management in major crops* (pp. 49–59). ICAR-VPKAS.

NATCOM. (2004). India's initial national communication to UNFCCC, Ministry of Environment & Forests, Government of India.

Neilson, R., & Boag, B. (1996). The predicted impact of possible climatic change on virus-vector nematodes in Great Britain. *European Journal of Plant Pathology, 102*, 193–199.

Niblack, T. L. (2005). Soybean cyst nematode management reconsidered. *Plant Disease, 89*(10), 1020–1026.

Nielsen, U. N., Wall, D. H., Adams, B. J., & Virginia, R. A. (2011). Antarctic nematode communities: Observed and predicted responses to climate change. *Polar Biology, 34*, 1701–1711.

Panday, S. C., Bisht, K. K. S., Srivastava, A. K., & Gupta, H. S. (2009). Climate change in mid Himalaya. In *4th world congress on conservation agriculture, innovations for improving efficiency, equity and environment* (417 p.), *New Delhi*.

Panday, S. C., Singh, R. D., & Srivastava, A. K. (2003). Change in climate and its consequence on crop management and production. In *6th agricultural science congress* (203 p). Multi-Enterprise Systems for Viable Agriculture. IISS, Bhopal.

Pandi, G. G. P., Chander, S., Singh, M. P., & Pathak, H. (2018). Impact of elevated CO_2 and temperature on brown planthopper population in rice ecosystem. *Proceedings of the National Academy of Sciences, India Section B: Biological Sciences, 88*(1), 57–64.

Papadopoulou, J., & Triantaphyllou, A. C. (1982). Sex differentiation in *Meloidogyne incognita* and anatomical evidence of sex reversal. *Journal of Nematology, 14*, 549–566.

Paroda, R. S., & Kumar, P. (2000). Food production and demand in South Asia. *Agricultural Economics Research Review, 13*, 1–24.

Patel C. (2016). *Preliminary studies on seasonal incidence and growth parameters of Helicoverpa armigera (Hubner) on wheat at Pantnagar* (108 p). Thesis M.Sc., G.B. Pant University of Agriculture and Technology, Pantnagar.

Prasad, J. S., & Somasekhar, N. (2009). *Nematode pests of rice: Diagnosis and management* (p. 29). Technical Bulletin No. 38, Directorate of Rice Research (ICAR), Hyderabad.

Prasad, P., Savadi, S., Bhardwaj, S. C., & Gupta, P. K. (2020). The progress of leaf rust research in wheat. *Fungal Biology, 124*(6), 537–550. https://doi.org/10.1016/j.funbio.2020.02.013

Rajashekara, H., Mishra, K. K., & Pattanayak, A. (2019). Blast disease in hill crops: Symptoms, identification and management. In J. Stanley, K. K. Mishra, A. R. N. S. Subbanna, H. Rajashekara, & A. Pattanayak (Eds.), *Integrated pest management in major crops* (pp. 38–48). ICAR-VPKAS.

Raji, P., Sumiya, K. V., Dhanya, S., Remya, K., & Narayanankutty, M. C. (2016). Screening of rice varieties and in vitro evaluation of botanicals against false smut pathogen, *Ustilaginoidea virens*. *International Journal of Agricultural Science Research, 6*(2), 79–86.

Rao, G. D., & Sinha, S. K. (1994). Impact of climatic change on simulated wheat production in India. In C. Rosenzweig & I. Iglesias (Eds.), *Implications of climate change for international agriculture: Crop modelling study* (pp. 1–10). EPA.

Rebetez, M., & Dobbertin, M. (2004). Climate change may already threaten Scots pine stands in the Swiss Alps. *Theoretical and Applied Climatology, 79*, 1–9.

Renton, A. (2009). Suffering the science: Climate change, people and poverty. Oxfam briefing No. 130. Oxford: Oxfam International.

Rogers, H. H., Runion, G. B., & Krupa, S. V. (1994). Plant responses to atmospheric CO_2 enrichment with an emphasis on roots and the rhizosphere. *Environmental Pollution, 83*, 155–189.

Rosenzweig, C., & Liverman, D. (1992). Predicted effects of climate change on agriculture: A comparison of temperate and tropical regions. In S. K. Dalam, & S. K. Majumdar (Ed.), *Dalam global climate change: Implications, challenges, and mitigation measures* (pp. 342–361). Pennsylvania: The Pennsylvania Academy of Sciences.

Ruess, L., Michelsen, A., Schmidt, I. K., & Jonassonm, S. (1999). Simulated climate change affecting microorganisms, nematode density and biodiversity in subarctic soils. *Plant and Soil, 212*, 63–73.

Samra, J. S., & Singh, G. (2002). *Drought management strategies* (68 p). Indian Council of Agricultural Research.

Sanghera, G. S., Ahanger, M. A., Kashayp, S. C., Bhat, Z. A., Rather, A. G., & Parray, G. A. (2012). False smut of rice (*Ustilaginoidea virens*) under temperate agro-climatic condition of Kashmir, India. *Elixir Bio Tech, 49*, 9827–9830.

Saseendran, S. A., Singh, K. K., Rathore, L. S., Singh, S. V., & Sinha, S. K. (2000). Effects of climate change on rice production in the tropical humid climate of Kerala, India. *Clim Chang, 44*, 495–514.

Savary, S., Castilla, N. P., Elazegui, F. A., McLaren, C. G., Ynalvez, M. A., & Teng, P. S. (1995). Direct and indirect effects of nitrogen supply and disease source structure on rice sheath blight spread. *Phytopathology, 85*, 959–965.

Schild, A. (2008). ICIMOD's position on climate change and mountain systems: The case of the Hindu Kush-Himalayas. *Mountain Research and Development, 28*(3), 328–331.

Shrestha, U. B., Gautam, S., & Bawa, K. S. (2012). Widespread climate change in the Himalayas and associated changes in local ecosystems. *PLoS ONE, 7*(5), e36741.

Singh, A. K., Ngachan, S. V., Meghalaya, U., Rawal, R., Singh, K. K., & Sathapathy, S. (2010). *Food security and climate change in Himalayan region: Status and strategies for livelihood improvement*. www.bhutanclimatesummit.org.bt. Assessed October 14, 2012.

Singh, P., Mazumdar, P., Harikrishna, J. A., & Babu, S. (2019). Sheath blight of rice: A review and identification of priorities for future research. *Planta, 250*, 1387–1407.

Somasekhar, N., & Prasad, J. S. (2009). Root-knot nematode *Meloidogyne graminicola*—An emerging threat to rice cultivation. *DRR Newsletter, 7*(4), 3–4.

Stanley, J., Aditya, J. P., Bhatt, J. C., Pandey, S. C., & Agrawal, P. K. (2009). Change in pest spectrum of rice under changing climatic situations in NW Himalayan hills. In *National conference on pest biodiversity in rice and their management under changed climate, CRRI, Cuttack, India.*

Stoeckli, S., Hirschi, M., Spirig, C., Calanca, P., Rotach, M. W., & Samietz, J. (2012). Impact of climate change on voltinism and prospective diapause induction of a global pest insect–*Cydia pomonella* (L.). *PLOS ONE*, *7*(4), 1–9.

Sushil, S. N., Pant, S. K., & Bhatt, J. C. (2004). Light trap catches of white grub and its relation with climatic factors. *Ann Plant Prot Sci, 12*, 254–256.

Tzortzakakis, E. A., & Trudgill, D. L. (2005). A comparative study of the thermal time requirements for embryogenesis in *Meloidogyne javanica* and *M. incognita. Nematology, 7*, 313–315.

UCCC. (2011). *Uttarakhand centre for climate change.* Interim Report. Kumaon University, Nainital. www.ucccku.com. Assessed November, 14 2011.

Upreti, D. K. (2010). Himalayan lichens as an indicator of climate change. In *International workshop on mountain biodiversity and impacts of climate change with special reference to Himalayan biodiversity Hotspot, G. B. PIHED, Kosi-Katarmal, Uttarakhand, pp. 90–95.*

Vaish, S. S., Bilal, A. S., & Prakash, K. (2011). First documentation on status of barley diseases from the high altitude cold arid Trans-Himalayan Ladakh region of India. *Crop Protection, 30*, 1129–1137.

Weaver, H. J., Hawdon, J. M., & Hoberg, E. P. (2010). Soiltransmitted helminthiases: Implications of climate change and human behavior. *Trends in Parasitology, 26*, 574–581.

Xu, J., Grumbine, R., Shrestha, A., Eriksson, M., & Yang, X. (2009). The melting Himalayas: Cascading effects of climate change on water, biodiversity, and livelihoods. *Conservation Biology, 23*, 520–530.

Yeates, G. W., Newton, P. C. D., & Ross, D. J. (1999). Response of soil nematode fauna to naturally elevated CO_2 levels influenced by soil pattern. *Nematology, 1*, 285–293.

Yeates, G. W., Newton, P. C. D., & Ross, D. J. (2003). Significant changes in soil microfauna in grazed pasture under elevated carbon dioxide. *Biology and Fertility of Soils, 38*, 319–326.

Yeates, G. W., Tate, K. R., & Newton, P. C. D. (1997). Response of the fauna of a grassland soil to doubling of atmospheric carbon dioxide concentration. *Biology and Fertility of Soils, 25*, 307–315.

Chapter 5
Understanding Wheat Thermo-Tolerance Mechanisms for Enhanced Sustainable Production

Mawuli Kwamla Azameti and Jasdeep C. Padaria

5.1 Introduction

Wheat (*Triticum aestivum* L.) provides many necessary nutrients to about 30% of people globally (Mayer et al., 2014). It is estimated that by the year 2050, developing countries must increase wheat production by 77% in order to meet the global wheat requirement (Sharma et al., 2015b). Various components of the weather continue to change. Among them, the increasing global temperature poses the major threat to sustainable crop production. Recent analysis carried out by scientific institutes such as NASA's Goddard Institute for Space Studies (GISS) gives the indication that there has been an increase of up to ~ 0.8 °C in the average global temperature since 1880. Lorenz et al. (2019) reported that 66.7% of the global warming occurred at a rate of 0.15–0.20 °C per decade, while Hansen et al. (2012) put the rate at 0.18 °C every decade. By the end of the twenty-first century, it is expected that global temperatures will have risen by more than 2 °C (IPCC, 2014).The threshold temperature of vegetative growth in wheat is 20–30 °C (Kobza & Edwards, 1987), while that of reproductive growth was 15 °C (Wardlaw et al., 1989). Every degree Celsius above this optimum temperature for reproductive growth leads to a 3–4% reduction in yield (Wardlaw et al., 1989). Consequently, there is detrimental pressure on wheat production and its quality due to the change in climate accompanied by the increasing frequency of extreme high temperature (Qi et al., 2016). Globally, heat stress, due to climate change, drastically affects crop productivity which results in major losses in yield. Wheat production is predicted to face more damaging effects from the rise in temperature (Tripathi et al., 2016). Enhancing our knowledge on the molecular

M. K. Azameti
Indian Agricultural Research Institute, Pusa, New Delhi 110012, India

J. C. Padaria (✉)
National Institute for Plant Biotechnology, Pusa, New Delhi 110012, India
e-mail: jasdeep_kaur64@yahoo.co.in

S. Sheraz Mahdi et al. (eds.), *Adapting to Climate Change in Agriculture-Theories and Practices*, https://doi.org/10.1007/978-3-031-28142-6_5

impacts of heat stress and the use of biotechnological interventions to safeguard its sustainable production under changing global climate should be considered as an essential component of enhancing sustainability in wheat production. The identification and modulation of heat shock proteins (HSPs) have significantly enhanced our knowledge on heat stress response (HSR) (Altschuler & Mascarenhas, 1982). To better enhance the development of more efficient thermo-tolerance in crops, it is imperative to delve more into the molecular response of crops to heat stress (Yeh et al., 2012).

5.2 Heat Stress

According to scientific standards, heat stress is caused by temperatures higher than 10–15 °C optimal temperature range (Larkindale et al., 2007). The situation caused by the heat stress becomes worse with an increase in soil temperature due to the rise in air temperature. This is a hindrance to sustainable crop production globally (Gourdji et al., 2013). Living organisms, including plants, are generally grouped into three according to their level of sensitivity to temperature. They are (a) Psychrophiles: which prefer low temperatures between 0 and 10 °C. They are considered as heat sensitive; (b) Mesophyles: which grow well between the temperature range of 10 and 30 °C and are regarded as moderately tolerant to heat stress; and (c) Thermophyles which are considered to be heat tolerant. They tolerate high temperature, even above 65 °C (Źróbek-Sokolnik, 2012) (Fig. 5.1).

The extent to which heat stress affects plants depends on the rate of temperature change, its intensity, and duration. As a survival mechanism, plants exhibit necessary responses to adapt to abiotic changes. A plant is said to be tolerant to heat when it can grow and produce during heat stress (Huang & Xu, 2008).

5.3 Plant Response to Heat Stress

Heat stress disrupts a number of plant activities which lead to changes in their morphology and physiology. These changes reduce the rate of development leading to loss of yield (Grant et al., 2011). Plants are sessile species; as a result, unfavourable environment, including high temperature affects their survival which can lead to death (Lobell and Field, 2007). The heat stress on wheat (plants) elicits various responses which can be grouped into four main domains; biochemical response, morphological response, physiological response, and molecular response. These collectively cause deleterious damages to the plants.

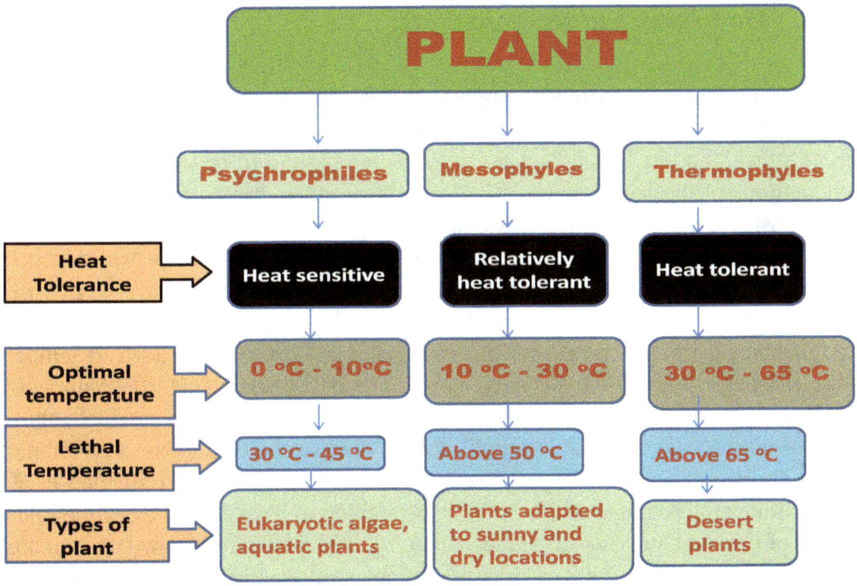

Fig. 5.1 Plant classification on the basis of their level of tolerance to temperature

5.3.1 Biochemical Response

Heat stress affects various proteins involved in numerous biochemical reactions in plants. These biochemical reactions play significant roles in plant's survival. High temperatures disrupt the functions of these temperature-sensitive proteins, causing irreversible denaturation of same (Howarth, 2005). Extremely high temperatures cause adverse cellular injuries and immediate cell death after a brief period of time (Howarth, 2005; Schoffl et al., 1999). Several major changes occur at the sub-cellular level due to extreme temperature. These include increase in lipid membrane fluidity, decrease in membrane integrity, inactivation of enzymes, protein denaturation and aggregation (Howarth, 2005). These changes cumulatively affect plants' existence negatively (Lipiec et al., 2013). Studies show that during heat stress, cell cycle and cell division are also significantly affected through a number of processes such as altering the organization of microtubules, and phragmoplast microtubules elongation (Smertenko et al., 1997). The chief constituent of wheat is starch which consists of amylose and amylopectin. Starch quality is mostly determined by amylose content. Therefore, variations in amylose content affects starch characteristics. Heat stress raises amylose levels and the ratio of amylose to amylopectin (Sharma et al. 2015a).

5.3.2 Morphological Responses

The major impact of extreme temperature on the morphology of plants is the hindrance to germination of seed (Hossain et al., 2013). The extent to which heat stress affects the reproductive stage of crops is reported to be more severe in the production of wheat (Nawaz et al., 2013). Heat stress has significant effects on various aspects of plants morphology, including plant germination, development and growth, yield and quality of seed. High temperature leads to growth retardation in plants (Źróbek-Sokolnik, 2012). During the reproductive phase, a unit-degree rise in average temperature causes a major loss of yield in wheat (Yu et al., 2014b). The rate of plant germination or seedling emergence is linearly correlated to temperature (Rizhsky et al., 2004; Roberts, 1988). The percentage of wheat germination increases with an increase in temperature up to the threshold, and then reduces above the threshold (Roberts & Summerfield, 1987). Plants subjected to elevated temperatures during sowing time experience a reduction in their height (Ahamed et al., 2010). Temperature range of 28–30 °C was reported to have caused a decline in the period of germination, days to anthesis booting, and the overall growth duration of plant (Yamamoto et al., 2008). Under heat stress, seed germination is also hampered (Essemine et al., 2007), and harvest index is decreased in wheat (Lukac et al., 2011). Hedhly et al. (2009) indicated that anthers produced under elevated temperature for 3 days during anthesis were found to be abnormal in their structures and are also non-functional. Increased day/night temperatures of 32/22 °C greatly decreased the rate of grain-filling in wheat plants (Song et al., 2015). Unlike the vegetative stage where high temperature during the day only affects leaf gas exchange processes, plant exposure to high temperatures under reproductive stage results in floral bud abortion and opened flowers (Young et al., 2004). Heat stress during meiosis is harmful to the early stages of gametogenesis (Ji et al., 2010). Alam et al., 2014 reported that the life cycle of wheat reduces under heat stress when compared to normal temperature situation. A temperature increase of even 1.5 °C leads to major decline in crop yields (Warland et al., 2006). Many studies involving various crops have confirmed yield reduction due to heat (Ahamed et al., 2010; Zhang et al., 2013). 4.1 to 10.0% yield reduction was reported by 1 °C rise in average temperature (Wang et al., 2012). Wheat grain was shown to have a large rise in grain starch and a decrease in dry matter aggregation under temperatures above 30 °C (Liu et al., 2011).

5.3.3 Physiological Responses

Plant physiology refers to the study of plant function and behaviour, encompassing all the dynamic processes that account for plants being alive. Plants' physiological responses to heat stress can best be explained under the following domains; photosynthesis, water relations, and osmotic adjustment.

Photosynthesis

Photosynthesis remains a very important physiological process that is heat-sensitive in photosynthetic plants. Heat stress causes a decline in the photosynthetic rate by decreasing leaf area expansion, and inhibiting photosynthetic machinery (Mathur et al., 2014). Heat stress in wheat mostly has negative effects on stroma and thylakoid lamellae (Mathur et al., 2014). Photosystem II (PSII) is very sensitive to heat stress; which leads to significant negative impact on its activity (Morales et al., 2003). Under heat stress, plant stomata closes, negatively affecting the rate of photosynthesis (Ashraf & Hafeez, 2004). Heat-tolerant crops are more able to maintain exchange of leaf gas and to assimilate CO_2 under heat stress. Decline in the amount of soluble proteins and Rubisco binding proteins (RBP) in darkness are other factors influencing the photosynthetic rate (Sumesh et al., 2008). Rubisco activity declines under heat stress due to Rubisco activase inhibition and inherently faster Rubisco inactivation rates (Lea & Leegood, 1999).

Water Status

Water plays very significant roles in the physiology of plants, and the water status of plants is mostly influenced by heat stress is plant. Heat stress leads to water loss in plant tissue, resulting in growth retardation. For plants to survive under elevated temperature, the rate of transpiration and water transportation must be increased. Heat stress raises the threshold of the soil water content to the point where it decreases plant functions (Centritto et al., 2011). A temperature of 31 °C during flowering is regarded as the maximum to maintain water level of a plant (Atkinson & Urwin, 2012).

Osmotic Response

Globally, various osmotic stresses significantly affect the production of wheat (Oyiga et al., 2016). Plants also produce certain organic compounds called compatible osmolytes (Sakamoto & Murata, 2002), whose accumulation is likely to be connected to enhanced stress tolerance in plants. The synthesis of glycine betaine (GB) has been shown to shield cells' redox potential under heat condition (Li et al., 2011).

5.4 Molecular Responses

Several heat-sensitive enzymes play significant roles in the various metabolic pathways in plants. Under heat stress, these enzymes and metabolic pathways get activated, leading to the increase in ROS (e.g. singlet oxygen (O_2), superoxide radical

(O_2^-), hydrogen peroxide (H_2O_2) and hydroxyl radical (OH^-)), resulting in oxidative stress (Asada, 2006). These ROS are mostly produced in the PSI and PSII in chloroplasts, and also in peroxisomes and mitochondria (Soliman et al., 2011). To safeguard themselves from the harmful effects of ROS, plants have systems which help the detoxification of the ROS in order to maintain their growth and metabolism (Sairam & Tyagi, 2004).

5.5 Heat Stress Signal Perception and Transduction

Prior to the development of thermo-tolerance in plants, there are series of steps involving signal perception and transduction. Plants perceive stress signals from the environment and transmit them through a cascade of signal transduction, eliciting the accumulation of transcription factors which induce gene expression that helps in plants' adaptation to environmental challenges (Mirouze & Paszkowski, 2011). Signals are perceived at both structural and molecular levels (Dickinson et al. 2018; Sajid et al. 2018). Structurally, signals may be perceived by cell wall, plasma membrane, cytoskeleton, and organelles while the molecular level of signal perception involves proteins, DNA, RNA, and phospholipids. Membranes are known for their extraordinary sensitivity to environmental cues, and various evidence from different organisms shows that membranes probably contain sensory devices that help to detect specific signals which are transmitted to expression of appropriate genes (Los & Murata, 2004). The plasma membrane is therefore generally regarded as a major heat sensor, which significantly helps to form inherent and acquired thermo-tolerance in plants (Li, 2020). It is reported that ion channels (such as Ca^{2+} channels), enzymes (NADPH oxidase (NOX)) and nitric oxide synthase (NOS) may also play roles as heat stress sensors, by triggering signaling involving Ca^{2+}, ROS (mainly H_2O_2 and nitric oxide (NO), which form signaling network involved in thermo-tolerance in plants (Demidchik et al. 2018). Apart from the plasma membrane and other membrane-associated ion channels, some organelles (mitochondria and chloroplast) also play roles in heat stress perception, which trigger ROS signaling cascades and induce thermo-tolerance through the regulation of transcription and metabolic networks (Qu et al. 2013). Dickinson et al. (2018) indicated that chloroplast can also perceive heat stress during the day, and they subsequently activate an unknown light-dependent chloroplast signaling which helps to induce the expression of heat shock factors (HSFs) and heat shock proteins (HSPs) involved in thermo-tolerance. However, during the night, absence of light halts the chloroplast-to-nucleus signaling, reducing the expression of HSF and HSP, and ultimately reducing the rate of plant survival under heat stress (Dickinson et al. 2018). Hormones are also believed to function in signal perception. Previous reports indicated that plant hormones play active roles in plant response to heat stress (Ahammed et al. 2014; Xia et al. 2015). Currently, all major hormones are implicated in aiding plants in their response to heat

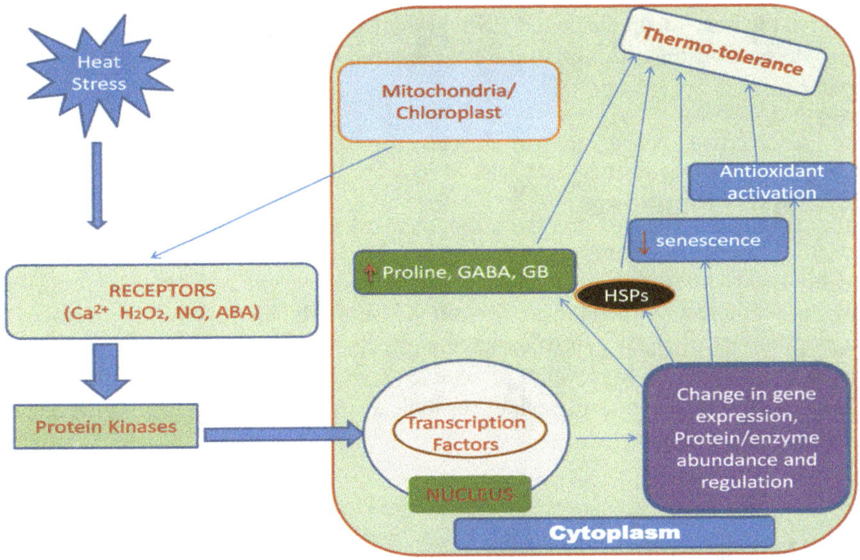

Fig. 5.2 Heat stress signalling pathway in plants. Perception of heat stress signal occurs at both structural and molecular level. This activates signals by hydrogen peroxide (H_2O_2), calcium (Ca^{2+}), plant hormone e.g., abscisic acid (ABA), and nitric oxide (NO). Relevant protein kinases are then synthesized as a result of these signals, which results into changes in expression of appropriate genes. This leads to alterations in plant metabolism, antioxidant activation and synthesis, heat shock protein, Osmo protectant and solute accumulation, and senescence reduction, and ultimately resulting to increased tolerance to heat stress

stress (Mittler et al. 2012; Xia et al. 2015; Zhou et al. 2014). Heat perception triggers a number of signaling pathways which involve Ca^{2+}, H_2O_2, NO, and hormones (Fig. 5.2).

5.6 Ca^{2+}Signaling

One very important signalling cascade in heat stress is Calcium signalling. It is the most important component of abiotic stress signalling in various situations, including heat stress. Ca^{2+} functions in structural, metabolite, and signalling mechanisms of plants' thermo-tolerance (Konrad et al. 2018). Heat stress triggers a specific transient inflow of Ca^{2+} across the plasma membrane (Wu & Jinn, 2010). Inflow of Ca^{2+} is the initial step involved in the signalling pathways to trigger plants' response to heat stress (Wang & Huang, 2017). Ca^{2+} enters the cell via Ca^{2+}-permeable channels and then elicits downstream responses. Afterwards, Ca^{2+} signalling is transmitted in sequential manner which further activates the formation of thermo-tolerance of plants (Campbell, 2018; Costa et al. 2018). Ca^{2+} ions are also involved in activating heat shock factors (HSFs). In the cell, the HSFs activated by the heat stress identify and bind to heat-shock elements (HSEs) (von Koskull-Doring et al., 2007). During heat stress, the time of influx of the heat-induced Ca^{2+} has been proven to be very important.

5.7 H_2O_2 Signaling

At high concentration H_2O_2 acts as a cytotoxic agent but performs the role of a messenger molecule when the concentration is less (Mohanta et al. 2018). Although higher amount of H_2O_2 causes damaging effects, it also acts in several protective pathways to influence the level of protective genes expression under abiotic stresses (Neill et al., 2002). Increasing evidences indicate that H_2O_2 influences various plant physiological processes.

5.8 NO Signalling

A number of researches indicated that NO builds up in various plant species under heat stress (Leshem, 2001; Yu et al., 2014a). It is suggested that one very crucial mechanism of plants' response to heat stress is the emission of heat-induced NO (Bouchard & Yamasaki, 2008; Yu et al. 2014a). In plants, the synthesis of NO involves enzymatic and non-enzymatic pathways. NO is instrumental in regulating stomatal closure (Chen et al. 2016). While some studies suggested that accumulation of NO under heat stress is independent of enzymes (Hancock, 2012), others reported involvement of enzymes, such as L-arginine (Wodala et al., 2008) and Nitrate Reductase (Siddiqui et al., 2017).

5.9 Plant Hormone

Hormones play a number of roles in signalling network and regulation of different plant systems. Our understanding of the roles of hormones is increased with our enhanced understanding of molecular biology technologies. Several hormones that were hitherto suggested to function only in plants growth and development are now suggested to function in plant response to heat stress (Dobra et al., 2015). Plant hormones activate a cascade of phosphorylation. Furthermore, several plant hormones such as JA, SA, and ABA interact with each other to form a defence network against extreme environmental conditions (Wani et al., 2016).

5.10 Plants Adaptation to Heat Stress

Plants must evolve mechanisms to react to high-temperature environments because they are sessile. Wheat adapts to high temperatures through a variety of pathways, including molecular, cellular, metabolic, physiological, and whole-plant changes (Kumar et al., 2013). Wheat adaptation to heat stress can best be described under avoidance mechanisms and tolerance mechanisms.

5.10.1 Avoidance Mechanism

Changes in plant architecture contribute significantly to avoiding heat stress. Under heat stress, plants show various morpho-phenological changes and short-term avoidance or acclimation mechanisms for surviving. The following are few examples of short-term mechanisms; rolling of leaf or change in leaf orientation, and increased transpiration. Increased leaf stomata and hair density, as well as greater vessels, are long-term adaptation mechanisms to heat stress (Srivastava et al., 2012). Adams et al. (2001) indicated that some crops adopt an important escape mechanism by maturing earlier under high temperature, leading to small yield losses. Unlike plants with larger leaves, plants such as wheat whose leaves are smaller have the potential of easily avoiding the heat stress. Plants also adopt other mechanisms to avoid heat stress. These include intensive transpiration to lower leaf temperature up to 6–10 °C (Fitter & Hay, 2002), and leaf rolling to enhance efficient use of water (Sarieva et al., 2010).

5.10.2 Tolerance Mechanisms

Plants show tolerance to heat stress through two main mechanisms; intrinsic and acquired. Under intrinsic mechanism, plants possess the natural capability to withstand extreme temperatures above the threshold. This is referred to as basal thermo-tolerance. Acquired thermo-tolerance is when plants survive under elevated temperatures within a short period of exposure to such temperatures (Song et al. 2012). There is a direct relationship between increase in antioxidative capacity and thermo-tolerance (Babu and Devraj, 2008). Antioxidant metabolites including AsA, GSH, tocopherol, and carotene protect plants from oxidative stress (Sairam et al., 2000). Chauhan (2005) reported a rise in GSH levels in wheat genotypes under heat stress. Almeselmani et al. (2009) reported that the antioxidant defence mechanism is actively involved in the thermotolerance in wheat genotypes. The damaging impacts of heat stress on crops can be lessened by the help of molecular chaperones, most importantly HSPs. HSPs are divided into various categories depending on the scale of their molecular mass (Table 5.1).

Table 5.1 Classification and functions of HSPs (*Source* Jee 2016)

Classification	Role
HSP10	Folding of mitochondrial protein
HSP40	Chaperoning intermediate filament
HSP60	Assembling of protein by the formation of hetero-oligomeric protein complexes
HSP70	Aids in the folding of protein
HSP90	Assisting in the folding of myosin and the formation of sarcomeres
HSP100	Participated in the refolding of aggregates
HSP110	Aiding the immune system

Osmolytes are significantly involved in alleviating heat stress by osmotic adjustment in plants (Li, 2020). Osmolytes have also been implicated in preserving membrane integrity, stabilizing proteins, reducing toxicity, and shielding cellular elements (Hasanuzzaman et al., 2019) (Fig. 5.3).

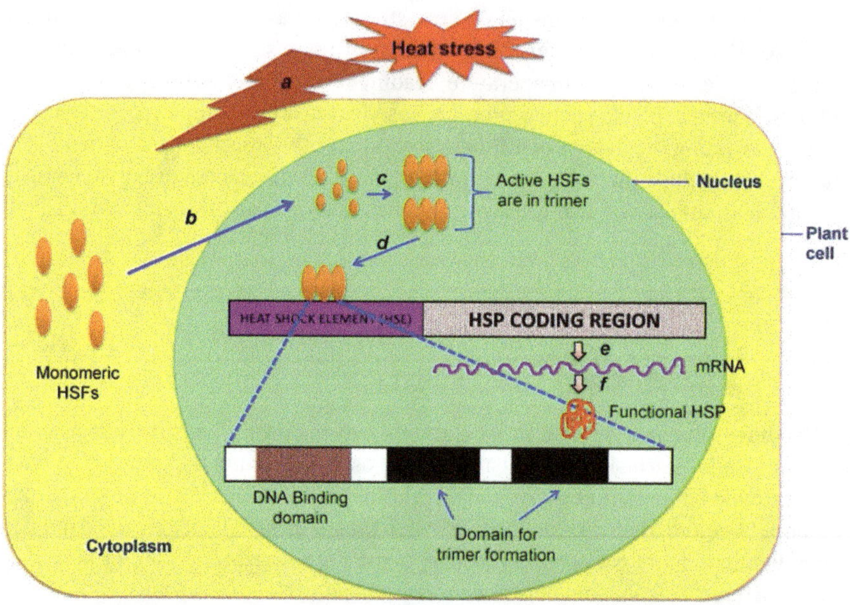

Fig. 5.3 Molecular regulatory mechanism of heat shock proteins (*Source* Hasanuzzaman et al., 2013). **a** heat stress signals are perceived, **b** heat shock factors (HSFs) enter the nucleus; **c** In the nucleus, HSF monomers become active by forming trimers; **d** This binds to the respective heat shock element (HSE), leading to **e** transcription and **f** translation to produce functional HSP leading to enhanced thermotolerance

5.11 Molecular and Biotechnological Approaches of Heat Stress Tolerance in Plants

5.11.1 Genetic Engineering and Transgenic Approaches

Heat stress's negative consequences are reduced by breeding crop plants that have a higher tolerance to heat by employing genetic engineering and transgenic strategies (Rodríguez et al., 2005). This approach employs the transfer of genes responsible for stress tolerance to a desired plants to enhance thermo-tolerance (Zheng et al. 2012). However, the complex status of wheat genome renders genetic modification in wheat difficult. Subjecting plants to longer period of heat stress increases the synthesis of the elongation factor (EF-Tu) which functions in thermo-tolerance. Fu et al. (2012) reported that when this EF-Tu was constitutively expressed in transgenic wheat, it resulted to an increased protection of the leaf proteins from aggregation, and increased photosynthesis. Consequently, the higher the amount of EF-Tu accumulated in wheat genotype, the better the tolerance to heat stress (Ristic et al., 2008).

5.11.2 Contribution of "Omics" Approaches in Enhancing Thermotolerance

Molecular proof for thermo-tolerance in plants is based on DNA. The growing global demand for higher crop yields necessitates the development of novel species with improved tolerance to climate change. This is an enormous task because it involves the use of complex information. Different processes including transcription, translation, and post-translation, as well as signalling pathways involved in response of plants to extreme environmental conditions, have been established thanks to the advancement of "omics" technologies (e.g. genomics, transcriptomics, proteomics, and metabolomics) (Aprile et al., 2009). Researchers have used the traditional transcriptomic method to analyze gene expression modifications under stress situations in crops (Qin et al., 2008). Recently, advanced omics strategies have been used to determine the effects of heat stress in wheat (Comastri et al. (2018); Narayanan et al., 2018). Protein profiling under heat stress conditions serves as an approach of identifying stress-responsive proteins which induce thermo-tolerance (Priya et al., 2019). Most HSPs are molecular chaperones and their functions under heat stress involve stabilization of protein and signal transduction (Arce et al., 2018). Subjecting crops to increased temperature induces synthesis and accumulation of HSPs which is regarded as one of the most necessary mechanisms of adaptation to heat stress (Keller and Simm, 2018). The analysis of direct gene products, which also vary depending on the degree of gene regulation, is one feature of proteomics. Various posttranslational modifications (PTMs) to proteins also function in adaptive responses to heat stress.

Other omic technologies used to enhance thermotolerance are *Metabolomics and lipidomics*. Thomason et al (2018) investigated metabolism reprogramming in wheat during heat stress and identified the metabolites that reduced in amount the most during heat stress: Drummondol, anthranilate, dimethylmaleate, galactoglycerol, guanine, and glycerone.

5.12 Breeding for Thermotolerance

Given the challenge of climate change to global food supply, agriculture would need to double its current crop production levels to reach food sufficiency (Sedeek et al., 2019). One approach to achieve this goal is to develop new breeding techniques for thermo-tolerance in crops.

5.12.1 Conventional Breeding Approach

Thermotolerance is a complicated quantitative trait and its inheritance is regulated by a number of genes/QTLs. Identification of genes responsible for thermotolerance through QTL mapping has not been very effective due to the large nature of QTL regions and its low marker density. With the advent of genotyping-by-sequencing (GBS) with an increased number of markers (Spindel & Iwata, 2018), researchers can now easily generate high-resolution genetic and precise QTL mappings (Bhat et al., 2016). GBS also aids in the identification of candidate genes that regulate quantitative traits.

Several genomic regions linked to heat resistance in wheat are mapped through the help of QTL analysis, which is often paired with GWAS and GBS. Heat tolerance in crops can be improved by targeted breeding activities (Reynolds et al., 2011).

Physiological characteristics can be calculated using a genetics approach in a model to improve wheat thermotolerance. In warmer climates, spring wheat breeding for abiotic stress tolerance yields greater genetic benefits (Gourdji et al., 2013).

5.12.2 New Breeding Techniques

Genetic engineering by biotechnology and other new breeding techniques (NBTs) is a method for improving crop adaptability that offers new possibilities. However, its application is limited due to increasing public concerns and legislation complexities. Wheat rice, and different other crops, have undergone genetic modifications for accelerated thermotolerance (Qi et al., 2011; Xue et al., 2015). The advent of genome sequencing technologies has led to the determination of the genome sequences of most crops. These genomic sequences, coupled with the available genome modification technologies have increased the chances of breeding crops for improved traits (Jaganathan et al., 2018). Genome editing (also called gene editing) technologies provide the opportunity to modify an organism's DNA to achieve a desired purpose. Several approaches to genome editing have been developed. Clustered Regularly Interspaced Palindromic Repeats (CRISPR)/CRISPR-associated protein 9 (Cas9), the most popular genome editing procedure, modifies a genome in a programmed manner and has been employed in different crops such as rice and wheat.

5.13 Conclusion

Heat stress, induced by an ever-changing climate, is a significant hindrance to long-term crop production. Due to the sensitive nature of wheat to elevated temperature especially at the reproductive and grain-filling stages, heat stress causes enormous yield loss in wheat. To safeguard the sustainable production of wheat in view of the changing climate, it is necessary to study the fundamental processes of thermotolerance and the emerging technologies to enhance the same. Molecular approaches help to reveal mechanisms of plants' response to heat stress. These technologies can help to produce thermo-tolerant crops that can withstand heat stress.

References

Adams, S., Cockshull, K., & Cave, C. (2001). Effect of temperature on the growth and development of tomato fruits. *Annals of Botany, 88*, 869–877.

Ahamed, K. U., Nahar, K., Fujita, M., & Hasanuzzaman, M. (2010). Variation in plant growth, tiller dynamics and yield components of wheat (Triticum aestivum L.) due to high temperature stress. *Advances in Agriculture & Botanics, 2*, 213–224.

Ahammed, G., Xia, X., Li, X., Shi, K., Yu, J., & Zhou, Y. (2014). Role of brassinosteroid in plant adaptation to abiotic stresses and its interplay with other hormones. *Current Protein and Peptide Science, 16*(5), 462–473.

Alam, M., Bodruzzaman, M., Hossain, M., & Sadekuzzaman, M. (2014). Growth performance of spring wheat under heat stress conditions. *International Journal of Agricultural Research, 4*(6), 91–103.

Almeselmani, M., Deshmukh, P. S., & Sairam, R. K. (2009). High temperature stress tolerance in wheat genotypes: Role of antioxidant defence enzymes. *Acta Agronomica Hungarica, 57*, 1–14.

Altschuler, M., & Mascarenhas, J. P. (1982). Heat shock proteins and effects of heat shock in plants. *Plant Molecular Biology, 1*, 103–115.

Aprile, A., Mastrangelo, A. M., De Leonardis, A. M., Galiba, G., Roncaglia, E., Ferrari, F., De Bellis, L., Turchi, L., Giuliano, G., & Cattivelli, L. (2009). Transcriptional profiling in response to terminal drought stress reveals differential responses along the wheat genome. *BMC Genomics, 10*, 279.

Arce, D., Spetale, F., Krsticevic, F., Cacchiarelli, P., Las, R. J., Ponce, S., Pratta, G., & Tapia, E. (2018). Regulatory motifs found in the small heat shock protein (sHSP) gene family in tomato. *BMC Genomics, 19*, 860.

Asada, K. (2006). Production and scavenging of reactive oxygen species in chloroplasts and their functions. *Plant Physiology, 141*, 391–396.
Ashraf, M., & Hafeez, M. (2004). Thermotolerance of pearl millet and maize at early growth stages: Growth and nutrient relations. *Biologi Plantarum, 48*, 81–86.
Atkinson, N. J., & Urwin, P. E. (2012). The interaction of plant biotic and abiotic stresses: From genes to the field. *Journal of Experimental Botany, 63*, 3523–3543.
Babu, N. R., & Devraj, V. R. (2008). High temperature and salt stress response in French bean (Phaseolus vulgaris). *Australian Journal of Crop Science, 2*, 40–48.
Bhat, J. A., Ali, S., Salgotra, R. K., et al. (2016). Genomic selection in the era of next generation sequencing for complex traits in plant breeding. *Frontiers in Genetics, 7*, 221.
Bouchard, J. N., & Yamasaki, H. (2008). Heat stress stimulates nitric oxide production in Symbiodinium microadriaticum: A possible linkage between nitric oxide and the coral bleaching phenomenon. *Plant and Cell Physiology, 49*, 641–652.
Campbell, A. K. (2018). *Fundamentals of intracellular calcium*. John Wiley and Sons Ltd.
Centritto, M., Brilli, F., Fodale, R., & Loreto, F. (2011). Different sensitivity of isoprene emission, respiration and photosynthesis to high growth temperature coupled with drought stress in black poplar (Populus nigra) saplings. *Tree Physiology, 31*, 275–286.
Chauhan, S. (June, 2005). *Physiological and molecular basis of heat tolerance with emphasis on oxidative stress metabolism in wheat*. Ph.D. Thesis, HNB Garhwal University; Srinagar, Uttaranchal, India.
Chen, Z. H., Wang, Y. Z., Wang, J. W., Babla, M., Zhao, C., García-Mata, C., et al. (2016). Nitrate reductase mutation alters potassium nutrition as well as nitric oxide-mediated control of guard cell ion channels in Arabidopsis. *New Phytologist, 209*, 1456–1469.
Comastri, A., Janni, M., Simmonds, J., Uauy, C., Pignone, D., Nguyen, H. T., & Marmiroli, N. (2018). Heat in wheat: Exploit reverse genetic techniques to discover new alleles within the Triticum durum sHsp26 family. *Frontiers in Plant Science, 9*, 1337.
Costa, A., Navazio, L., & Szabo, I. (2018). The contribution of organelles to plant intracellular calcium signalling. *Journal of Experimental Botany, 69*, 4175–4193.
Demidchik, V., Maathuis, F., & Voitsekhovskaja, O. (2018). Unravelling the plant signalling machinery: An update on the cellular and genetic basis of plant signal transduction. *Functional Plant Biology, 45*, 1–8.
Dickinson, P. J., Kumar, M., Martinho, C., Yoo, S. J., Lan, H., Artavanis, G., Charoensawan, V., Schottler, M. A., Bock, R., Jaeger, K. E., & Wigge, P. A. (2018). Chloroplast signaling gates thermotolerance in Arabidopsis. *Cell Reports, 22*, 1657–1665.
Dobra, J., Cerny, M., Storchova, H., Dobrev, P., Skalak, J., Jedelsky, P. L., Luksanova, H., Gaudinova, A., Pesek, B., Malbeck, J., Vanek, T., Brzobohaty, B., & Vankova, R. (2015). The impact of heat stress targeting on the hormonal and transcriptomic response in Arabidopsis. *Plant Science International Journal of Experimental Plant Biology, 231*, 52–61.
Essemine, J., Ammar, S., Jbir, N., & Bouzid, S. (2007). Sensitivity of two wheat species's seeds (Triticum durum, Variety Karim and Triticum aestivum, Variety Salambo) to heat constraint during germination. *Pakistan Journal of Biological Science, 10*, 3762–3768.
Fitter, A. H., & Hay, R. (2002). *Environmental physiology of plants* (3rd ed.). Academic Press.
Fu, J., Momclovic, I., & Prasad, V. (2012). Molecular bases and improvement of heat tolerance in crop plants. In S. Josipovic, & E. Ludwig (Eds.), *Heat stress: Causes. Prevention and treatments* (pp. 185–214). Nova Science.
Gourdji, S. M., Mathews, K. L., Reynolds, M., Crossa, J., & Lobell, D. B. (2013). An assessment of wheat yield sensitivity and breeding gains in hot environments. *Proceedings of the Royal Society B: Biological Sciences, 280*, 20122190.
Grant, R. F., Kimball, B. A., Conley, M. M., White, J. W., Wall, G. W., & Ottman, M. J. (2011). Controlled warming effects on wheat growth and yield: Field measurements and modeling. *Agronomy Journal, 103*(6), 1742–1754.
Hancock, J. T. (2012). NO synthase? Generation of nitric oxide in plants. *Period Biology, 114*, 19–24.

Hansen, J., Sato, M., & Ruedy, R. (2012). Perception of climate change. *Proceedings of the National Academy of Sciences of the United States of America, 109*, E2415–E2423.
Hasanuzzaman, M., Anee, T. I., Bhuiyan, T. F., Nahar, K., & Fujita, M. (2019). Emerging role of osmolytes in enhancing abiotic stress tolerance in rice. In *Advances in rice research for abiotic stress tolerance* (pp. 677–708). Elsevier.
Hasanuzzaman, M., Nahar, K., Alam, M., Roychowdhury, R., & Fujita, M. (2013). Physiological, biochemical, and molecular mechanisms of heat stress tolerance in plants. *International Journal of Molecular Sciences, 14*(5), 9643–9684.
Hedhly, A., Hormaza, J. I., & Herrero, M. (2009). Global warming and sexual plant reproduction. *Trends in Plant Science, 14*, 30–36.
Hossain, A., Sarker, M. A. Z., Saifuzzaman, M., Teixeira da Silva, J. A., Lozovskaya, M. V., & Akhter, M. M. (2013). Evaluation of growth, yield, relative performance and heat susceptibility of eight wheat (*Triticum aestivum* L.) genotypes grown under heat stress. *International Journal of Plant Production, 7*, 615–636.
Howarth, C. J. (2005). Genetic improvements of tolerance to high temperature. In M. Ashraf, & P. J. C. E. Harris (Eds.), *Abiotic stresses: plant resistance through breeding and molecular approaches*. Howarth Press Inc.
Huang, B., & Xu, C. (2008). Identification and characterization of proteins associated with plant tolerance to heat stress. *Journal of Integrative Plant Biology, 50*, 1230–1237.
IPCC (Intergovernmental Panel on Climate Change). (2014). Summary for policymakers. In: T. F. Stocker, D. Qin, G. K. Plattner, M. Tignor, S. K. Allen, J. Boschung, A. Nauels, Y. Xia, V. Bex, & P. M. Midgley (Eds.), Climate change 2013: The physical science basis. Contribution of Working Group I to the Fifth Assessment Report of the Intergovernmental Panel on Climate Change. Cambridge University Press.
Jaganathan, D., Ramasamy, K., Sellamuthu, G., Jayabalan, S., & Venkataraman, G. (2018). CRISPR for crop improvement: An update review. *Frontiers in Plant Science, 9*, 985.
Jee, H. (2016). Size dependent classification of heat shock proteins: A mini-review. *Journal of Exercise Rehabilitation, 12*(4), 255–259.
Ji, X., Shiran, B., Wan, J., Lewis, D. C., Jenkins, C. L. D., Condon, A. G., et al. (2010). Importance of pre-anthesis anther sink strength for maintenance of grain number during reproductive stage water stress in wheat. *Plant, Cell and Environment, 33*, 926–942.
Keller, M., Simm, S., & SPOT-ITN Consortium. (2018). The coupling of transcriptome and proteome adaptation during development and heat stress response of tomato pollen. *BMC Genomics, 19*, 447.
Kobza, J., & Edwards, G. E. (1987). Influences of leaf temperature on photosynthetic carbon metabolism in wheat. *Plant Physiology, 83*, 69–74.
Konrad, K. R., Maierhofer, T., & Hedrich, R. (2018). Spatio-temporal aspects of Ca^{2+} signalling: Lessons from guard cells and pollen tubes. *Journal of Experimental Botany, 69*, 4195–4214.
Kumar, S., Thakur, P., Kaushal, N., Malik, J. A., Gaur, P., & Nayyar, H. (2013). Effect of varying high temperatures during reproductive growth on reproductive function, oxidative stress and seed yield in chickpea genotypes differing in heat sensitivity. *Archives of Agronomy & Soil Science, 59*, 823–843.
Larkindale, J., Mishkind, M., & Vierling, E. (2007). Plant responses to high temperature. In M. A. Jenks & P. M. Hasegawa (Eds.), *Plant abiotic stress* (pp. 100–144). Blackwell.
Lea, P. J., & Leegood, R. C. (1999). Plant biochemistry and molecular biology (p. 384). John Wiley and Sons.
Leshem, Y. Y. (2001). Nitric oxide in plants. In *Occurrence, function and use*. Kulwer Academic Publishers.
Li, S., Li, F., Wang, J., Zhang, W. E. N., Meng, Q., Chen, T. H. H., & Yang, X. (2011). Glycinebetaine enhances the tolerance of tomato plants to high temperature during germination of seeds and growth of seedlings. *Plant, Cell and Environment, 34*(11), 1931–1943.

Li, Z. G. (2020). Mechanisms of plant adaptation and tolerance to heat stress. In: M. Hasanuzzaman (Eds.), *Plant ecophysiology and adaptation under climate change: Mechanisms and perspectives II*. Springer.

Lipiec, J., Doussan, C., Nosalewicz, A., & Kondracka, K. (2013). Effect of drought and heat stresses on plant growth and yield: A review. *International Agrophysics, 27*, 463–477.

Liu, P., Guo, W., Jiang, Z., Pu, H., Feng, C., Zhu, X., Peng, Y., Kuang, A., & Little, C. R. (2011). Effects of high temperature after anthesis on starch granules in grains of wheat (*Triticum aestivum* L.). *Journal of Agricultural Science, 149*, 159–169.

Lobell, D. B., & Field, C. B. (2007). Global scale climate–Crop yield relationships and the impacts of recent warming. *Environmental Research Letters, 2*.

Lorenz, R., Stalhandske, Z., & Fischer, E. M. (2019). Detection of a climate change signal in extreme heat, heat stress, and cold in Europe from observations. *Geophysical Research Letters, 46*, 8363–8374.

Los, D. A., & Murata, N. (2004). Membrane fluidity and its roles in the perception of environmental signals. *Biochimica et Biophysica Acta, 1666*, 142–157.

Lukac, M., Gooding, M. J., Griffiths, S., & Jones, H. E. (2011). Asynchronous flowering and within-plant flowering diversity in wheat and the implications for crop resilience to heat. *Annals of Botany, 109*, 843–850.

Mathur, S., Agrawal, D., & Jajoo, A. (2014). Photosynthesis: Response to high temperature stress. *Journal of Photochemistry and Photobiology B: Biology, 137*, 116–126.

Mayer, K. F., Rogers, J., Dolezel, J., Poznaik, C., Eversole, K., & Feuillet, C. (2014). A chromosome based draft sequence of the hexaploid bread wheat (Triticum aestivum) genome. *Science, 18*, 345.

Mirouze, M., & Paszkowski, J. (2011). Epigenetic contribution to stress adaptation in plants. *Current Opinion in Plant Biology, 14*, 267–274.

Mittler, R., Finka, A., & Goloubinoff, P. (2012). How do plants feel the heat? *Trends in Biochemical Sciences, 37*, 118–125.

Mohanta, T. K., Bashir, T., Hashem, A., Abd_Allah, E. F., Khan, A. L., & Al-Harrasi, A. S. (2018). Early events in plant abiotic stress signaling: Interplay between calcium, reactive oxygen species and phytohormones. *Journal of Plant Growth Regulation, 37*, 1033–1049.

Morales D., Rodriguez P., Dellamico J., Nicolas E., Torrecillas A. and Sanchez-Blanco M.J. (2003). High-temperature preconditioning and thermal shock imposition affects water relations, gas exchange and root hydraulic conductivity in tomato. Biologia Plantarum, 47: 203–208.

Narayanan, S., Prasad, P. V. V., & Welti, R. (2018). Alterations in wheat pollen lipidome during high day and night temperature stress. *Plant, Cell & Environment, 41*, 1749–1761.

Nawaz, A., Farooq, M., Cheema, S. A., & Wahid, A. (2013). Differential response of wheat cultivars to terminal heat stress. *International Journal of Agriculture and Biology, 15*, 1354–1358.

Neill, S. J., Desikan, R., Clarke, A., Hurst, R. D., & Hancock, J. T. (2002). Hydrogen peroxide and nitric oxide as signalling molecules in plants. *Journal of Experimental Botany, 53*, 1237–1247.

Oyiga, B. C., Sharma, R. C., Shen, J., Baum, M., Ogbonnaya, F. C., & Léon, J. (2016). Identification and characterization of salt tolerance of wheat germplasm using a multivariable screening approach. *Journal of Agronomy and Crop Science, 202*, 472–485.

Priya, M., Dhanker, O. P., Siddique, K. H. M., HanumanthaRao, B., Nair, R. M., Pandey, S., Singh, S., Varshney, R. K., Prasad, P. V. V., & Nayyar, H. (2019). Drought and heat stress-related proteins: An update about their functional relevance in imparting stress tolerance in agricultural crops. *Theoretical and Applied Genetics, 132*, 1607–1638.

Qi, X., Xu, W., Zhang, J., Guo, R., Zhao, M., Hu, L., Wang, H., Dong, H., & Li, Y. (2016). Physiological characteristics and metabolomics of transgenic wheat containing the maize C4 phosphoenolpyruvate carboxylase (PEPC) gene under high temperature stress. *Protoplasma, 254*(2), 1017–1030.

Qi, Y., Wang, H., Zou, Y., Liu, C., Liu, Y., Wang, Y., & Zhang, W. (2011). Over-expression of mitochondrial heat shock protein 70 suppresses programmed cell death in rice. *FEBS Letters, 585*, 231–239.

Qin, D., Wu, H., Peng, H., Yao, Y., Ni, Z., Li, Z., Zhou, C., & Sun, Q. (2008). Heat stress-responsive transcriptome analysis in heat susceptible and tolerant wheat (Triticum aestivum L.) by using wheat genome array. *BMC Genomics, 9*, 432.

Qu, A. L., Ding, Y. F., Jiang, Q., & Zhu, C. (2013). Molecular mechanisms of the plant heat stress response. *Biochemical and Biophysical Research Communications, 432*, 203–207.

Reynolds, M., Bonnett, D., Chapman, S. C., Furbank, R. T., Manès, Y., Mather, D. E., & Parry, M. A. (2011). Raising yield potential of wheat. I. Overview of a consortium approach and breeding strategies. *Journal of Experimental Botany, 62*, 439–452.

Ristic, Z., Bukovnik, U., Momcilovic, I., Fu, J., & Prasad, P. V. V. (2008). Heat-induced accumulation of chloroplast protein synthesis elongation factor, EF-Tu, in winter wheat. *Journal of Plant Physiology, 165*, 192–202.

Rizhsky, L., Liang, H. J., Shuman, J., Shulaev, V., Davletova, S., et al. (2004). When defense pathways collide. The response of Arabidopsis to a combination of drought and heat stress. *Plant Physiology, 134*, 1683–1696.

Roberts, E. B. (1988). What we've learned: Managing invention and innovation. *Research Technology Management, 31*, 11–29.

Roberts, E. H., & Summerfield, R. J. (1987). Measurements and prediction of flowering in annual crops. In J. G. Atherton (Ed.), *Manipulation of flowering* (pp. 17–50). Butterworth.

Rodríguez, M., Canales, E., & Borrás-Hidalgo, O. (2005). Molecular aspects of abiotic stress in plants. *Biotecnología Aplicada, 22*, 1–10.

Sairam, R. K., & Tyagi, A. (2004). Physiology and molecular biology of salinity stress tolerance in plants. *Current Science, 86*, 407–421.

Sairam, R. K., Srivastava, G. C., & Saxena, D. C. (2000). Increased antioxidant activity under elevated temperature: A mechanism of heat stress tolerance in wheat genotypes. *Biologia Plantarum, 43*, 245–251.

Sajid, M., Rashid, B., Ali, Q., & Husnain, T. (2018). Mechanisms of heat sensing and responses in plants. It is not all about Ca^{2+} ions. *Biologia Plantarum, 62*, 409–420.

Sakamoto, A., & Murata, N. (2002). The role of glycine betaine in the protection of plants from stress: Clues from transgenic plants. *Plant, Cell and Environment, 25*(2), 163–171.

Sarieva, G. E., Kenzhebaeva, S. S., & Lichtenthaler, H. K. (2010). Adaptation potential of photosynthesis in wheat cultivars with a capability of leaf rolling under high temperature conditions. *Russian Journal of Plant Physiology, 57*(1), 28–36.

Schoffl, F., Prandl, R., & Reindl, A. (1999). Molecular responses to heat stress. In: K. Shinozaki, & K. Yamaguchi-Shinozaki (Eds.), *Molecular responses to cold, drought, heat and salt stress in higher plants* (pp. 81–98). R.G. Landes.

Sedeek, K. E. M., Mahas, A., & Mahfouz, M. (2019). Plant genome engineering for targeted improvement of crop traits. *Frontiers in Plant Science, 10*, 114.

Sharma, D., Mamrutha, H. M., Gupta, V. K., Tiwari, R., & Singh, R. (2015a). Association of SSCP variants of HSP genes with physiological and yield traits under Heat stress in wheat. *Research on Crop, 16*(1), 139–146.

Sharma, I., Tyagi, B. S., Singh, G., Venkatesh, K., & Gupta, O. P. (2015b). Enhancing wheat production—A global perspective. *Indian Journal of Agricultural Science, 85*, 3–13.

Siddiqui, M., Alamri, S. A., Mutahhar, Y. Y., Al-Khaishany, M. A., Al-Qutami, H. M., & Nasir Khan, M. A. (2017). Nitric Oxide and calcium induced physiobiochemical changes in tomato (Solanum Lycopersicum) plant under heat stress. *Fresenius Environmental Bulletin, 26*, 1663–1672.

Smertenko, A., Draber, P., Viklicky, V., & Opatrny, Z. (1997). Heat stress affects the organization of microtubules and cell division in Nicotiana tabacum cells. *Plant, Cell and Environment, 20*, 1534–1542.

Soliman, W. S., Fujimori, M., Tase, K., & Sugiyama, S. I. (2011). Oxidative stress and physiological damage under prolonged heat stress in C3 grass Lolium perenne. *Grassland Science, 57*, 101–106.

Song, L., Jiang, Y., Zhao, H., & Hou, M. (2012). Acquired thermotolerance in plants. *Plant Cell, Tissue and Organ Culture (PCTOC), 111*, 265–276.

Song, W. F., Zhao, L. J., Zhang, X. M., Zhang, Y. M., Li, J. L., Zhang, L. L., Song, Q. J., Zhao, H. B., Zhang, Y. B., Zhang, C. L., Xin, W. L., Sun, L. F., & Xiao, Z. M. (2015). Effect of timing of heat stress during grain filling in two wheat varieties under moderate and very high temperature. *Indian Journal of Genetics, 75*, 121–124.

Spindel, J., & Iwata, H. (2018). Genomic selection in rice breeding. In: T. Sasaki, & M. Ashikari (Eds.), *Rice genomics, genetics and breeding* (pp. 473–496). Springer Singapore.

Srivastava, S., Pathak, A. D., Gupta, P. S., Shrivastava, A. K., & Srivastava, A. K. (2012). Hydrogen peroxide-scavenging enzymes impart tolerance to high temperature induced oxidative stress in sugarcane. *Journal of Environmental Biology, 33*, 657–661.

Sumesh, K. V., Sharma-Natu, P., & Ghildiyal, M. C. (2008). Starch synthase activity and heat shock protein in relation to thermal tolerance of developing wheat grains. *Biologia Plantarum, 52*, 749–753.

Thomason, K., Babar, M. A., Erickson, J. E., Mulvaney, M., Beecher, C., & MacDonald, G. (2018). Comparative physiological and metabolomics analysis of wheat (*Triticum aestivum* L.) following post-anthesis heat stress. *PLoS One 13*, e0197919.

Tripathi, A., Tripathi, D. K., Chauhan, D. K., Kumar, N., & Singh, G. S. (2016). Paradigms of climate change impacts on some major food sources of the world: A review on current knowledge and future prospect. *Agriculture, Ecosystems & Environment, 216*, 356–373.

von Koskull-Doring, P., Scharf, K. D., & Nover, L. (2007). The diversity of plant heat stress transcription factors. *Trends in Plant Science, 12*, 452–457.

Wang, X., & Huang, B. (2017). Lipid-and calcium-signaling regulation of HsfA2c-mediated heat tolerance in tall fescue. *Environmental and Experimental Botany, 136*, 59–67.

Wang, X., Cai, J., Liu, F., Jin, M., Yu, H., Jiang, D., Wollenweber, B., Dai, T., & Cao, W. (2012). Pre-anthesis high temperature acclimation alleviates the negative effects of post-anthesis heat stress on stem stored carbohydrates remobilization and grain starch accumulation in wheat. *Journal of Cereal Science, 55*, 331–336.

Wani, S. H., Kumar, V., Shriram, V., & Sah, S. K. (2016). Phytohormones and their metabolic engineering for abiotic stress tolerance in crop plants. *The Crop Journal, 4*, 162–176.

Wardlaw, I. F., Dawson, I. A., Munibi, P., & Fewster, R. (1989). The tolerance of wheat to high temperatures during reproductive growth. I. Survey procedures and general response patterns. *Australian Journal of Agricultural Research, 40*, 965–980.

Warland, J. S., McDonald, M. R., & McKeown, A. M. (2006). Annual yields of five crops in the family Brassicacae in southern Ontario in relation to weather and climate. *Canadian Journal of Plant Science, 86*, 1209–1215.

Wodala, B., Deák, Z., Vass, I., Erdei, L., Altorjay, I., & Horváth, F. (2008). In vivo target sites of nitric oxide in photosynthetic electron transport as studied by chlorophyll fluorescence in pea leaves. *Plant Physiology, 146*, 1920–1927.

Wu, H. C., & Jinn, T. L. (2010). Ethylesterase activity and cytosolic Ca^{2+} oscillation are crucial for plant thermotolerance. *Plant Signaling & Behavior, 5*, 1252–1256.

Xia, X. J., Zhou, Y. H., Shi, K., Zhou, J., Foyer, C. H., & Yu, J. Q. (2015). Interplay between reactive oxygen species and hormones in the control of plant development and stress tolerance. *Journal of Experimental Botany, 66*(10), 2839–2856.

Xue, G. P., Drenth, J., & McIntyre, C. L. (2015). TaHsfA6f is a transcriptional activator that regulates a suite of heat stress protection genes in wheat (Triticum aestivum L.) including previously unknown Hsf targets. *Journal of Experimental Botany, 66*, 1025–1039.

Yamamoto, Y., Aminaka, R., Yoshioka, M., Khatoon, M., Komayama, K., Takenaka, D., Yamashita, A., Nijo, N., Inagawa, K., Morita, N., et al. (2008). Quality control of photosystem II: Impact of light and heat stresses. *Photosynthesis Research, 98*, 589–608.

Yeh, C.-H., Kaplinsky, N. J., Hu, C., & Charng, Y. (2012). Some like it hot, some like it warm: Phenotyping to explore thermotolerance diversity. *Plant Science, 195*, 10–23.

Young, L. W., Wilen, R. W., & Bonham-Smith, P. C. (2004). High temperature stress of Brassica napus during flowering reduces micro- and megagametophyte fertility, induces fruit abortion, and disrupts seed production. *Journal of Experimental Botany, 55*, 485–495.

Yu, M., Lamattina, L., Spoel, S. H., & Loake, G. J. (2014a). Nitric oxide function in plant biology: A redox cue in deconvolution. *New Phytologist, 202*, 1142–1156.

Yu, Q., Li, L., Luo, Q., Eamus, D., Xu, S., Chen, C., Wang, E., Liu, J., & Nielsen, D. C. (2014b). Year patterns of climate impact on wheat yields. *International Journal of Climatology, 34*, 518–528.

Zhang, X., Cai, J., Wollenweber, B., Liu, F., Dai, T., Cao, W., & Jiang, D. (2013). Multiple heat and drought events affect grain yield and accumulations of high molecular weight glutenin subunits and glutenin macropolymers in wheat. *Journal of Cereal Science, 57*, 134–140.

Zheng, B., Chenu, K., Dreccer, M. F., & Chapman, S. C. (2012). Breeding for the future: what are the potential impacts of future frost and heat events on sowing and flowering time requirements for Australian bread wheat (*Triticum aestivium*) varieties? *Global Change Biology, 18*, 2899–2914.

Zhou, J., Wang, J., Li, X., Xia, X. J., Zhou, Y. H., Shi, K., Chen, Z., & Yu, J. Q. (2014). H_2O_2 mediates the crosstalk of brassinosteroid and abscisic acid in tomato responses to heat and oxidative stresses. *Journal of Experimental Botany, 65*(15), 4371–4383.

Źróbek-Sokolnik, A. (2012). Temperature stress and responses of plants. In: P. Ahmad & M. N. V. Prasad (Eds.), *Environmental adaptations and stress tolerance of plants in the era of climate change* (pp. 113–134). Springer.

Chapter 6
Role of Neglected Potential Crops in Climate Resilient Sustainable Agriculture

Mamta Singh, Supriya Babasaheb Aglawe, Chandana Behera, R. Gowthami, Jyotika Purohit, Vikender Kaur, and Rashmi Yadav

6.1 Introduction

Climate change is one of the most defined threats to the sustainable agricultural systems. Every year the impact of environmental disasters such as heavy rains, high temperatures, erratic rainfall, extreme cold, storms and earthquakes etc. cause great losses to the natural resources. Since agriculture sector is purely dependent on environmental circumstances, it can be considered as one of the most affected sectors due to climate change. The threats to the agriculture may be grouped into two major categories. The one which is caused by the external factors that fall outside control of human endeavors and agricultural research practices i.e., that we do not have any control over them e.g., natural disasters, and the other is the threat which is caused due to inefficient agriculture practices and research focus as influence by short term achievements. Apart from the climate disaster the other threat to the

M. Singh (✉) · V. Kaur · R. Yadav
Germplasm Evaluation Division, ICAR-NBPGR, New Delhi 110012, India
e-mail: mamta.singh@icar.gov.in

S. B. Aglawe
School of Agriculture, SR University, Warangal, Telangana 506371, India

C. Behera
OUAT, Bhawanipatna, Odisha 766001, India

R. Gowthami
Tissue Culture and Cryopreservation Unit, ICAR-NBPGR, New Delhi 110012, India
e-mail: Gowthami.R@icar.gov.in

J. Purohit
Department of Plant Pathology, C.P. College of Agriculture, S.D. Agricultural University, Sardarkrushinagar, Gujarat 385506, India

S. Sheraz Mahdi et al. (eds.), *Adapting to Climate Change in Agriculture-Theories and Practices*, https://doi.org/10.1007/978-3-031-28142-6_6

agriculture sector is to fulfill the food and nutritional security requirements of ever-growing human population which intensifies the volume of threat to the agriculture and pose greatest threat to the sector. Given the increasing intensities and frequencies of natural disasters coupled with serving to the human needs, there is an urgent need to debate on strengthening the resilience of agriculture system. India ranks second largest in terms of agricultural land holding with 20 diverse agroclimatic regions and 157.35 million hectares of cultivated land (DES, 2014). As per second advance estimates of production in 2019–20 is 291.95 million tones for food grains, 34.19 million tonnes for oilseeds, 34.89 million bales for cotton, 9.81 million bales for jute and mesta and 353.85 million tonnes for sugarcane. Most contributing crops to the total agricultural production are only few major crops such as rice, wheat, maize, major pulses and oilseeds. Although India is self-sufficient now in terms of food production, the condition was quite painful before independence. To alleviate extreme poverty and malnutrition green revolution was initiated with the objective to feed millions of people in India. During the period of green revolution agriculture production enjoyed the high production of rice and wheat due to fertilizer responsive high yielding varieties. Due to their rapid popularization and adoption in most of the growing states the superior varieties of these major food crops replaced most of the local food crops. Major crops being cultivated during the era of green revolution were rice, wheat, maize, major millets, sorghum and barley (Hall, 1964; USDA, 1963). In fact, the production of rice and millets was more than rest of the crops combined altogether. But, after the commencement of green revolution, the crops that were once consumed in every household lost their value and started to be grown as a source of fodder crops in just a few decades.

6.2 Climate Change and Food Systems

Climate change has particularly affected food security in dryland areas and will be increasingly affected in future if sustainable alternatives are not sought after. Global warming has negative impacts on period of cold accumulation which is important to get good harvest in many fruits and vegetables. However, secured mitigation systems and judicious technological interventions can promise a safe and secured ecosystem with diversity and sustainability. Scientific assessment and modelling of food patterns, trades, agricultural production and demands should be undertaken. To increase the credibility of assessment all socioeconomic and local needs should be taken into consideration. Seeing at the increasing risk of climate change on health and nutritional aspects diversity of foods in the human diets is much emphasized. Climate change has its effects on global health concerns such as reduction in the available seafood protein due to acidification, temperature rise and warming in ocean and overfishing. The major food crops viz., rice and wheat (C3 grains) have been reported to accumulate less concentration of zinc and iron when cultivated in high CO_2 concentrations (Myers et al., 2014) and legumes have low protein content which may have serious health consequences. Further, it leads to accumulation of more of sugars and starch in C3 plants that exacerbate the prevalence of obesity and "hidden hunger". Thus, climate change mitigation measures may rely on the crops that can produce more with less water requirement, cause less greenhouse gas emission, climate smart, survive droughts, erratic rainfalls, resistant to diseases and insect pests, have multipurpose applications and ability to survive on marginal lands. Proper policies, markets, institutional supports, and governance will help adaptation and mitigation throughout the food system in the situation of climate change (Mbow et al., 2019).

6.3 Genetic Potential of Orphan Crops to Mitigate Different Abiotic Stresses

Today's agriculture is facing an enormous challenge to feed ever increasing world population. The expected world population is expected to reach 9 billion by 2050 (Tyczewska et al., 2018). Ensuring food security for such a huge population, global food production should be doubled (Ray et al., 2013). With the escalating problem of global warming and climate change, abiotic stresses are becoming major threat and putting additional pressure on agriculture. Orphan crops have potential to address this question. Orphan crops are generally grown at small scale in marginal land and different climatic conditions. Most of the time they are grown in adverse growth conditions such as drought, high temperature, cold, salinity, acidic soils etc. As a result, many orphan crops are not only better survived under extreme climatic and soil conditions but also evolved tolerance mechanism toward these stresses. Adoption of such tolerant orphan crops for abiotic stress prone area is one of the important strategies to achieve sustainable crop production (Fig. 6.1) (Tables 6.1 and 6.2). Nevertheless, tolerance mechanism can be studied further at molecular level and exploited for crop improvement.

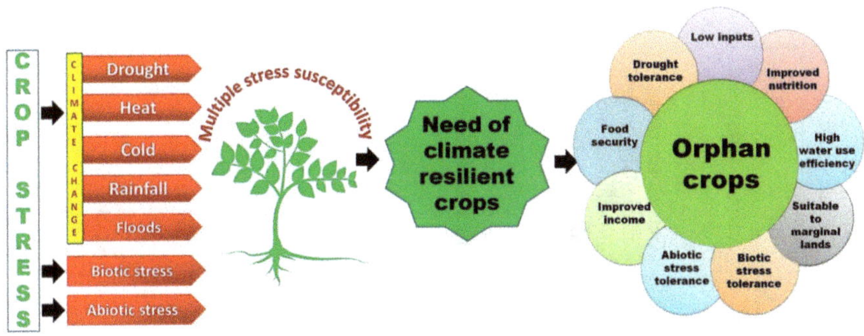

Fig. 6.1 Importance of underutilized crops as climate resilient crops

Table 6.1 Potential of some underutilized crops towards food/nutritional security and sustainable agriculture system

Crop	Climate resilient feature	Food usage	Health benefit and nutritional potentials	Agronomic and horticultural uses	Industrial uses	Medicinal usage
Linseed	Adapted to different environments and agro-ecologies. Genetic resources are available with drought tolerance, salinity and alkalinity tolerance	Soaked seeds, Oil in cooking, Ground flax seeds used as mix in breakfast cereal-cookies, bread or muffins	Flax seeds are a rich source of lignans. High omega-3 alpha linolenic acid (55–57%) concentration found in linseed	As a break crop for cereal production it fits well into crop rotations. Can be taken as intercrop with other pulses	Valued for its natural textile fiber (linen). paper, wax, nutraceutical, straw, thread, rope and other packaging materials, cigarette paper, currency notes	Known for its nutraceutical properties, anticancer property. Oil is used in anti-hypercholesterolemic drugs for cardiovascular diseases
Sesame	Crop of hot dry tropics, drought resistant	Seeds are used in sweetmeats and confections such as "rewari" and "gazak". Seed may be eaten fried or mixed with sugar	Seeds have 20 to 25% protein and around 50% oil. The seeds are recommended as protein supplements for wheat or corn flour	Its oil is used in insecticidal preparations, sesame oil and sesamin concentrates are used for their synergistic effect in pyrethrum insecticide	Lower grade oils are used in soap and paint industries and in lubricant and illuminant	Its oil is highly prized in medicine as a carrier or suspending agent for antibiotics, vitamins and steroid hormones

(continued)

Table 6.1 (continued)

Crop	Climate resilient feature	Food usage	Health benefit and nutritional potentials	Agronomic and horticultural uses	Industrial uses	Medicinal usage
Safflower	It has the potential to grow with less rainfall, high temperatures, drought, and salinity (Hussain et al., 2016) than other major oilseed crops such as sunflower soybeans and canola	High quality vegetable oil is extracted from seeds. Tender leaves and shoots are used as potherb and salad locally. Dried safflower petals are used as herbal tea	Oil rich in polyunsaturated fatty acids which helps in reducing cholesterol level in blood. Fresh leaves of safflower are good source of fiber, vitamins minerals, and antioxidants	–	Oil is used in cosmetics industry. The oil is particularly used with white paints, as it does not have the yellow tint	Plant exhibit many medicinal properties to cure diseases like hypertension, cardiovascular problems, spondylosis, arthritis and fertility disorders
Niger	Grown predominantly under rainfed conditions. It can tolerate slight alkalinity and salinity also	Seed contains 37–47% oil, that has pleasant odour and used for culinary purposes. Niger seed cake is used as feed for milch cattle	The seeds contain 20% protein. Fatty acid composition of oil has high linoleic acid and omega-3 fatty acids	Niger meal with 30% protein and 17% crude fiber can be used as a good manure	Oil is used in paint manufacturing and soft soap making, lighting and for lubrication. Being good fragrance absorbent, its oil is used in perfume industry	Oil is used for anointing the body, has anti-inflammatory properties, help dementia and insomnia

(continued)

Table 6.1 (continued)

Crop	Climate resilient feature	Food usage	Health benefit and nutritional potentials	Agronomic and horticultural uses	Industrial uses	Medicinal usage
Perilla (*Perilla frutescens*)	The plant can be grown in varying soil conditions viz acidic, neutral and basic (alkaline) soils	Perilla seeds are used as spice. Oil obtained from seeds is used as cooking oil. Leaves are used as garnish or as vegetable	Perilla seeds are good source of, fiber, fatty acids, dietary minerals such as calcium, iron, protein, niacin and thiamine. Leaves are rich in Vit A, riboflavin and Vit C	Very well suited in mixed cropping practice with Amaranthus, *Elusine coracana* and and various cultivated legumes crops	Seeds are used as bird food and for human consumption too. Seed oil can be used as fuel and drying oil in paint, varnish, printing ink and waterproof coatings	Its foliage part is distilled to get essential oil for flavoring. Perilla aldehyde is known to be 2000 times as sweet as Sugar (Guenther, 1949). Plant is used for treating chest stuffiness, colds, cough, vomiting, abdominal pain, Constipations etc
Amaranth	Moderately heat tolerant, high photosynthetic activity due to C4 cycle, has good water use efficiency in high temperatures and radiation intensities, Also it's a low risk crop for small-scale farmers that require relatively low inputs	Used as grain amaranth and leafy vegetable The gluten-free high-protein seed can be popped like popcorn. Grains are grounded as amaranth flour. Grains are used to make ladoos and sattu	Amaranth grain is relatively higher in protein content (14–19%), high lysine and methionine, which are limiting in cereals and legumes respectively. Lipid and starch content are lower than most of major cereals	Reduced insect population when taken as intercrop. e.g., cucumber beetles and fruit flies population gets lowered with amaranth establishment (Pitan & Esan, 2014)	Gaining its importance as nutritious superfood in supermarkets. Amaranth oil is used in cosmetic industries for making hair oil. Used in infant/weaning food formulations, for making alcoholic beverages	A fluid extract obtained from the plant acts as an astringent internally to treat mouths and throats ulcerate

(continued)

Table 6.1 (continued)

Crop	Climate resilient feature	Food usage	Health benefit and nutritional potentials	Agronomic and horticultural uses	Industrial uses	Medicinal usage
Buckwheat (*Fagopyrum* spp.)	A potential crop under rainfed and degraded conditions. Buckwheat is not particularly drought tolerant but its short growing season (70 to 90 days) may allow it to avoid droughts and thrives in poor soils	Used in buckwheat tea or processed into groats, flour, and noodles, tender shoots are eaten as leafy vegetables, grains used for human consumption and livestock feed preparation	Because of its well-balanced amino acid profile, the protein in buckwheat is very high quality. It is gluten-free and therefore suitable for people with gluten intolerance	Due to speedy growth within short duration it is a good cover crop for suppressing weeds. Also attracts insects with its abundant blossoms which is good for pollination	In food industry used to make varieties of food products such as noodles, buckwheat honey, buckwheat flour, buckwheat groats etc.	Improved blood sugar control, suppressing gallstone formation, and reducing the risk of colon cancer in animals. Green leaves used medicinally to promote blood circulation
Job's tear (*Coixlacryma-jobi*)	Grows well in areas where corn and rice do not perform well in higher areas	Its grains are widely consumed as cereal grains. A thick consistency drink known as Job's tears tea is prepared from powdered grains of Job's tears	Grains are gluten free, contain thiamine, riboflavin, niacin and ascorbic acid, 13.05 g protein)	Can be intercropped with maize or sorghum as a supplemental grain or as alternative crop if main crop fails, it promotes resilience of farms in the warm, humid tropics	In Asian supermarkets Job's tears sold as Chinese pearl barley, used in brewing industry and an excellent mixture in bakery products	Used to strengthen the spleen, treatment of urination, arthritis, diarrhea and draining the pus, decreasing liver fat accumulation, decreasing osteoporosis etc.

(continued)

Table 6.1 (continued)

Crop	Climate resilient feature	Food usage	Health benefit and nutritional potentials	Agronomic and horticultural uses	Industrial uses	Medicinal usage
Quinoa (*Chenopodium quinoa.*)	This crop possesses exceptionally superior ability to survive under extreme soil water deficit due to its inherent low water consumption and ability to resume its photosynthetic activity by maintaining its leaf area after long duration drought and high temperature. Also shows remarkable adaptations to varying soil salinization	*Chenopodium quinoa* is an herbaceous annual plant in the amaranth family which is primarily known for its seeds that are rich in protein, dietary fiber, Vitamin B, and dietary minerals. It is a non-grass pseudocereal	Seeds provide all the essential amino acids required by the human body. Quinoa protein can supply histidine (180%), isoleucine (274%), lysine (338%), methionine + cysteine (212%), phenylalanine + tyrosine (320%), threonine (331%), tryptophan (228%) and valine (323%). This is more than the nutritional requirements for an adult (Jaikishun et al., 2019)	Their survival in wide ranging agroecological conditions make them more growth conducive and give appreciable yields. Quinoa has the ability of excellent phytodesalinization, has potential to remove heavy metals from the saline soil	All flour industry products can be prepared from flour and whole grains of quinoa. Flour is used as food supplement in the flour industry to prepare gluten-free products. Starch from quinoa is used for production of aerosol, self-copy paper, pulps, foods for desserts, good source of excipients in plastics industry, anti-off-set powders and talcs	Leaves, stems and grains are used for healing wounds, toothache, reducing inflammation and treating urinary infections. Embryo of the quinoa seed has high protein (45%) that can be added to infant feed for rapid nutritional recovery

(continued)

Table 6.1 (continued)

Crop	Climate resilient feature	Food usage	Health benefit and nutritional potentials	Agronomic and horticultural uses	Industrial uses	Medicinal usage
Rice bean (*Vigna umbellata*)	A crop with potential for food security. Source of genes for biotic and abiotic stress tolerance including drought, soil acidity, storage pest and aluminum toxicity tolerance	Used as nutritionally rich food and fodder crop. Primarily consumed as a dal, roasted whole grain, added to make deep-fried snacks	Nutrient dense crop which can address micro-nutrient deficiencies in affected areas, contain high methionine, niacin, riboflavin and thiamine. Considered as considered as a low-fat food	Adapted to in low-input production systems, satisfy the need of a multi-tier agriculture	Flour may be added as food supplement in bakery products to enrich the nutritional values	–
Adzuki bean (*Vigna angularis*)	Adapted to human-disturbed habitats and natural establishment of weedy forms are found due to derivatives of hybrids between cultivars and wild forms which may serve as useful gene donor. The crop is grown as rainfed and withstand drought	Consumed as sweet beans, serves as a filling in Japanese sweets, eaten sprouted or boiled in a hot, tea-like drink	Like most beans, this bean also loaded with protein, fiber, beneficial plant compounds, complex carbs, antioxidants and minerals	Adapted to in low-input production systems, satisfy the need of a multi-tier agriculture	Due to its high concentration of saponins, a natural foaming agent that helps cleanse pores, remove dead skin cells, and draw out impurities, the beans are used in exfoliating soap making industries	Consumed to improve gut health where the fiber in the beans serve as food for good gut bacteria, known to be taken in diabetes, Molybdenum in the beans serve to detoxify liver. Known to improve bone and skin health

(continued)

Table 6.1 (continued)

Crop	Climate resilient feature	Food usage	Health benefit and nutritional potentials	Agronomic and horticultural uses	Industrial uses	Medicinal usage
Faba bean (*Vicia faba*)	Faba bean inclusions in crop rotation results in reduced CO_2 emissions, improved soil physical properties like porosity, bulk density and water holding capacity of soil	Both dry and fresh seeds of faba bean can be consumed	They contain high protein (up to 35% in dry seeds), are a good source of mineral nutrients such as K, Ca, Mg, Fe, and Zn, vitamins, and numerous bioactive compounds	Highly efficient in the symbiotic fixation of atmospheric nitrogen and enhance agricultural sustainability	As a sustainable dietary choice faba bean may partially replace meat consumption for protein	Accumulates a large amount of L-Dopa in its various parts which has its importance in treating Parkinson's disease and hormonal imbalances
Winged bean (*Psophocarpus tetragonolobus*)	Due to their biological nitrogen-fixation they can support low-input farming and improve soil conditions	All parts are edible. Known as "One Species Supermarket" for its nutrient-dense green pods, flowers, immature seeds, nutty tubers, leaves, and mature seeds	Mature winged bean seeds contain on average around 15–18% fat and 30–37% crude protein, and favorable amino acid and micronutrient composition. Protein content and fatty acid profiles compare favorably with soybean and chickpea	Crops like rice and maize taken after winged bean require less nitrogen application and give more yield. Improves soil fertility and benefits soil physical properties	Winged bean seeds contain 15.0–20.4% oil content that makes it a potential oilseeds crop. Can be used for fortification of wheat bread	Leaf extracts are used for smallpox treatments, leaves and seed are eaten to cure skin sores, green leafy vegetable rich in vitamin A that benefits eye vision, increase immunity and fight against cold. Cure sprains and inflammation since it consists of considerable amount of manganese

(continued)

Table 6.1 (continued)

Crop	Climate resilient feature	Food usage	Health benefit and nutritional potentials	Agronomic and horticultural uses	Industrial uses	Medicinal usage
Basella	This can be easily grown as kitchen garden crop which does not require much care and maintenance, thus ensures nutritional security	Used as a leafy vegetable. Leaves are used to make snacks and delicacies. It is also consumed mixed with dal, chutney, soup, curry etc. in almost all parts of India	Leaves are rich source of Vitamin A, C, calcium and iron, folic acid. Essential amino acids such as arginine, leucine, isoleucine, lysine, threonine and tryptophan are found. Credited with antioxidant properties. Its mucilage is known for hypoglycemic activities	Grown as kitchen garden crop	Leaf and fruit are the source of basella. It has potential bioactivity of Phyto molecules found in basella makes it potential opportunity for food processing industry. The betalains can act as alternative for synthetic colourants in the food industry (Khan, 2015)	A fine paste of its leaves and stem can be applied on burn wounds to remove the marks on skin. In Ayurveda, basella is recommended to cure anemia and dysentery. It's a good laxative, and the cooked roots are used in the treatment of diarrhea (Kumar et al., 2013)

(continued)

Table 6.1 (continued)

Crop	Climate resilient feature	Food usage	Health benefit and nutritional potentials	Agronomic and horticultural uses	Industrial uses	Medicinal usage
Drumstick tree	The drumstick plant is fast growing and very well adapted to sustain in adverse growth conditions where most of the plants are unable to grow, to require at least 400 mm of rain per annum	Leaves and flowers are eaten as salad, cooked vegetables, added to soups and sauces or used to make tea. Young, tender pods are used as vegetables and pickle. Root bark is used as a condiment	High antioxidant content. Moringa oil has 1–2% of essential fatty acids especially omega 3 and omega 6	Windbreak or living fence, as a support for climbers, can be intercropped with a range of vegetables, leaves and twigs also serve as forage for livestock, fast-growing tree that adapts well to hot, semi-arid regions, controlling soil and wind erosion	Seed contains about 40% non-drying, rancidity resistant oil that can be used in used for cooking, lubrication, and in the cosmetic industry Ground moringa seeds are used to purify drinking water	Extract of leaves and seeds of possess marked analgesic activity, beneficial in glandular swellings, root extract exhibit significant anti-inflammatory activity

(continued)

Table 6.1 (continued)

Crop	Climate resilient feature	Food usage	Health benefit and nutritional potentials	Agronomic and horticultural uses	Industrial uses	Medicinal usage
Paradise tree (*Simarouba glauca*)	Tree root system are very well formed in the soil, preventing soil erosion. Provides evergreen canopy that protects soil fauna and flora. Improves groundwater level and reduce greenhouse effects. Large-scale planting in wastelands help wasteland reclamation	Seeds produce an edible oil. The fruits closely resemble olive and are sold in local markets	–	This tree is an example of a low input cost technology that may give assured returns and livelihood improvement in rural areas. Being a shade tolerant tree, it may be grown as under-storey tree, particularly under the canopy of large fruit trees	The wood of this tree is insect resistant, thus, used in quality furniture making, toys making, pulp preparation for paper making and matches. Also used as biofuel, lubricants, soaps, detergents, varnishes, pharmaceuticals and cosmetics	Extracts from this tree are proposed to have potent anticancer properties (which needs further experimental proof). The bark, leaves and roots are used for anemia, diarrhea, hemorrhages, intestinal worms, malaria etc.

(continued)

Table 6.1 (continued)

Crop	Climate resilient feature	Food usage	Health benefit and nutritional potentials	Agronomic and horticultural uses	Industrial uses	Medicinal usage
Kankoda (*Momordica dioica*)	Since pollination and fertilization are adversely impacted by the climate changes, development of parthenocarpic Kankoda may serve as a potent nutritious alternative for cucurbitaceous vegetables	Fruits are used as vegetable. The root is tuberous and cooked as a vegetable. Tender shoots and leaves are used as greens locally	Per 100 g edible fruit consists of moisture (84.1%), carbohydrate (7.7 g), protein (3.1 g), fat (3.1 g), fiber (3.0 g) and minerals (1.1 g). Contains ascorbic acid, carotene, thiamin, riboflavinand niacin (Internet source)	Can be allowed to climb on treetops. Harvesting starts about 50–60 days after sowing	Various parts of Kankoda are used as traditional herbal remedies to cure diabetes mellitus. Also known for its anti-malarial and anti-ulcer potential	It has commercial value and exported and used locally. A semi-drying oil is extracted from the seed
Jojoba (*Simmondsia chinensis*)	As a low water requiring crop Jojoba has a great scope to be cultivated in the relatively hot weather of desert, low fertility soil and water scarce regions. It has a great ability to withstand the high salt in the soil	Jojoba can be used as a source of palatable forage for livestock. Once detoxified, oil meal resulting from oil extraction may be used as livestock feed	Jojoba oil is edible but non-caloric and non-digestible and pass out of the intestine without being absorbed in the body (Internet source) because of this the oil has no nutritious value	Can be cultivated from seeds or cuttings. Can be used for afforestation in rainfed arid areas	Jojoba oil is considered good for skin and hair and, used for treating acne, eczema, and psoriasis	Can be cultivated in desert lands as an emerging biofuel crop and be used as a replacement of petroleum. Jojoba oil is used as an additive in many cosmetic products. The oil is outstandingly stable on high temperatures and never becomes rancid (Bassam, 2010)

(continued)

Table 6.1 (continued)

Crop	Climate resilient feature	Food usage	Health benefit and nutritional potentials	Agronomic and horticultural uses	Industrial uses	Medicinal usage
Guayule (*Parthenium argentatum*)	Plant is an evergreen woody shrub that thrives well in rocky and limestone desert areas in hot weather. Its extensive root system makes it resistant to drought	Edible usage is not known	–	As a biofuel guayule can be cultivated in areas where food crops cultivation is a struggle	The plant can source hypoallergic natural rubber, non-toxic adhesives, terpenes, ethanol and other specialty chemicals. Also, may serve as economically viable biofuel crop	–
Jatropha (*Jatropha curcas*)	Its cultivation decreases greenhouse gas emission by replacing fossil diesel with biofuels. Contributes to diversifying household income and national fuel security. Also, it is a carbon neutral crop. It is a drought tolerant perennial crop	–	–	Successfully used to develop agroforestry cropping system. Can be grown on wastelands. Its biomass feedstock is used to produce biogas and high-quality organic fertilizer	Potential bioenergy fuel crop in biofuel markets. Traditionally used for the manufacture of candles and soap, as lamp oil	Plant contains various phytochemicals that may contribute to extraction of antimicrobial compounds for making medicinal drugs. However, proper toxicological profiling is needed before exploiting jatropha as a medicinal plant

Table 6.2 Abiotic stress resistant/tolerant features of some orphan crops as potential trait donor

Common name	Botanical name	Trait of interest	References
Water use efficient/drought tolerant orphan crops			
Bambara groundnut	*Vigna subterranea*	Drought tolerant	National Research Council (2006), Rosell and Holmer (2007)
Baobab	*Adansonia digitata*	Drought tolerant	Rosell and Holmer (2007)
Cassava	*Manihot esculentum*	Drought tolerant	National Research Council (2006), Ceballos et al. (2004)
Cowpea	*Vigna unguiculata*	Drought tolerant	Rosell and Holmer (2007)
Enset	*Enseteventricosum*	Drought tolerant	Williams and Haq (2000), Brandt (1997)
Foxtail millet	*Setariaitalica*	Drought tolerant	Ketema (1997)
Grass pea	*Lathyrus sativus*	Extremely drought tolerant	Campell (1997)
Little millet	*Panicum sumatrense*	Drought tolerant	Ketema (1997)
Pearl millet	*Pennisetum glaucum*	Drought tolerant	IPGRI (2004)
Proso millet	*Panicum miliaceum*	Drought tolerant	Ketema (1997)
Tef	*Eragrostis tef*	Tolerant to abiotic stresses	Heslop-Harrison and Schwarzacher (2007), Getinet and Sharma (1996)
Yam	*Dioscorea* spp.	Drought tolerant	Williams et al. (2000)
Bambara groundnut	*Vigna subterranea*	Drought tolerance	National Research Council (2006)
Ethiopian mustard	*Brassica carinata*	Drought tolerance	Getinet et al. (1996)
Common bean	*Phaseolus vulgaris*	High WUE, Deeper and vigorous roots	Ghanbari et al. (2013), Polania et al. (2017)
Faba bean	*Vicia faba*	Root growth	Belachew et al. (2018)
Wild wheat	*Aegilops tauschii*	Drought tolerance	Ieshisa and Takumi (2012)
Barley	*Hordeum spontaneum*	Drought tolerant	Ivandic et al. (2000)
Amaranthus	*Amaranthus hypochondriacus*	Drought tolerant	González-Rodríguez et al. (2019), Massange-Sanchez et al. (2016)
Quinoa	*Chenopodium quinoa*	Drought tolerance	Pulvento et al. (2010), Hinojosa et al. (2018)

(continued)

Table 6.2 (continued)

Common name	Botanical name	Trait of interest	References
Salinity/alkalinity tolerant orphan crops			
Amaranthus	*Amaranthus cruentus* L.	Salt tolerance	Huerta-Ocampo et al. (2014), Massange-Sanchez et al. (2016)
Quinoa	*Chenopodium quinoa*	Salt tolerance	Adolf et al. (2013), Ruiz et al. (2016)
Buckwheat	*Fagopyrum esculentum*	Salt tolerance	Wu et al. (2017)
Barley	*Hordeum vulgare* L.	Salt tolerance	Hassan et al. (1970)
Guar	*Cyamopsis tetragonoloba* (L). Taub	Salt tolerance	Francois et al. (1990)
Kenaf	*Hibiscus cannabinus* L.	Salt tolerance	Francois et al. (1992)
Channel millet	*Echinochloaturnerana* (Domin) J.M. Black	Salt tolerance	Shannon et al. (1981)
Oats	*Avena sativa* L.	Salt tolerance	Mishra and Shitole (1986)
Rye	*Secale cereale* L.	Salt tolerance	Francois et al. (1989)
Sugar beet	*Beta vulgaris* L.	Salt tolerance	Bower et al. (1954)
Triticale	*X Triticosecale* Wittmack	Salt tolerance	Francois et al. (1988)
Durum wheat	*T. turgidum* L. var. *durum* Desf	Salt tolerance	Francois et al. (1986)
Bermudagrass	*Cynodondactylon* (L.) Pers	Salt tolerance	Bernstein and Francois (1962), Langdale and Thomas (1971)
Kallargrass	*Leptochloafusca* (L.)	Salt tolerance	Sandhu et al. (1981)
Oats	*Avena sativa* L.	Salt tolerance	Mishra and Shitole (1986), USSL
Rye	*Secale cereal* L.	Salt tolerance	Francois et al. (1989)
Wildrye	*Elymus angustus* Trin	Salt tolerance	McElgunn and Lawrence (1973)
Asparagus	*Asparagus officinalis* L.	Salt tolerance	Francois (1987)
Jojoba	*Simmondsia chinensis*	Salt tolerance	Tal et al. (1979), Yermanos et al. (1967)
Date-palm	*Phoenix dactylifera* L.	Salt tolerance	Furr and Armstrong, (1962), Furr and Ream (1968)
Oat	*Avena sativa* L.	Alkalinity tolerance	Foy et al. (1987)
Triticale	*X Triticosecale* Wittmack	Alkalinity tolerance	Beck et al. (2020)

(continued)

Table 6.2 (continued)

Common name	Botanical name	Trait of interest	References
Narrow leaf lupine	*Lupinus angustifolius*	Alkalinity tolerance	Kettel et al. (2003)
Heat tolerant orphan crops			
Pearl millet	*Pennisetum glaucum*	Heat tolerance	Yadav et al. (2016)
Cow pea	*Vigna unguiculata*	Heat tolerance	Sanginga et al. (2000), Ehlers et al. (2000), Ehlers and Hall (1998)
Sorghum	*Sorghum bicolor*	Heat tolerance	Prasad et al. (2006)
Common bean	*Phaseolus vulgaris*	Heat tolerance	Porch (2006)
Quinoa	*Chenopodium quinoa*	Heat tolerance	Yang et al. (2016), Becker et al. (2017)
Metal toxicity tolerant orphan crops			
Buckwheat	*Fagopyrum esculentum*	Tolerance to Al toxicity	Reyna-Llorens et al. (2015), Lei et al. (2017)
Squash	*Cucurbita pepo*	Tolerance to Al toxicity	Ahn et al., (2001)
Cranberry	*Vaccinium macrocarpon*	Tolerance to Al toxicity	Osaki et al. (1997)
Malabar melastome	*Melastomamalabathricum*	Tolerance to Al toxicity	
French hydrangea	*Hydrangea macrophylla*	Tolerance to Al toxicity	
Duck weed	*Lemna minor* L.	Hyper accumulation of heavy meatless	Prasad and Freitas (2003)
Waterlogging tolerant orphan crops			
Kodo millet	*Paspalum scrobiculatum*	Tolerant to flooding condition	Ketema (1997)
Niger	*Guizotiaabyssinica* Cass	Grows best on poorly drained heavy clay soils	Alemaw and Wold (1995)
Tef	*Eragrostis tef* (Zucc.)	Grow well on poorly drained soils	Assefa et al. (2015)

6.3.1 *Water Use Efficient/Drought Tolerant Orphan Crops*

Out of total cultivable land of the world, negligible land is under irrigation. Replacing the drought susceptible staple food crops with drought tolerant orphan crops is one of the good options. Many orphan crops have potential to be used as staple food crop and, at the same time they exhibit good amount of tolerance towards moisture stress. Orphan crops, including millets and some of the legumes show very good tolerance or resistance towards drought. Among millets especially pearl millet (*Pennisetum glaucum*), foxtail millet (*Setaria italic*), proso millet (*Panicum miliaceum*), little millet (*Panicum sumatrense*), tef (*Eragrostis tef*) and fonio (*Digitariaexilis*) have tolerance towards moisture stress as they are mainly cultivated in dry land area (Haussmann et al., 2012; Tadele, 2016). They have inherent tolerance for water deficit. They possess traits like long roots, increased water-use-efficient (WUE), waxy leaves etc. which help them to survive under water deficient conditions. By the virtue of short life cycle, African rice (*Oryza glaberrima*) completes its lifecycle early and escape from the drought (Linares, 2002). Orphan legumes such as faba bean (*Vicia faba*) developed traits like increased root length, increased root volume and fibrous root system which are the promising traits for drought tolerance (Duc et al., 2015; Hall, 2012; Khan et al., 2010). It is proven fact that several orphan crops grow at low water due to high WUE (Chibarabada et al., 2017).

6.3.2 Heat Stress Tolerant Orphan Crops

For any plant species there occurs particular range of temperature which is suitable for its growth and called as optimal temperature. Temperature more than optimal temperatures described as heat stress that adversely affects plant growth, development and ultimately yield. Heat stress is another major threat to agriculture and sustainable crop production. Though it occurs all over the globe, it is more prevalent in particular parts of the world such as certain part of Africa, Asia and Middle East. In the scenario of global warming and climate change risk of heat stress is increasing day by day. Rise of temperature by 3–4 °C can decreases yield up to the extent of 15–35%, particularly in Asia and Africa (Ortiz et al., 2008). The adverse effects of the global warming on agriculture could be minimized by developing heat tolerant cultivar or by adopting crops pertaining resistance/tolerance towards high temperature. Many orphan crops can sustain higher temperature due to their inherent heat tolerance. These crops possess traits viz., transcriptional cooling, reflecting solar radiation, altering membrane lipid composition, leaf shading extensive rooting system, early maturity with short life cycle, high rate of photosynthesis, stomatal conductance which are valuable traits for heat stress breeding (Lehman & Engelke, 1993; Wahid, 2007; Yadav et al., 2007).

6.3.3 Salinity/Alkalinity Tolerant Orphan Crops

Soil salinity is becoming big crisis for irrigated agriculture and the agriculture form the costal lands. It is estimated that 20% of the total cultivable and 33% of irrigated agriculture lands are affected by salinity problem worldwide. Moreover, it is predicted that salinity rate is increasing 10% per year and by 2050, more than 50% of the agriculture land would turn saline (Jamil et al., 2011). Raising sea levels, poor quality of irrigation water, poor cultural practices, and surface evaporation are some of the reasons for soil salinity. Salinity not only affects crop production but also degrades soils. Salinity affects almost all the stages of crop growth viz., germination, vegetative and reproductive stages (Akbarimoghaddam et al., 2011). Though soil alkalinity is not much severe problem, soil salinity in some parts of world is quite severe. Nearly 40% of the arable land is having problem of acidic soils (Gale, 2002). Acidic soils generally possess excess Al and Mn which affect crop growth and production. Increasing problem of salinity and alkalinity can be overcome by growing crops tolerant to salt stress. Many orphan crops have developed good mechanism to cope up with high salinity. Adoption of such salt tolerant crops in salty soils and coastal area will be one of the important steps to bring these areas under cultivation and to ensure global food security. Plant tolerant to excess Al and Mn can grow well on alkaline soils.

6.3.4 Cold Tolerant Crops

Cold stress is a major abiotic stress and severe in many parts of world particularly northern hemisphere. It is more common in hilly and mountainous regions. Lower temperature affects crop growth and productivity and ultimately decreases crop yield (Xin & Browse, 2001). Many tropical crop plants are susceptible to lower temperature and shows cold injuries such as a necrosis, chlorosis and reduction in growth whereas low temperature tolerant crops can able to grow under cold temperatures also. There are two groups of cold tolerant crops, chilling tolerant (0–15 °C) and freeing tolerant (> 0 °C) plants. Orphan crops such as carrot (*Daucus carota* subsp. sativus), fava bean (*Vicia faba*), turnip (*Brassica rapa* subsp. rapa), kale (*Brassica oleracea* var. sabellica), collard (*Brassica oleracea* var. viridis) etc. are cold tolerant and good choice for the area where cold stress is frequent and severe.

6.3.5 Metal Toxicity Tolerant Orphan Crops

Agricultural lands are contaminated with heavy metals like Lead, Copper, Cadmium, Zinc, Cobalt, Nickel, Chromium etc. in many parts of the world. This is due to anthropological activities such as water pollution with heavy metals, industrial waste, and irrigation of land with contaminated water and excess use of fertilizers etc. (Passariello et al., 2002). Whereas in some of the agricultural lands there is increased concentration of one or more than one micronutrient such as aluminum (Al), iron (Fe), magnesium (Mg) etc. based on their parent rock and climatic conditions. Metal toxicity is causing serious threat for crop growth and production. Metals when they present excess in soil, they turn toxic to plant and hamper physiological, cellular, biochemical and molecular processes. Some of the crops have evolved tolerance mechanism against metal toxicity. Detoxification of excess metal is mainly based on chelation and sub-cellular compartmentalization (Yadav, 2010). Orphan crop buck wheat (*Fagopyrum esculentum*) (Lei et al., 2017; Reyna-Llorens et al., 2015) and squash (*Cucurbita pepo*) (Ahn et al., 2001) shows good tolerance level toward excess Al in soil. Woody plant species such as *Vaccinium macrocarpon*, *Melastomama labathricum* and *Hydrangea macrophylla* (Osaki et al., 1997) and timber plant species *Melaleuca cajuputi* (Maejima & Watanabe, 2014) also shows Al tolerance. Moreover, many other orphan crops possess good tolerance to excess metals and heavy metals.

6.3.6 Waterlogging Tolerant Orphan Crops

There are two types of submergence viz., water logging complete submergence. In water logging root and some lower part of the plant is under water whereas in submergence whole plant is under water. Water logging is also known as anoxia, soil saturation, hypoxia etc. Water logging condition slowdowns the exchange of gases which hamper normal growth of the plant (Colmer & Greenway, 2011). Some of the orphan crops are well adapted to water logging and marshy condition and evolved good mechanism to overcome difficulties associated with water logging condition. Tef *Eragrostis tef* (Zucc.) Trotter is an African cereal orphan crop which possesses good tolerance to submergence and can grow well on poorly drained soils (Assefa et al., 2015).Kodo millet (*Paspalum scrobiculatum*) and niger (*Guizotiaabyssinica* Cass) can grow on poorly drained heavy soils (Alemaw & Wold, 1995; Ketema, 1997). Some of the orphan fruit crops also show tolerance to waterlogging conditions and can grow well in wet and saturated soils e.g., lowbush cranberry, fox grape, American red raspberry, taro etc.

6.4 Genetic Potential of Orphan Crops to Mitigate Different Biotic Stresses

It is evident from earlier literature that biodiversity of orphan crops in agroecosystem can reduce the impact of insect-pests, diseases and weeds with various mechanisms viz., resource dilution, stimulo-deterrent diversion, disruption of spatial cycle and temporal cycle, allelopathic effects, soil suppressiveness, crop physiological resistance, conservation of natural enemies (Ratanadass et al. 2012). The other important pathway identified for induction the plant's defense reaction, contributing to phenotypewhich is evolved by plants at different levels, such as perceptionof cellular signals and transduction that induces expression of specialized sub-sets of defense related genes (Fraire-Velázquez et al., 2011). As per the plant TF database, some important TFs/genes have been reported in Table 6.3. In the microbe-associated molecular pattern-triggered immunity (MAMP-triggered immunity), effector-triggered immunity (ETI), or system acquired resistance (SAR), plant WRKYs have been implied (Javed et al., 2020). Buckwheat has high allelopathic potential. Rutin, the main flavonoid of buckwheat, which plays significant role in smothering of annual weeds (Tsuzuki & Dong, 2003).

Table 6.3 Genes/TFs involved in orphan crops against biotic stresses

Potential crop	TF/Gene identified	Target crops	Resistance towards	References
Buck wheat	*AP2/ERF* transcription factor	Arabidopsis	Disease	Liu et al. (2019)
Grain Amaranthus	*Ah24* gene	Tobacco Arabidopsis	Insect herbivory	Massange-Sanchez et al. (2015)
Adzuki bean	WRKY transcription factor family	–	Stem rot and other bean diseases	Srivastava et al. (2018)
Faba bean	*Af1, Af2, Af3, Af4, Af5, Af6, Af7*	–	Ascochyta blight	Rashid et al. (1991)
Faba bean	*Fr1, Fr2, Fr3*	–	Rust (*Uromyces viciae-fabae*)	Conner and Bernier (1982)
Faba bean	*Uvf1*	–	Rust (*Uromyces viciae-fabae*)	Avila et al. (2003)
Faba bean	*Oc2, Oc3, Oc4, Oc5*	–	Orobanche	Diaz-Ruiz et al. (2010)
Faba bean	*Of1, Of2, Oc1, Oc2, Oc3, Oc4, Oc5*	–	Orobanche	Gutierrez et al. (2013)
Winged bean	Protease Inhibitors (WBTI 1 to 7) Ranging from Kunitz typeto Bowman–Birk type serine proteinase inhibitors	–	*Helicoverpa*	Giri et al. (2003)
Gayule	Trans-cinnamic acid; 3,4-dihydroxyhydrocinnamic acid; and 3-hydropropionic acid with latex/rubber transferases	–	Nematodes/ Insects	Cespedes et al. (2001)
Jatropha	*NBS-LRR* gene	–	*Disease*	Sood et al. (2014)
Jatropha	*JcPR-10a* gene	–	*Disease*	Agarwal et al. (2014)

6.5 Strategic Approaches to Gain Attention for Orphan Crops

Orphan crops play a vital role in the advent of drastic climate change, as they can establish themselves to the abiotic stresses such as harsh environment, moisture deficit etc. So orphan crops play an important role in the income generation and healthier diets to urban poor farmers and consumers. They also have an important role in global food and nutrition security in the developing countries. Despite many beneficial aspects there are several challenges which are responsible for low productivity of orphan crops.

Regarding this, it is necessary to develop new strategies and approaches to increase the crop productivity in orphan crops to gain more attention. So, it is important to do several activities that have been discussed.

6.5.1 Awareness and Knowledge Enhancement

There are several ways to create the awareness for orphan crops.

- International events that supported orphan crops or underutilized crops also creates awareness for orphan crops.
- The establishment of the International Year of Quinoa (IYQ-2013), aimed to increase awareness and knowledge about quinoa and focused on food security (FAO, 2013).
- Disseminate the information about the orphan crops by conducting training at local communities and execute information campaigns to generate the message for nutritional status and benefits of orphan crops.
- The Lugar center provides information regarding the major recent analysis on global food security where it focuses on a range of scientific journals, research, and scholarly reviews.
- Besides original research works and reports in scientific journals and research, some journals dedicated to special issue for underutilized crops. Among them, were 'Future Smart Food', in eight countries in South and Southeast Asia, the countries included in the report are Bhutan, Cambodia, Bangladesh, Lao PDR, Nepal, Myanmar, Viet Nam and West Bengal in India. The crops selection was based on assuring nutrient dense, healthy diets, climate-resilient and economically adapted and viable crops. The crops included, six root crops, ten cereals crops and six tuber crops, nine fruit, nine pulses and vegetable crops, and five nuts and spices (Tadele, 2019).

6.5.2 Identifying End-Use or Specific Importance of Orphan Species

Since the Green Revolution most of the scientific study focused on major and staple crops like rice, wheat, maize, green gram, soybean, mustard, ground nut etc. Several policies inclined towards major crops. Despite the number of advantages of underutilized or orphan crops, these are always neglected by the scientists, researchers, policy makers, traders and government. For these reasons, countries' capacity for research on NUS is generally weak (Rudebjer et al., 2013).

6.5.3 Strategic Approaches of Orphan Crops

Orphan or underutilized crops are multipurpose crops. There is an urgent need of strategic approaches for the development and growth of orphan crops. These underutilized/neglected/orphan crops are overlooked despite having a few special traits (e.g., abiotic stress resistance gene, high nutritive value, low glycaemic indices and high concentrations of micronutrients).

6.5.4 Advanced Research

Due to the importance of underutilized/orphan crops across the world, there is an urgent need for a paradigm shift from the present scenario of neglected crops into sustainable cultivation, exploitation, conservation, evaluation and utilization of different types of species of orphan crops. orphan crops are an important crop in a global solution for food and nutrition insecurity. Crop improvement techniques, conventional or classical breeding methods such as selection, hybridization, mutation etc. and molecular or biotechnological methods such as marker assisted breeding, transgenic are being used in orphan crops. TILLING and EcoTILLING methods do not require prior knowledge of the genome sequence except for the gene of interest. When high throughput phenotyping (Phenomics) coupled with high throughput genotyping empowers the recovery of desirable trait data, that will assist in breeding of orphan crops. Genome editing tools such as CRISPR/Cas9 have been applied for the modification of undesirable traits in cassava and few other selected orphan crops. The examples of advanced breeding tools on underutilized or orphan crops are represented in Table 6.4.

Table 6.4 Examples of conventional and advanced breeding tools on underutilized or orphan crops

Technique	Crop	Improved trait/ features	References
Selection	Tef	Soil acidity	Abate et al. (2013)
Hybridization (intra- and inter specific)	Triticale	Soil acidity	Kim et al. (2001)
Mutation breeding	Chickpea, cassava	High yield, blight resistant, early maturity and in cassava it lowers the insulin level, which prevents quick spikes in glucose contents	Ceballos et al. (2008)
Marker assisted breeding	Foxtail millet (*Setariaitalica*)	Provides fundamental resources for genetics research and genetic improvement in foxtail millet	Jia et al. (2013)
Marker-assisted Genotype by sequencing	Finger millet, cassava, chickpea, cowpea, pearl millets etc.	Developed an unbelievably valuable dataset of several thousand SNPs segregating within each accession of finger millet	Kumar et al. (2016)
Genome wide association study	Cassava, finger millet	Identified association loci and candidate genes, and their tissue-specific expression in cassava	Zhang et al. (2018), Li et al. (2018)
TILLING (Targeting Induced Local Lesions IN Genomes	Cassava, tef, chickpea, pearl millet	Drought tolerant for cassava	Esfeld et al. (2013)
Genome editing (CRISPR/Cas9)	Cassava, ground cherry (*Physalis pruinosa*), tef	Modified undesired traits, improvement on plant architecture, flower production and fruit size	Mollins (2017), Lemmon et al. (2018), Bull et al. (2018)
Omics tools (transcriptomics and proteomics)	Tef, chickpea, finger millet	Drought and nutrition-related genes	Cannarozzi et al. (2014)
Speed breeding	Cassava, finger millet, chickpea	Rapid creation of new crop varieties	Chiurugwi et al. (2018)

6.5.5 Conservation of Germplasm

The germplasm or biodiversity of orphan or underutilized crops are protected by both in-situ and ex-situ conservation. Wherever there is less chance of conservation at ex situ then we must follow the on-farm conservation of orphan crops. It empowers local farmers, particularly women, and strengthens the cultural identity of local communities.

6.5.6 Effective Policy Formulation

In the present scenario there is a limited number of legal protections for orphan or underutilized crops. Different types of national and international policies can meet the challenges of orphan or underutilized crops such as effective use and conservation, better access to the international market. Some of the important policies are

- Incentives for growing and conserving underutilized or orphan crops ex situ and on farm.
- Incorporate underutilized or orphan crops under international legal framework for germplasm conservation, genetic improvement and sharing of germplasm for research.
- Generate a supportive policy environment at national level such as, provide the subsidies for growing and marketing of underutilized or orphan crops, organize the education programmes to promote the use and nutritional value of underutilized or orphan crops.
- Create awareness about the plant breeders' rights and farmers rights among the local communities to protect the crops which are uniform, distinct and stable.
- Worldwide in many areas a new concept now being tested is called payments for agrobiodiversity conservation services (PACS) is an incentive provided by the governments.

6.5.7 Partnership and Network

There is an urgent need for improving orphan or underutilized crops by accelerating the strategic harmony among existing national, regional and international networks and collaborative platforms. Networking of orphan or underutilized crops helps to broaden the base of agriculture and assimilate them into sustainable adoption to meet the global food, nutrition and income generation of local people, farmers at the national or international level.

6.6 Global Research on Underutilized Crops

Though underutilized crops are being a part of livelihood of many local communities across the globe, but there was less emphasis as given for research, conservation and popularization of these crops (Table 6.5). Already the underutilized crops attain the attention of global scientific community, policy and decision makers, governments, industries etc. to plan and impose strategies, research and development, implementation of policies etc. (Khoury et al., 2014; Padulosi et al., 2011, 2013). Across the globe several institutes are working on promotion and mainstreaming of underutilized crops (https://www.cgiar.org/).

6.7 Future Opportunities and Priorities for Orphan Species Under Climate Change

From ancient man to the today's modern agriculture, the currently cultivated main crops are the results of selective preference. This in the increased genetic uniformity due to the cultivation of selected crops and populations lead to loss of plant diversity specially the underutilized crops (Khoury et al., 2014). Fortunately, these neglected crops still exist in the natural habitats, forests, villages etc. and still locals are depending on these region specific local underutilized crops in their day-to-day life for both food and other needs (Mayes et al., 2012). These crops have been used to meet the multiple demands viz., food and other uses among small or marginal farmers in rural areas. In recent years, few underutilized crops have also entered the export and finding its place in the world market (Akinnifesi et al., 2008). Despite with many uses, these crops are facing few challenges to get attention by the larger communities due to social, agronomic, environmental, policy and economic challenges (Fig. 6.2) i.e., meager competitiveness with main crops, limited or lack of germplasm availability lack of information with respect to crop production, diversity, life cycle, breeding strategies, lack of national policies for supporting research and development, lack of interest from farmers, commercial growers, producers, private sectors, post-harvest processing companies, markets, consumers etc. Prioritization of orphan crops is region and country specific; hence it is necessary to do thorough research to streamline few underutilized crops (Fig. 6.3). At outset mainstreaming of underutilized crops helps to attain multiple securities among the poor and marginal farmers, better utilization of marginal and wastelands, less use of chemicals during cultivation, improved livelihood, and most importantly helps to cope up with climate change effects (Mabhaudhi et al., 2019).

Table 6.5 Summary of global activities on underutilized crops

Year	Activity
1975	US National Academy of Sciences (NAS) issued a report on "Underutilized Tropical Plants with Promising Economic Value". Later NAS issued further reports
1987	Establishment of International Centre for Underutilized Crops (ICUC) at Sri Lanka
1990s	A total of 24 monographs on individual underutilized species by IPGRI and two volumes of ICUC on specific species including many crops have been issued
1992	Recognition has been given to underutilized crops to make agriculture farming more sustainable in EARTH Summit of The United Nations Conference on Environment and Development (UNCED)
1996	First Global Plan of Action (GPA) for the Conservation and Sustainable Use of Plant Genetic Resources for Food and Agriculture (PGRFA) adopted by 150 countries in 1996 have incorporated an Activity 12 for promoting development and commercialization of underutilized crops
1999	At International workshop in Chennai, India, Consultative Group of International Agricultural Research (CGIAR) recognized the importance of neglected and underutilized crops in food security and in poverty alleviation of rural poor
2002	Establishment of Global Facilitation Unit of Underutilized Species (GFU) of the Global Forum on Agricultural Research (GFAR) and was housed within Biodiversity International, Rome, Italy
2004	International Treaty on Plant Genetic Resources for Food and Agriculture
2008	Crops for the Future (CFF) establishment
2008	International Symposium on Underutilized Plants for Food Security, Nutrition, Income and Sustainable Development held in Arusha, Tanzania
2011	Second GPA for Plant Genetic Resources for Food and Agriculture
2011	Establishment of Crops for the Future Research Centre (CFFRC) in Malaysia
2011	II International Symposium on Underutilized Plant Species: Crops for the Future—Beyond Food Security held in Kuala, Lumpur, Malaysia
2012	Cordoba Declaration result of the International Seminar "Crops for the XXI Century" held in Córdoba, Spain emphasized the importance of underutilized and promising crops for the international arena
2013	III International Conference on Neglected and Underutilized species held in Accra, Ghana
2015	III International Symposium on Underutilized Plant Species held in Tamil Nadu, India

6.8 Conclusion

Each plant on this earth may serve some special purpose. The plants that were once used mere for subsistence at the time of domestication, became major food crops of today's modern agricultural food systems. This phenomenon indicates that identification and selection of critical trait that has potential benefit in any aspect to the humanity and ecosystem, may turn a trait of great importance. In this respect, it is quite important that whatever biodiversity available on earth should be protected, because once eroded or vanished, it will not be able to be recovered. There is an

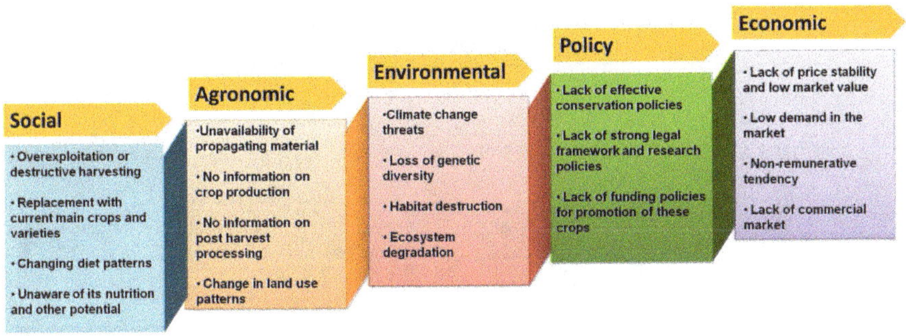

Fig. 6.2 General challenges in mainstreaming underutilized crops

urgent need to seriously look for other food sources available through plant systems, because climate change is affecting not only the major crops but also the crops of so-called minor importance. For a sustainable system it is very important to protect and utilize overall diversity in the agriculture. Mainstreaming minor crops is one of the most viable and ecofriendly options that brings multiple securities in terms of food security, nutritional security, security against climate threats, livelihood security, income security etc. Strong and effective policies are required, and potential minor crops need to be strategically researched, conserved, protected, documented, promoted and utilized so that the burden on agriculture sector for providing quality diet can be minimized specially when we talk about the most burning climate change related issues.

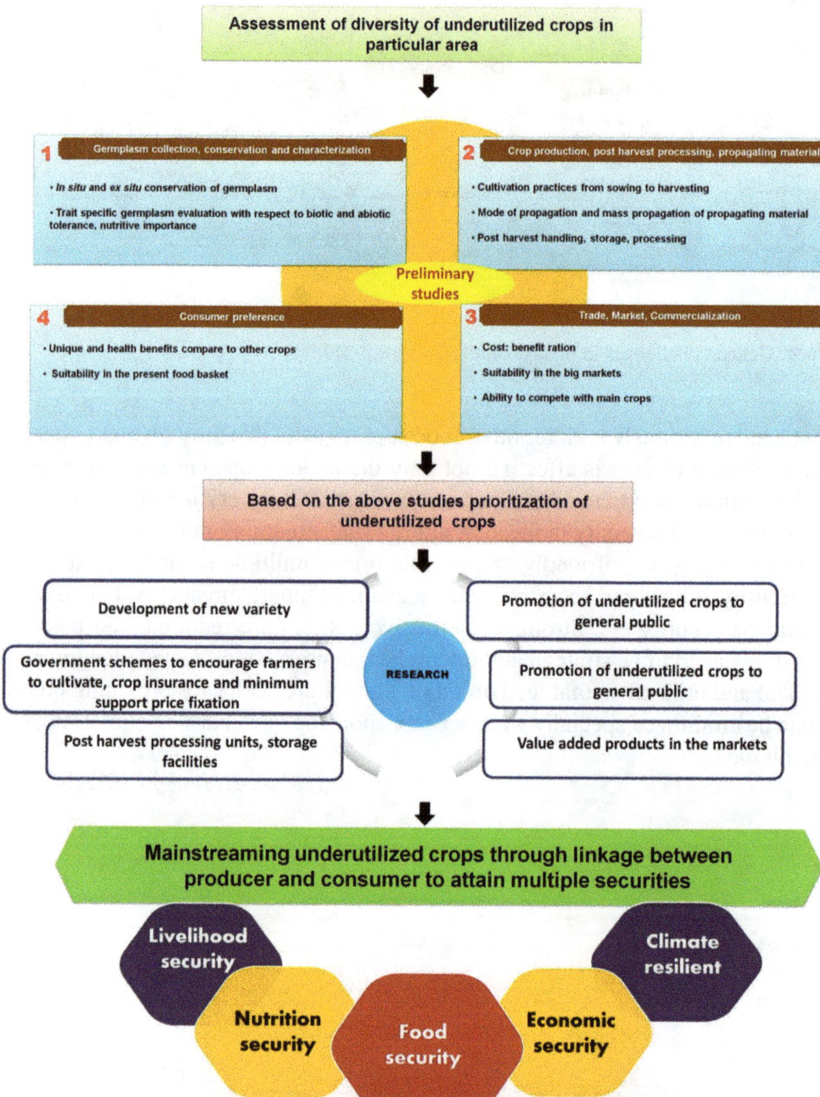

Fig. 6.3 Generalized model for prioritization and mainstreaming of underutilized crops under climate change

References

Abate, E., Hussein, S., Laing, M., & Mengistu, F. (2013). Quantitative responses of tef [*Eragrostis tef* (Zucc.) Trotter] and weeping love grass [*Eragrostiscurvula* (Schrad.) Nees] varieties to acid soil. *Australian Journal of Crop Science, 7*(12), 1854–1860.

Adolf, V. I., Jacobsen, S.-E., & Shabala, S. (2013). Salt tolerance mechanisms in quinoa (*Chenopodium quinoa* Willd.). *Environmental and Experimental Botany, 92*, 43–54.

Agarwal, P., Bhatt, V., Singh, R., Das, M., Sopory, S. K., & Chikara, J. (2014). Pathogenesis-related gene, JcPR-10a from *Jatropha curcas* exhibit RNase and antifungal activity. *Molecular Biotechnology, 54*, 412–425. https://doi.org/10.1007/s12033-012-9579-7

Ahn, S. J., Sivaguru, M., Osawa, H., Chung, G. C., & Matsumoto, H. (2001). Aluminum inhibits the H+-ATPase activity by permanently altering the plasma membrane surface potentials in squash roots. *Plant Physiology, 126*, 1381–1390.

Akbarimoghaddam, H., Galavi, M., Ghanbari, A., & Panjehkeh, N. (2011). Salinity effects on seed germination and seedling growth of bread wheat cultivars. *Trakia Journal of Science, 9*(1), 43–50.

Akinnifesi, F. K., Leake, R. R. B., Ajayi, O. C., Sileshi, G., Tchoundjeu, Z., Matakala, P., & Kwesiga, F. R. (2008) Indigenous fruit trees in the tropics: Domestication, utilization and commercialization (p. 438). CAB International Publishing.

Alemaw, G., & Wold, A. T. (1995). An agronomic and seed-quality evaluation of noug (*Guizoti-aabyssinica* Cass.) germplasm in Ethiopia. *Plant Breeding, 114*(4), 375–376.

Ali, S., Abbas, Z., Rizwan, M., Zaheer, I. Z., Yavaş, I., Ünay, A., Mohamed, M., & Abdel et al. (2020). Application of floating aquatic plants in phytoremediation of heavy metals polluted water: a review. *Sustainability, 12*(5), 1927. https://doi.org/10.3390/su12051927

Assefa, K., Cannarozzi, G., & Girma, D. (2015). Genetic diversity in tef [*Eragrostis tef* (Zucc.) Trotter]. *Frontiers in Plant Science, 6*, 177.

Avila, C. M., Sillero, J. C., Rubiales, D., Moreno, M. T., Torres, A. M. (2003). Identification of RAPD markers linked to the Uvf-1 gene conferring hypersensitive resistance against rust (*Uromyces viciae-fabae*) in *Vicia faba* L. *Theory of Applied Genetics, 107*, 353–358.

Bassam, E. N. (2010). *Handbook of bioenergy crops: a complete reference to species, development and applications*. Earthscan.

Beck, P., Jennings, J., & Rogers, J. (2020). Management of pastures in the upper south: The I-30 and I-40 corridors. In *Management strategies for sustainable cattle production in southern pastures* (pp. 189–226). Academic Press.

Becker, V. I., Goessling, J. W., Duarte, B., Caçador, I., Liu, F., Rosenqvist, E., & Jacobsen, S. E. (2017). Combined effects of soil salinity and high temperature on photosynthesis and growth of quinoa plants (*Chenopodium quinoa*). *Functional Plant Biology, 44*, 665–678.

Belachew, K. Y., Nagel, K. A., Fiorani, F., & Stoddard, F. L. (2018). Diversity in root growth responses to moisture deficit in young faba bean (*Vicia faba* L.) plants. *Peer Journal, 6*, e4401.

Berg, A., de Noblet-Ducoudré, N., Sultan, B., Lengaigne, M., & Guimberteau, M. (2013). Projections of climate change impacts on potential C4 crop productivity over tropical regions. *Agricultural and Forest Meteorology, 170*, 89–102.

Bernstein, L., & Francois, L. E. (1962). The salt tolerance of bermudagrass—NK37. In *United States salinity report to collaborators* (pp. 37–38).

Bower, C. A., Moodie, C. D., Orth, P., & Gschwend, F. B. (1954). Correlation of sugar beet yields with chemical properties of a saline-alkali soil. *Soil Science, 77*, 443–451.

Brandt, S. A. (1997). The "Tree Against Hunger": Enset-based agricultural system in Ethiopia. American Association for the Advancement of Science.

Bull, S. E., Seung, D., Chanez, C., Mehta, D., Kuon, J. E., Truernit, E., Hochmuth, A., Zurkirchen, I., Zeeman, S. C., Gruissem, W., & Vanderschuren, H. (2018). Accelerated ex situ breeding of GBSS- and PTST1-edited cassava for modified starch. *Science Advances*. https://doi.org/10.1126/sciadv.aat6086

Campell, C. G. (1997). Grass pea *(Lathyrus sativus L.)*. International Plant Genetic Resources Institute.

Cannarozzi, G., Plaza-Wuthrich, S., & Esfeld, K. (2014). Genome and transcriptome sequencing identifies breeding targets in the orphan crop tef (Eragrostis tef). *BMC Genomics, 15*(1), article no. 581.

Ceballos, H., Iglesias, C. A., Perez, J. C., & Dixon, A. G. O. (2004). Cassava breeding: Opportunities and challenges. *Plant Molecular Biology, 56*, 503–516.

Ceballos, H., Sanchez, T., Denyer, K., Tofino, A. P., Rosero, E. A., Dufour, D., Smith, A., Morante, N., Perez, J. C., & Fahy, B. (2008). Induction and identification of a small-granule, high-amylose mutant in cassava (*Manihot esculenta* Crantz). *Journal of Agriculture and Food Chemistry, 56*, 7215–7222.

Céspedes, C. L., Martínez-Vázquez, M., Calderón, J. S., Salazar, J. R., & Aranda, E. (2001). Insect growth regulatory activity of some extracts and compounds from *Parthenium argentatum* on fall armyworm *Spodoptera frugiperda*. *Zeitschrift Für Naturforschung, 56*, 95–105.

Chibarabada, T. P., Modi, A. T., & Mabhaudhi, T. (2017). Expounding the value of grain legumes in the semi-and arid tropics. *Sustainability, 9*(1), 60.

Chiurugwi, T., Kemp, S., Powell, W., & Hickey, L. T. (2018). Speed breeding orphan crops. *Theoretical and Applied Genetics*. https://doi.org/10.1007/s00122-018-3202-7

Colmer, T. D., & Greenway, H. (2011). Ion transport in seminal and adventitious roots of cereals during O_2 deficiency. *Journal of Experimental Botany, 62*, 39–57.

Conner, R. L., & Bernier, C. C. (1982). Inheritance of rust resistance in inbred lines of *Vicia faba*. *Phytopathology, 72*, 1555–1557.

Díaz-Ruiz, R., Torres, A. M., Satovic, Z., Gutierrez, M. V., Cubero, J. I., & Román, B. (2010). Validation of QTLs for *Orobanche crenata* resistance in faba bean (*Vicia faba* L.) across environments and generations. *Theoretical and Applied Genetics, 120*(5), 909–919.

Directorate of Economics and Statistics (DES), Ministry of Agriculture, India. (2014). https://eands.dacnet.nic.in/PDF/Glance-2016.pdf

Duc, G., Agrama, H., Bao, S., Berger, J., Bourion, V., De Ron, A. M., Gowda, C. L. L., Mikic, A., Millot, D., & Singh, K. B. (2015). Breeding annual grain legumes for sustainable agriculture: New methods to approach complex traits and target new cultivar ideotypes. *CRC Critical Reviews in Plant Sciences, 34*, 381–411.

Ehlers, J. D., & Hall, A. E. (1998). Heat tolerance of contrasting cowpea lines in short and long days. *Field Crops Research, 55*, 11–21.

Ehlers, J. D., Hall, A. E., Patel, P. N., Roberts, P. A., & Mathews, W. C. (2000). Registration of california blackeye 27. *Crop Science, 40*, 854–855.

Esfeld, K., Plaza-Wüthrich, S., & Tadele, Z. (2013). TILLING as a high throughput technique of tef improvement. In K. Assefa, S. Chanyalew, & Z. Tadele (Eds.), *Achievements and prospects of tef improvement* (pp. 53–65). EIAR-University of Bern.

FAO. (2012, December 10). "Neglected crops need a rethink—Can help world face the food security challenges of the future, says Graziano da Silva". *FAO Media Centre*. Retrieved December 17, 2012.

FAO. (2013, February 20). "Official launch of the International Year of Quinoa" (PDF). Retrieved August 28, 2013.

Foy, C. D., Smith, D. H., Jr., & Briggle, L. W. (1987). Tolerances of oat cultivars to an acid soil high in exchangeable aluminum. *Journal of Plant Nutrition, 10*(9–16), 163–1174.

Fraire-Velázquez, S., Rodríguez-Guerra, R., & Sánchez-Calderón, L. (2011). Abiotic and biotic stress response crosstalk in plants. In *Abiotic stress response in plants—Physiological, biochemical and genetic perspectives* (pp. 3–26). Intech Open.

Francois, L. E. (1987). Salinity effects on asparagus yield and vegetative growth. *Journal of American Society for Horticultural Science, 112*, 432–436.

Francois, L. E. (1994). Growth, seed yield, and oil content of canola grown under saline conditions. *Agronomy Journal, 86*, 233–237.

Francois, L. E., Donovan, T. J., & Maas, E. V. (1992). Yield, vegetative growth, and fiber length of kenaf grown on saline soil. *Agronomy Journal, 84*, 592–598.

Francois, L. E., Donovan, T. J., Lorenz, K., & Maas, E. V. (1989). Salinity effects on rye grain yield, quality, vegetative growth, and emergence. *Agronomy Journal, 81*, 707–712.

Francois, L. E., Donovan, T. J., Maas, E. V., & Rubenthaler, G. L. (1988). Effect of salinity on grain yield and quality, vegetative growth, and germination of triticale. *Agronomy Journal, 80*, 642–647.

Francois, L. E., Maas, E. V., Donovan, T. J., & Youngs, V. L. (1986). Effect of salinity on grain yield and quality, vegetative growth, and germination of semi-dwarf and durum wheat. *Agronomy Journal, 78*, 1053–1058.

Francois, L. E., Donovan, T. J., & Maas, E. V. (1990). Salinity effects on emergence, vegetative growth, and seed yield of guar. *Agronomy Journal, 82*, 587–592.

Fried, H. G., Narayanan, S., & Fallen, B. (2019). Evaluation of soybean (*Glycine max* L. Merr.) genotypes for yield, water use efficiency, and root traits. *PLoS ONE, 14*, e0212700.

Furr, J. R., & Armstrong, W. W., Jr. (1962). A test of mature Halawy and Medjool date palms for salt tolerance. *Date Growers Institute Rpt, 39*, 11–13.

Furr, J. R., & Ream, C. L. (1968). Salinity effects on growth and salt uptake of seedlings of the date, *Phoenix dactylifera* L. *Proceedings of the American Society for Horticultural Science, 92*, 268–273.

Gale, M. (2002). *Applications of molecular biology and genomics to genetic enhancement of crop tolerance to abiotic stress: A discussion document*. Food and Agriculture Organization of the United Nations.

Gaur, P. M., Jukanti, A. K., & Varshney, R. K. (2012). Impact of genomic technologies on chickpea breeding strategies. *Agronomy, 2*, 199–221.

Getinet, A., Rakow, G., & Downey, R. K. (1996). Agronomic performance and seed quality of Ethiopian mustard in Saskatchewan. *Canadian Journal of Plant Science, 76*, 387–392.

Getinet & Sharma, S. M. (1996). *Guizotiaabyssinica (L. f.) Cass*. Institute of Plant Genetics and Crop Plant Research, Gatersleben/International Plant Genetic Resources Institute.

Ghanbari, A. A., Shakiba, M. R., Toorchi, M., & Akbar, A. (2013). Morpho-physiological responses of common bean leaf to water deficit stress. *European Journal of Experimental Biology, 3*, 487–492.

Giri, A. P., Harsulkar, A. M., Ku Maurice, S. B., Gupta, V. S., Deshpande, V. V., Ranjekar, P. K., & Franceschi, V. R. (2003). Identification of potent inhibitors of *Helicoverpaarmigera* gut proteinases from winged bean seeds. *Phytochemistry, 63*(5), 523–532.

González-Rodríguez, T., Ismael, C. H., Jonathan, A. B., Enrique, R. C., Norma, M. G., Erika, M. M., Mercedes, G. L. P., Jorge, M. T., & John, D. F. (2019). Identification of factors linked to higher water-deficit stress tolerance in *Amaranthus hypochondriacus* compared to other grain amaranths and A. hybridus, their shared ancestor. *Plants, 8*(7), 239.

Guenther, E. (1949). *The essential oils* (Vol. 3). D. Van Nostrand.

Gutierrez, N., Palomino, C., & Satovic, Z. (2013). QTLs for *Orobanche* spp. resistance in faba bean: Identification and validation across different environments. *Molecular Breeding, 32*, 909–922.

Hall, A. E. (2012). Phenotyping cowpeas for adaptation to drought. *Frontiers in Physiology, 3*, 155.

Hall, W. F. (1964). *Agriculture in India* (p. 3). Regional Analysis Division, Economic Research Service, United States Department of Agriculture. https://archive.org/details/agricultureinind64hall

Hassan, N. A. K., Drew, J. V., Knudsen, D., & Olson, R. A. (1970). Influence of soil salinity on production of dry matter and uptake and distribution of nutrients in barley and corn. I. Barley (*Hordeum vulgare*). *Agronomy Journal, 62*, 43–45.

Haussmann, B. I. G., Rattunde, H. F., Weltzien-Rattunde, E., Traor´e, P. S. C., vomBrocke, K., & Parzies, H. K. (2012). Breeding strategies for adaptation of pearl millet and sorghum to climate variability and change in West Africa. *Journal of Agronomy and Crop Science, 198*(5), 327–339.

Heslop-Harrison, J. S., & Schwarzacher, T. (2007). Domestication, genomics and the future for banana. *Annals of Botany, 100*(5), 1073–1084.

Hinojosa, L., González, J. A., Barrios-Masias, F. H., Fuentes, F., & Murphy, K. M. (2018). Quinoa abiotic stress responses: A review. *Plants, 7*(4), 106. https://www.tenthacrefarm.com/perennial-crops-wet-soil/

Huerta-Ocampo, J. A., Barrera-Pacheco, A., Mendoza-Hernández, C. S., Espitia-Rangel, E., Mock, H. P., & Barba de la Rosa, A. P. (2014). Salt stress-induced alterations in the root proteome of *Amaranthus cruentus* L. *Journal of Proteome Research, 13*, 3607–3627.

Hussain, M. I., Lyra, D., Farooq, M., et al. (2016). Salt and drought stresses in safflower: A review. *Agronomy for Sustainable Development, 36*, 4. https://doi.org/10.1007/s13593-015-0344-8

Ieshisa, J. C., & Takumi, S. (2012). Variation in abscisic acid responsiveness of *Aegilopstauschii* and hexaploid wheat synthetics due to the D-genome diversity. *Genes and Genetic Systems, 87*, 9–18.

Inukai, S., Kock, K. H., & Bulyk, M. L. (2017). Transcription factor–DNA binding: Beyond binding site motifs. *Current Opinion in Genetics & Development, 43*, 110–119.

IPGRI. (2005). Promoting fonio production in West and Central Africa through germplasm management and improvement of post-harvest technology, in *Project Number: 8600-00100*, 18, p. 00100, International Plant Genetic Resources Institute.

Ivandic, V., Hackett, C. A., Zhang, Z. J., Staub, J. E., Nevo, E., Thomas, W. T. B., & Forster, B. P. (2000). Phenotypic responses of wild barley to experimentally imposed water stress. *Journal of Experimental Botany, 51*, 2021–2029.

Jaikishun, S., Li, W., Yang, Z., & Song, S. (2019). Quinoa: In perspective of global challenges. *Agronomy, 9*(4), 176. https://doi.org/10.3390/agronomy9040176

Jamil, A., Riaz, S., Ashraf, M., & Foolad, M. R. (2011). Gene expression profiling of plants under salt stress. *Critical Reviews in Plant Sciences, 30*(5), 435–458.

Javed, T., Shabbir, R., Ali, A., Afzal, I., Zaheer, U., & Gao, S. J. (2020). Transcription factors in plant stress responses: Challenges and potential for sugarcane improvement. *Plants, 9*, 491.

Jha, U. C., Chaturvedi, S. K., Bohra, A., Basu, P. S., Khan, M. S., & Barh, D. (2014). Abiotic stresses, constraints and improvement strategies in chickpea. *Plant Breeding, 133*, 163–178.

Jia, G., Huang, X., Zhi, H., Zhao, Y., Zhao, Q., Li, W., Chai, Y., Yang, L., Liu, K., Lu, H., et al. (2013). A haplotype map of genomic variations and genome-wide association studies of agronomic traits in foxtail millet (*Setariaitalica*). *Nature Genetics, 45*, 957–961.

Ketema, S. (1997). *Tef, Eragrostis tef (Zucc.) Trotter*, vol. 52, Institute of Plant Genetics and Crop Plant Research, Gatersleben/International Plant Genetic Resources Institute.

Kettel, K. F., Tuck, B., Payne, W., Chen, C., Machado, S., & Karow, R. (2003). Narrow-leaf lupin.

Khan, H. R., Paull, J. G., Siddique, K. H. M., & Stoddard, F. L. (2010). Faba bean breeding for drought-affected environments: A physiological and agronomic perspective. *Field Crops Research, 115*, 279–286.

Khan, M. I. (2015). Plant betalains: Safety, antioxidant activity, clinical efficacy, and bioavailability. *Comprehensive Reviews in Food Science and Food Safety, 15*(2), 316–330. https://doi.org/10.1111/1541-4337.12185

Khoury, C. K., Bjorkman, A. D., Dempewolf, H., Ramirez-Villegas, J., Guarino, L., Jarvis, A., Rieseberg, L. H., & Struik, C. P. (2014). Increasing homogeneity in global food supplies and the implications for food security. *Proceedings of the National Academy of Sciences, 111*(11), 4001–4006.

Kim, B. Y., Baier, A. C., Somers, D. J., & Gustafson, J. P. (2001). Aluminum tolerance in triticale, wheat, andrye. *Euphytica, 120*, 329–337.

Kumar, A., Sharma, D., Tiwari, A., Jaiswal, J. P., Singh, N. K., & Sood, S. (2016). Genotyping-by-sequencing analysis for determining population structure of finger millet germplasm of diverse origins. *Plant Genome*. https://doi.org/10.3835/plantgenome2015.07.0058

Kumar, S., Prasad, A. K., Iyer, S. V., & Vaidya, S. K. (2013). Systematic pharmacognostical, phytochemical and pharmacological review on an ethno medicinal plant *Basella Alba* L. *Journal of Pharmacognosy and Phytotherapy, 5*, 53–58.

Langdale, G. W., & Thomas, J. R. (1971). Soil salinity effects on absorption of nitrogen, phosphorus, and protein synthesis by coastal bermudagrass. *Agronomy Journal, 63*, 708–711.

Lehman, V. G., & Engelke, M. C. (1993). Heritability of creeping bentgrass shoots water content under soil dehydration and elevated temperatures. *Crop Science, 33*, 1061–1066.

Lei, G. J., Yokosho, K., & Yamaji, N. (2017). Functional characterization of two half-size ABC transporter genes in aluminium-accumulating buckwheat. *New Phytologist, 215*(3), 1080–1089.

Lemmon, Z. H., Reem, N. T., Dalrymple, J., Soyk, S., Swartwood, K. E., Rodriguez-Leal, D., Van Eck, J., & Lippman, Z. B. (2018). Rapid improvement of domestication traits in an orphan crop by genome editing. *Nature Plants, 4*(10), 766–770. https://doi.org/10.1038/s41477-018-0259-x

Li, Y. L., Ruperao, P., Batley, J., Edwards, D., Khan, T., Colmer, T. D., Pang, J. Y., Siddique, K. H. M., & Sutton, T. (2018). Investigating drought tolerance in chickpea using genome-wide association mapping and genomic selection based on whole-genome resequencing data. *Frontiers in Plant Science, 9*, 190. https://doi.org/10.3389/fpls.2018.00190

Linares, O. F. (2002). African rice (*Oryza glaberrima*): History and future potential. *Proceedings of the National Academy of Sciences of the United States of America, 99*(25), 16360–16365.

Liu, M., Sun, W., Ma, Z., Zheng, T., Huang, L., Wu, Q., Zhao, G., Tang, Z., Bu, T., Li, C., & Chen, H. (2019). Genome-wide investigation of the AP2/ERF gene family in tartary buckwheat (*FagopyumTataricum*). *BMC Plant Biology, 19*, 1–19.

Mabhaudhi, T., Chimonyo, V. G. P., Hlahla, S., Massawe, F., Mayes, S., Nhamo, L., & Modi, A. T. (2019). Prospects of orphan crops in climate change. *Planta, 250*, 695–708. https://doi.org/10.1007/s00425-019-03129-y

Maejima, E., & Watanabe, T. (2014). Proportion of phospholipids in the plasma membrane is an important factor in Al tolerance. *Plant Signaling & Behavior, 9*, e29277.

Massange-Sanchez, J. A., Palmeros-Suarez, P. A., Martinez-Gallardo, N. A., Castrillon-Arbelaez, P. A., Avilés-Arnaut, H., Alatorre-Cobos, F., Tiessen, A., & Délano-Frier, J. P. (2015). The novel and taxonomically restricted *Ah24* gene from grain amaranth (*Amaranthus hypochondriacus*) has a dual role in development and defense. *Frontiers in Plant Science, 6*, 602. https://doi.org/10.3389/fpls.2015.00602

Massange-Sanchez, J. A., Paola, A. P., Eduardo, E. R., Isaac, R. A., Lino, S. S., Norma, A. M. G., Fulgencio, A. C., Axel, T., & John, P. D. F. (2016). Overexpression of grain amaranth (*Amaranthus hypochondriacus*) AhERF or AhDOF transcription factors in Arabidopsis thaliana increases water deficit-and salt-stress tolerance, respectively, via contrasting stress-amelioration mechanisms. *PLoS ONE, 11*(10), e0164280.

Mayes, S., Massawe, F. J., Alderson, P. G., Roberts, J. A., Azam-Ali, S. N., & Hermann, M. (2012). The potential for underutilized crops to improve security of food production. *Journal of Experimental Botany, 63*(3), 1075–1079.

Mbow, C., Rosenzweig, C., Barioni, L.G., Benton, T.G., Herrero, M., Krishnapillai, M., Liwenga, E., Pradhan, P., Rivera-Ferre, M. G., Sapkota, T., Tubiello, F. N., & Xu, Y. (2019). Food security. In P. R. Shukla, J. Skea, E. Calvo Buendia, V. Masson-Delmotte, H. O. Pörtner, D. C. Roberts, P. Zhai, R. Slade, S. Connors, R. van Diemen, M. Ferrat, E. Haughey, S. Luz, S. Neogi, M. Pathak, J. Petzold, J. Portugal Pereira, P. Vyas, E. Huntley, K. Kissick, M. Belkacemi, J. Malley (Eds.), *Climate Change and Land: An IPCC special report on climate change, desertification, land degradation, sustainable land management, food security, and greenhouse gas fluxes in terrestrial ecosystems*. In Press.

McElgunn, J. D., & Lawrence, T. (1973). Salinity tolerance of Altai wild ryegrass and other forage grasses. *Canadian Journal of Plant Science, 53*, 303–307.

Mishra, D. N., & Shitole, M. G. (1986). Growth and yield of oat (*Avena sativa* L.) cv. Kent under Na_2SO_4 dominated saline soil. *Geobios, 13*, 253–257.

Mollins, J. (2017). Smallholder farmers to gain from targeted CRISPRCas9 crop breeding. https://www.cimmyt.org/smallholder-farmers-to-gain-from-targeted-crispr-cas9-breeding/. Accessed January 9, 2019.

Myers, S., Zanobetti, A., Kloog, I., Huybers, P., Leakey, A., Bloom, A., Carlisle, E., Dietterich, L., Fitzgerald, G., Hasegawa, T., Holbrook, N., Nelson, R., Ottman, M., Raboy, V., Sakai, H., Sartor, K., Schwartz, J., Seneweera, S., Tausz, M., & Usui, Y. (2014). Increasing CO_2 threatens human nutrition. Nature, advance online publication.

National Research Council. (2006). *Lost crops of Africa: Volume II: Vegetables*. The National Academies Press. https://doi.org/10.17226/11763

Ortiz, R., Sayre, K. D., Govaerts, B., Gupta, R., Subbarao, G. V., Ban, T., Hodson, D., Dixon, J. M., Ortiz-Monasterio, J. I., & Reynolds, M. (2008). Climate change: Can wheat beat the heat? *Agriculture, Ecosystems and Environment, 126*, 46–58.

Osaki, M., Watanabe, T., & Tadano, T. (1997). Beneficial effect of aluminum on growth of plants adapted to low pH soils. *Soil Science and Plant Nutrition, 43*, 551–563.

Padulosi, S., Heywood, V., Hunter, D., & Jarvis, A. (2011). Underutilized species and climate change: current status and outlook. In *Crop adaptation to climate change* (1st edn., pp. 507–521). Wiley.

Padulosi, S., Thompson, J., & Rudebjer, P. (2013). *Fighting poverty, hunger and malnutrition with neglected and underutilized species (NUS): Needs, challenges and the way forward*. Bioversity International.

Passariello, B., Giuliano, V., Quaresima, S., Barbaro, M., Caroli, S., Forte, G., Garelli, G., & Iavicoli, I. (2002). Evaluation of the environmental contamination at an abandoned mining site. *Microchemical Journal, 73*, 245–250.

Pitan, O. O. R., & Esan, E. O. (2014). Intercropping cucumber with amaranth (*Amaranthus cruentus* L.) to suppress populations of major insect pests of cucumber (*Cucumis sativus* L.). *Archives of Phytopathology and Plant Protection, 47*(9), 1112–1119.

Polania, J., Rao, I. M., Cajiao, C., Grajales, M., Rivera, M., Velasquez, F., Raatz, B., & Beebe, S. E. (2017). Shoot and root traits contribute to drought resistance in recombinant inbred lines of MD 23–24_SEA 5 of common bean. *Frontiers in Plant Science, 8*.

Porch, T. G. (2006). Application of stress indices for heat tolerance screening of common bean. *Journal of Agronomy and Crop Science, 192*, 390–394.

Prasad, M. N. V., & Freitas, H. M. D. (2003). Metal hyperaccumulation in plants biodiversity prospecting for phytoremediation technology. *Electronic Journal of Biotechnology, 93*(1), 285–321.

Prasad, P. V. V., Boote, K. J., Jr., & Allen, L. H. (2006). Adverse high temperature effects on pollen viability, seed-set, seed yield and harvest index of grain-sorghum (*Sorghum bicolor* (L.) Moench) are more severe at elevated carbon dioxide due to higher tissue temperatures. *Agricultural and Forest Meteorology, 139*, 237–251.

Pulvento, C., Riccardi, M., Lavini, A., D'Andria, R., Iafelice, G., & Marconi, E. (2010). Field trial evaluation of two *Chenopodium quinoa* genotypes grown under rain-fed conditions in a typical Mediterranean environment in south Italy. *Journal of Agronomy and Crop Science, 196*, 407–411.

Rashid, K. Y., Bernier, C. C., & Conner, R. L. (1991). Genetics of Resistance in Faba BeanInbred Lines to Five Isolates of *Ascochyta Fabae*. *Canadian Journal of Plant Pathology, 13*(3), 218–225.

Ray, D. K., Mueller, N. D., West, P. C., & Foley, J. A. (2013). Yield trends are insufficient to double global crop production by 2050. *PLoS ONE, 8*(6), e66428. https://doi.org/10.1371/journal.pone.0066428

Reyna-Llorens, I., Corrales, C., & Poschenrieder (2015). Both aluminum and ABA induce the expression of an ABC-like transporter gene (*FeALS3*) in the Al-tolerant species *Fagopyrum esculentum*. *Environmental and Experimental Botany, 111*, 74–82.

Rosell, S., & Holmer, B. (2007). Rainfall change and its implications for Belg harvest in South Wollo, Ethiopia. *GeografiskaAnnaler: Series a, Physical Geography, 89*(4), 287–299.

Rudebjer, P., Chakeredza, S., Dansi, A., Ekaya, W., Ghezae, N., Aboagye, L.M., Kwapata, M., Njoroge, K., & Padulosi, S. (2013). Beyond commodity crops: strengthening young scientists' capacity for research on underutilized species in sub-Saharan Africa. In: F. Massawe, S. Mayes, & P. Alderson (Eds.), *Proceedings of the second international symposium on underutilized plant species: crops for the future—Beyond food security. Acta Horticulturae, 979*, 577–588. International Society for Horticultural Science.

Ruiz, K. B., Biondi, S., Martínez, E. A., Orsini, F., Antognoni, F., & Jacobsen, S.-E. (2016). Quinoa—A model crop for understanding salt-tolerance mechanisms in halophytes. *Plant Biosystems—International Journal Dealing with All Aspects of Plant Biology, 150*, 357–371.

Sandhu, G. R., Aslam, Z., Salim, M., Sattar, A., Qureshi, R. H., Ahmad, N., & Wyn Jones, R. G. (1981). The effect of salinity on the yield and composition of *Diplachnefusca* (Kallar grass). *Plant, Cell and Environment, 4*, 177–181.

Sanginga, N., Lyasse, O., & Singh, B. B. (2000). Phosphorus use efficiency and nitrogen balance of cowpea breeding lines in a low P soil of the derived savanna zone in West Africa. *Plant and Soil, 220*(1–2), 119–128.

Shannon, M. C., Wheeler, E. L., & Saunders, R. M. (1981). Salt tolerance of Australian channel millet. *Agronomy Journal, 73*, 830–832.

Soares, B. L., Ferreira, P. A. A., Oliveira-Longatti, S. M. D., Marra, L. M., Rufini, M., Andrade, M. J. B. D., & Moreira, F. M. D. S. (2014). Cowpea symbiotic efficiency, pH and aluminum tolerance in nitrogen-fixing bacteria. *Scientia Agricola, 71*(3), 171–180.

Sood, A., Jaiswal, V., Chanumolu, S. K., Malhotra, N., Pal, T., & Chauhan, R. S. (2014). Mining whole genomes and transcriptomes of Jatropha (*Jatrophacurcas*) and Castor bean (*Ricinuscommunis*) for NBS-LRR genes and defense response associated transcription factors. *Molecular Biology Reports, 41*, 7683–7695. https://doi.org/10.1007/s11033-014-3661-0

Srivastava, R., Kumar, S., & Kobayashi, Y. (2018). Comparative genome-wide analysis of WRKY transcription factors in two Asian legume crops: Adzuki bean and Mung bean. *Science and Reports, 8*, 16971.

Tadele, Z. (2016). In A. Shanker, & C. Shanker (Eds.), *Drought adaptation in millets, in abiotic and biotic stress in plants: recent advances and future perspectives* (pp. 639–662). InTech.

Tadele, Z. (2017). Raising crop productivity in Africa through intensification. *Agronomy, 7*, 22. https://doi.org/10.3390/agronomy7010022

Tadele, Z. (2019). Orphan crops: Their importance and the urgency of improvement. *Planta, 250*, 677–694. https://doi.org/10.1007/s00425-019-03210-6

Tal, M., Rosental, I., Abramovitz, R., & Forti, M. (1979). Salt tolerance in *Simmondsia chinensis*: Water balance and accumulation of chloride, sodium and proline under low and high salinity. *Annals of Botany, 43*, 701–708.

Tsuzuki, E., & Dong, Y. J. (2003). Buckwheat allelopathy: Use in weed management. *Allelopathy Journal, 12*, 1–11.

Tyczewska, A., Woźniak, E., Gracz, J., Kuczyński, J., & Twardowski, T. (2018). Towards food security: Current state and future prospects of agrobiotechnology. *Trends in Biotechnology, 36*(12), 1219–1229.

US Department of Agriculture—Economic Research Service, Regional Analysis Division. (1963). The 1964 Far East, Communist China, Oceania Agricultural Situation: Supplement No. 4 to the 1964 World Agricultural Situation (pp. 49–50).

Wahid, A. (2007). Physiological implications of metabolite biosynthesis for net assimilation and heat-stress tolerance of sugarcane (*Saccharum oicinarum*) sprouts. *Journal of Plant Research, 120*, 219–228.

Williams, J.T., & Haq, N. (2000). Global research on underutilized crops: an assessment of current activities and proposals for enhanced cooperation (p. 50). International Centre for Underutilized Crops (ICUC).

Wu, Q., Bai, X., & Zhao, W. (2017). De novo assembly and analysis of tartary buckwheat (*Fagopyrum tataricum* Garetn.) transcriptome discloses key regulators involved in salt-stress response. *Genes, 8*(10)

Xin, Z., &Browse, J. (2001). Cold comfort farm: The acclimation of plants to freezing temperatures. *Plant Cell & Environment, 23*, 893–902.

Yadav, A. K., Arya, R. K., Singh, M. K., Kumar, D., & Panchta, R. (2016). Heat tolerance in pearl millet: A Review. *Forage Research, 42*(2), 65–81.

Yadav, S. K. (2010). Heavy metals toxicity in plants: An overview on the role of glutathione and phytochelatins in heavy metal stress tolerance of plants. *South African Journal of Botany, 76*(2), 167–179.

Yadav, S. S., Redden, R., Chen, W., & Sharma, B. (Eds.). (2007). *Chickpea breeding and management* (pp. 474–496). CAB International.

Yang, A., Akhtar, S. S., Amjad, M., Iqbal, S., & Jacobsen, S. E. (2016). Growth and physiological responses of quinoa to drought and temperature stress. *Journal of Agronomy and Crop Science, 202*, 445–453.

Yermanos, D. M., Francois, L. E., & Tammadoni, T. (1967). Effects of soil salinity on the development of Jojoba. *Economic Botany, 21*, 69–80.

Zhang, S.K., Chen, X., Lu, C., Ye, J.Q., Zou, M.L., Lu, K.D., Feng, S.B., Pei, J.L., Liu, C., Zhou, X.C., Ma, P.A., Li, Z.G., Liu, C.J., Liao, Q., Xia, Z.Q., & Wang, W.Q. (2018). Genome-wide association studies of 11 agronomic traits in cassava (*Manihot esculenta* Crantz). *Frontiers in Plant Science*. https://doi.org/10.3389/fpls.2018.00503

Chapter 7
Exciting Journey: Our Past, Present and Future in the Horrific Light of Climate Change

Abhisek Saha

7.1 The Biological History of the Human Species

The human species started its survival someplace in the scope of 110,000 and 50,000 years earlier. More than 5,000,000 years 3,000,000 years back, a piece of these primates formed into Australopithecus. Around 2,000,000 years earlier, another piece of Australopithecus transformed into our ancestor, Homo Habilis, which suggests "handyman". Around one million years earlier came our very own closer forerunner, Homo Erectus. Around 100,000 years earlier, Homo Sapiens enters the scene, back in Africa (Ward, 2006).

7.2 In Just Two Seconds, an Extraordinary Change!

The globe has been ready for sustaining life for around three billion years prior and the capacity to sustain life for around 5 billion years later on. Subsequently, if we see the time of having the capacity to help life as twenty-four hours period, at that point humans are presently about 9'o clock in the initial segment of the day, individuals isolated from the chimps about a minute prior and the human species began to survive one to two seconds as well.

A. Saha (✉)
Department of Chemistry, Tufanganj College (Affliated to Coochbehar Panchanan Barma University), Cooch Behar, West Bengal 736160, India
e-mail: abhisek.nbu@gmail.com; abhisek@tufanganjmahavidyalaya.net

S. Sheraz Mahdi et al. (eds.), *Adapting to Climate Change in Agriculture-Theories and Practices*, https://doi.org/10.1007/978-3-031-28142-6_7

7.3 Why, and What Will Replace the Human Species?

Reliably, an enormous number of creature assortments stop to exist and a colossal number of fresh varieties are made. Will in this manner the human species stop existing and a considerable number of fresh creature assortments are made. Will thusly the humans stop to survive in two or three million years, like most various flora and fauna? Given this is valid, why, and what will displace it?

7.4 Development Develops Humans Started to Influence the Earth

We do not plummet from gorillas. We and primates were slid from basic predecessors. Chimps and gorillas and orangs and people all experienced a great many long stretches of progress to wind up our identity today. All through more than 90% of its set of experiences, Homo sapiens lived in little gatherings of searcher finders. They lived stationary lives in changeless settlements as ranchers. Agriculture had a noteworthy effect, people in progress to influence the earth as at no other time (Kating, 1988). Civic establishments rapidly emerged in old Egypt around 6000 BC, at the Indus River valleys, and in China. Beginning around 3000 BC, Hinduism, one of the majority seasoned religions still drilled today. Social development immediately outplaces biological advancement. In the fourteen century, the Renaissance began in Italy with progress in religion, craftsmanship, and science. European human advancement inciting sensible and mechanical bombshells. Change has proceeded at a high speed from the mid-1940s to now. The globe has been ready for sustaining life for around three billion years prior and the capacity to sustain life for around 5 billion years later on. Subsequently, if we see the time of having the capacity to help life as twenty-four hours period, at that point humans are presently about 9'o clock in the initial segment of the day, individuals isolated from the chimps about a minute prior and the human species began to survive one to two seconds as well (Richard & Roger, 1977).

7.5 Methodology Used for the Present Study

The proposed examination is founded on the official records and the logical methodology. For this, the fundamental sources, for instance, the files have been assembled. Also, the entire scope of writing on the likely destiny of people upon synthetic and catastrophic events has been broken down for the planning of the significant system. As for the social affair of fundamental data and essential references, the libraries and web have been used.

7.6 Innovative Improvement is Abuse for Natural Version Through a Dangerous Atmospheric Deviation?

In the present day, human headway has affected the outside of the planet. We have been expanding progressively more non-sustainable power sources like solid, liquid, and gaseous fuel. This emits all the carbon that had been taken care of the underground for a long time. This extra carbon dioxide gas is causing a general temperature change. Individuals have in like manner hacked down the colossal woodland trees that used to cover our planet. Plants and trees hold carbon dioxide and emit oxygen. Seven billion people of earth utilizing earth assets and conveying a mountain of wastes. Different engineered creations are liable for the destruction of the defensive ozone layer. Beating the rundown: Chlorofluorocarbons, human-created, harmless, and inactive in the troposphere level and the stratosphere level are photolyzed, discharging responsive chlorine gas iotas that chemically obliterate the ozone layer. A blend of lower temperatures and raised chlorine gas and bromine gas obsessions are responsible for the obliteration of the ozone layer in the upper stratosphere level, in this way, molding "hole" (Kerr, 1987).

High essentialness electromagnetic waves transmitted from the sun which is involved 100–400 nm frequency. UV outflow fuses UV-A, the less unsafe kind of UV radiation, with a frequency range between 315 and 400 nm, UV-B with a frequency of 280–315 nm, and UV-C which is primarily hazardous between 100 and 280 nm. UV-C can't arrive at our planet's surface owing to the stratospheric ozone layer's capacity to ingest it. Expanded UV-B presentation lessens, profitability by meddling with procedures of photosynthesis harms DNA, adjusts nitrogen digestion, restrains versatility. It assumes a basic job in the amphibian framework. Decomposers—retain broken down natural carbon elements and reuse over the earth, Most important makers—found at the focal point of the nourishment network. Inclined to UV-B strain represses development, meddles with systems for N_2 obsession and CO_2 obsession, elevated mortality. Versatile Strategy shows Pigmentation which retains over ninety percent of UV-B before infiltrates hereditary stuff. Macroalgae, Seagrasses are sessile which limited to development destinations and have assorted territories higher than tidal zones, intertidal zones, and few not at all offered to air. It additionally has adjusted to fluctuating sun-oriented introduction and ready to shield itself from unnecessary radiation utilizing systems of photoinhibition Mechanisms (electron transport) decline photosynthesis during exorbitant radiation. An excessive amount of ultra-violet light can bring about skin malignant growth, eye harm, for example, waterfalls, Immune framework harm, lessening in phytoplankton, harm to the different living things DNA (Neale et al., 1998). This has been found in Antarctic ice-fish that has a pressing need for shades to safeguard them from the ultra-violet light. DNA assimilates UV-B waves and alters the shape in DNA. Alteration in the DNA implies that chemicals can't "read" the DNA sequence and as a consequence transformation of cells or the cells passes on. Cells can be built up the capacity to fix the DNA sequence. Uncommon catalyst land at the harm site evacuates the harmed area of DNA and replaces it with the correct segments. DNA coding sequence to some degree strong to harm by UV-B (Britt, 2000).

Trials were done in higher Plants to decide whether expanded UV-B is a risk to earthly flora and discovered elevated UV-B presentation induces restraint of photosynthesis. A few scientists have inferred that it isn't valid. It is hard to expose UV-B Effects. Limitations under control consider remembering huge contrasts for temperature change, precipitation effect, soil character from one year to another and in different zones. UV-B influences covered by different worries of surface plants, for example, dry spell, Drought creates enormous decreases in photosynthesis and development concealing the impacts of UV-B. Water zeroed in on plants produces a transcending gathering of leaf flavonoids giving more conspicuous UV-B affirmation. UV-B radiation can change the hour of blooming just as the number of blossoms in specific species (Allen et al., 1998).

Overexposure may build the danger of non-melanoma, risky melanoma skin disease, elevated dangers of threatening melanoma from serious burns from the sun, smother the invulnerable framework, quicken maturing of the skin because of high presentation, cause an episode of extreme rash in reasonable cleaned individuals due to photoallergy. Assurance of the skin next to sun-powered energy utilizing skin-protecting creams with sun protection factor is fundamental. Introduction, to UV-B, builds the danger of cataracts. and the main source of visual impairment. It additionally causes pterygium (A wedge-molded development over the focal cornea). So Protection is required which incorporates shades with a cent percent UV block.

7.7 Carbon Footprint and Carbon Accounting

Every single human leaving a Carbon impression when they purchase nourishment and commodities the creation of the nourishment and products likewise transmitted a few amounts of Carbon-di-oxide. An individual's, nation's, or affiliation's carbon impression can be assessed by endeavored a GHG releases examination, a genuine presence cycle assessment, or other calculative activities connoted as carbon accounting. When the size of a Carbon Footprint is known, a methodology can be conceived to diminish it by innovative advancements, vitality proficiency enhancements, better procedure, item the executives, changed green open or private acquirement (GPP), carbon catch, utilization procedures, carbon counterbalancing (Rolert, 1998).

7.8 Sustainable Agriculture

The exemplary meaning of sustainability gathered from the Brundtland report that we should address the issues of the present without compromising the limit of individuals later on to address their issues. The fundamental issue covers farming exercises that can deliver food effectively and effortlessly and beneficially without corrupting common assets. The idea of maintainable agriculture is a perplexing one that fuses various other similarly significant variables. There are significant qualities of sustainable advancement applicable to agriculture that it is a unique procedure, a worldwide idea and it is a multidimensional wonder, incorporating financial, ecological, and social measurements. The idea of manageable advancement goes past the monetary development that is ordinarily estimated in Gross Domestic Product and considers the condition of assets and ecological execution of the economy. The objectives of sustainable agriculture are to produce sheltered, solid food and non-food items because of market requests for present and future, empower suitable jobs to be produced using supportable land the board, assessing installments for open advantages gave control within biophysical requirements, and adjust to other natural objectives Provide ecological enhancements. Economical Sustainable agriculture can be broken into three sections: monetary, characteristic, and social (Hader et al., 1998).

The effect of climate change and sustainable agriculture: Environmental change could make irreversible harm to land and water biological systems and loss of creative potential in horticulture which influence the agro-natural reasonableness of yields, which may prompt expanded nuisance and sickness pervasions and will disproportionately affect destitute individuals in rustic territories where employments of the lion's share rely upon farming.

Environmental degradation due to climate change and sustainable agriculture: Agriculture could be a semi-synthetic activity that's closely associated with climate. Property agriculture is closely related to climatic, ecological, and economic factors. It's a need inside the comprehension of the effects of temperature change is the degree that agriculture will be influenced. World temperature change is an urgent space of examination in common sciences and designing, and water system assumes a huge part during which temperature change could likewise be fundamentally fundamental for policymaking. Temperature change contains an indispensable effect on water accessibility and water system necessities. Temperature change contains a significant part inside the hydrological cycle and as an outcome of water assets frameworks. Warming may prompt changes in water accessibility and request, moreover as inside the dissemination of water assets, inside the construction and nature of water utilization. The look and style technique must be adequately flexible to incorporate the prospect of and reactions to a few potential environments sway. The most factors which will impact the cost of joining temperature change into the technique square measure the measure of the plan, the trustworthiness of the expectation. Popular agriculture has advanced by soul the downside hazard of those components through the water system, the usage of pesticides and composts, and the control of hereditary assets (Van Der Mei et al., 2003). A genuine worry inside the comprehension of the effects of temperature change is the degree that agriculture will be influenced. In this manner, inside the medium and long terms, the temperature change is an additional test that horticulture should look at in gathering public food needs. Temperature change effectively affects the hydrological cycle thus, on water assets frameworks. Warming may prompt changes in water accessibility and request, similarly as inside the dispersion of water assets, inside the construction and nature of water utilization, and bother clashes among water clients. Regardless, it's reasonable to accept that the main changes inside the hydrological cycle square measure expected for the snow-ruled bowls of Alpine Europe, while yearly streamflow is most likely going to diminish over the stream bowls inside the southern piece of the area. The effect of world warming on crop water needs plays an assignment of predominant significance in surveying water system wants. Inside the most recent decade, world vegetation models are built up that epitomize the definition of physiological cycles like synthetic change, breath, happening, and soil water consumption. This may not be along these lines inside the future due to warming, and the environmental marvel. Subsequently, designers and chiefs should reliably audit concocting standards, principles, activity rules emergency courses of action, and water the board arrangements. Vulnerabilities on, notwithstanding, the environment can change and how water system frameworks should be constrained to adjust to those progressions square measure issues that water specialists square measure constrained to deal with. The test is to spot short approaches to manage semi-perpetual vulnerabilities. The inquiry is the thing that the best course for a task is over succeeding fifty years or a ton of, but instead, what's the best heading for succeeding not many years, realizing that a reasonable supporting methodology can empower time to be told and adjust direction (Wrachien & Goli, 2015). The look and style technique must be adequately adaptable to incorporate the possibility of and reactions to a few potential environments sway. The most factors which will impact the cost of fusing temperature change into the technique square measure the measure of the plan, the trustworthiness of the expectation. the occasion of an exhaustive methodology that coordinates these elements into water system project decision, needs more investigation on the cycles administering environment changes, the effects of collected ozone harming substance on vegetation and spillover, the aftereffect of environment factors on crop water needs, and the effects of the environment on foundation execution (Kluson, 2001).

7.9 Green Economy

Green Economy targets making issues of diminishing natural dangers and biological shortages. A healthy green economy leads to sustainable development without degrading the environment. Green Economy has three primary columns which are monetary, social, and natural.

7.10 Montreal Protocol (1987)

The leading body of pros was confined to investigate substances responsible for the whole course of action, set up procedures that thwart future use of explicit sorts of engineered substances. They specified that the age and utilization of mixes contributing towards usage of ozone in the stratosphere were to be discarded constantly in 2000 (2005 for methyl chloroform).

7.11 Indications of Recovery?

There have been a couple of signs of recovery. In 1997 satellite showed a decay of two or three known ozone-exhausting gases. Satellite pictures give some dropping down of ozone mishap. Regardless, recovery is moderate. Specialists should see the Stations that action levels of ozone and surface radiation changes like the occasion speed of skin risky turn of events and falls. More evaluations to pick characteristic impacts (counting human) on UV radiation is required. Exploration on creams and analgesics and their suitability in frustrating skin ailments and subverting melanoma is additionally required. More impression of UV-related wickedness to different species living in high degrees is in like way gigantic.

A couple of features are up 'til now vague which consolidates current models that can't copy ozone variability definitely, speeds of future augmentations in ozone harming substances are not yet settled kept seeing of ozone and ozone-exhausting substances are head. Ozone layer recovery is expected by 2050. Depends upon the total expulsion of normal ozone-draining substances and replacements for HCFCs, methyl bromide are so far being looked for, and assessments of the new compound should proceed (www.space.com/54-earth-history composition-and-atmosphere.html).

Our children will live in a circumstance extensively progressively unpleasant with less drinking water, more tempests, more floods, even more, seismic tremors, rising sea levels, progressively unlawful specialists, all the more starving individuals. Billions should leave their homes because the sea will cover some piece of their country. A plenitude of 33% of the land, the surface has been adjusted by human activities and individuals use about 20% off as a rule fundamental creation The get-together of carbon dioxide in the air has reached out by near 30% since the beginning of the bleeding edge revolt. Environmental disasters are on the ascension. Around 70% of disasters are before long atmosphere-related. Generally, 40% of the universes cultivating zone is genuinely adulterated. We are undeniably a provider.

7.12 The Cataclysmic Event in the Light of Devastating 'Ice Age'

Another basic truth is that there have been five acknowledged ice ages in earth history, with the earth experiencing the quaternary ice age in the midst of right now. The earth is correct now in an interglacial time of the Quaternary ice age, with the last cold time of the quaternary having completed about 10,000 years back with the start of

Holocene. It is beginning and end some portion of a dependable, obvious cycle, a trademark cycle that benefits typically at standard interims. Likewise, since the last ice age was completed 11,500 years earlier (Randal & Wu, 1999).

7.13 Conclusion

The earth's environment isn't an indoor controller that can be arranged to accomplish the perfect temperature. The authentic fact that human usage and will full tainting of nature is and has been unsafe to our very endurance is something that should cause phenomenal stress for us all. Shortly, one anthropogenic end circumstance exists one broad-scale volcanism or different catastrophic natural change. Both ordinary causes have happened more than once in the geologic past and there is no inspiration to consider them fantastical on what's to come. The way that most by a wide margin of the species that have existed on earth has ended up being ended and thusly, human disposal would be certain.

What's so staggering about the class Homo? How might we make sense of how to manage with our little teeth and worthless fingernails, and not using any means a lot of concealing to shield us from the sun and the infection? Regardless, the main problem is that if we suffer will it be our formative fate? What's so unprecedented about the class Homo? How might we make sense of how to get by with our little teeth and inconsequential fingernails, and for no situation much conceal to shield us from the sun and the infection? What are the consequences of rising development in the coming ages? Or on the other hand, will mass devastation like a nuclear war or a space rock crash or an interplanetary war squashing and crash the whole human race?

A couple of assessments recommend that in another ten thousand years, individuals may have paid a genetic expense contingent upon development. Because of the maltreatment and over-dependence on gadgets to address each issue, we would come to look like restrained animals!! Individuals will continue progressing by methods for customary basic assurance over countless years and individuals will ceaselessly change into at any rate one new creature gathering. Men following a million years to be either unitarns, survivalistians, Numans, cybongs, or Astrans. Regardless of whether we will wrap up being one of these or advance into something different. No one but time can tell. Advancement is continuing. Environmental change is expected to critically influence both domesticated animals and harvest creation frameworks in most locales, even though there are a few nations that may truly advance modifying conditions. The changing environment is additionally tallied to source issues, like contamination, water deficiency, and soil quality corruption. The primary direct rural GHG discharges incorporate Nitrous oxide outflows from soils, composts, fertilizer, and pee from touching creatures; and Methane gas creation by ruminant creatures and from paddy rice development. Both of these gases have a fundamentally higher warming potential than Carbon dioxide. Governments should make certain the specification and engendering of fitting and current data on asset use capability and hazard management to help ranchers and other private specialists. Escalation admission to information and hazard management systems is a contribution to expanding the execution of maintainable and innovative unique practices.

References

Allen, D. J., Nogues, S., & Baker, N. (1998). Ozone depletion and increased UV-B radiation: Is there a real threat to photosynthesis? *Journal of Experimental Botany, 49*(328), 1775–1788.

Britt, A. B. (2000). Plant biology: An unbearable beating by light? *Nature, 406*, 30–31.

Hader, D. P., Kumar, H. D., Smith, R. C., & Worrest, R. C. (1998). Effects on aquatic ecosystems. *Journal of Photochemistry and Photobiology. B: Biology, 46*, 53–68.

Kating, J. F. (1988). Runaway and most greenhouse atmosphere and the evolution of earth and Venus. *Icanus, 74*(3), 472–494.

Kerr, R. (1987). Winds, pollutants drive ozone hole. *Science, 238*, 156–159.

Kluson Robert, A. (2001). *Sustainable agriculture definitions and concepts.* Agriculture/Natural Resource Extension.

Neale, P. J., Davis, R., & Cullen, J. (1998). Interactive effects of ozone depletion and vertical mixing on photosynthesis of Antarctic phytoplankton. *Nature, 392*, 585–589.

Randal, W. J., & Wu, F. (1999). Cooling of Arctic and Antarctic polar stratosphere due to depletion. *Journal of Climate, 12*, 1467–1479.

Richard, E. L. & Roger, L. (1977). *Origins: The emergence and evolution of our species and it is possible further*. ISBN 0-525-48013-7

Rolerts, N. (1998). *The Holocene: An environmental history* (2nd ed.). Willey Blackwell. ISBN 0631186387.

Van Der Mei, I. A., Ponsonby, A. L., Dwyer, T., Blizzard, L., Simmons, R., Taylor, B. V., Butzkueven, H., & Kilpatrick, T. (2003). Past exposure to sun, skin phenotype and risk of multiple sclerosis: A case-control study. *British Medical Journal, 327*, 316–322.

Ward, P. D. (2006). *Out of thin air, dinosaurs, birds, and earth's ancient atmosphere.* National Academies Press. ISBN 0309100615.

Wrachien, D. D., & Goli, B. (2015). Global warming effects on irrigation development and crop production: A worldwide view. *Agricultural Sciences, 6*, 734–747.

www.space.com/54-earth-history-composition-and-atmosphere.html. Accessed 16 April, 2020.

Chapter 8
Impact of Climate Change on Honeybees and Crop Production

Bhabani Mahankuda and Ruchira Tiwari

8.1 Diversity of Honeybee Population

There are 25,000 different types of bee species described globally and majority of them are solitary (Michener, 2007). All the identified species can be grouped under 9 families and 4000 genera (Aguiar et al., 2013). Amongst the families, the members of Apidae family viz., honeybee, bumble bee and carpenter bee are well studied. Whereas, the data on total estimate of faunal composition of honeybees in India is still lacking. Out of the total 9 families, six families are present in India (Pannure, 2016). The diversity of bee genera and species of these six families in India are presented in Table 8.1.

8.2 Pollination and Honey Bee

With every bite of food, we must thank a pollinator for their services to mankind. Pollination is one among the most crucial functions of ecosystem whereby plants and honeybees interacts resulting in sustainable crop production. Fertilization is the process of transfer of pollen from male anther to female stigma, later enabling a plant to produce seeds. Pollination is the process of mutual benefit to both plants and pollinators, whereby pollinators obtain pollen and nectar from plant as their food and transfer pollen from one plant to another, aiding them in pollination. Insects of many species viz. bees, moths, butterflies, wasps, midges, beetles and flowers visit

B. Mahankuda · R. Tiwari (✉)
Department of Entomology, College of Agriculture, GBPUA&T, Pant Nagar, Uttarakhand 263145, India
e-mail: ruchis03@rediffmail.com

B. Mahankuda
Division of Entomology, College of Agriculture, OUAT, Bhawanipatna, Odisha, India

S. Sheraz Mahdi et al. (eds.), *Adapting to Climate Change in Agriculture-Theories and Practices*, https://doi.org/10.1007/978-3-031-28142-6_8

Table 8.1 Diversity of honeybees in India

Name of the families	No of bee genera	No of bee species
Apidae	25	241
Megachilidae	28	237
Halictidae	27	194
Andrenidae	1	31
Colletidae	2	22
Melittidae	1	1

Source Gupta (2010), Ascher and Pickering (2010), Saini and Rathor (2012)

different flowers and pollinate them. Many cultivated food crops are entomophiles in nature i.e., dependant on insect for their pollination. Honeybees contribute to a major share in pollination (73%) amongst all the pollinators. Whereas, amongst the other pollinators flies and bats contributes to 19 and 6.5% of total pollination. Beetles and birds contribute less (5%) towards pollination in comparison with other pollinators. Similarly, only four percentage of total pollination was carried out by moths and pollinators individually (Abrol, 2009). Thirty five percent of global food production is aided by animal pollination through production of fruits, vegetables and seeds in 87 major crop plants across the world (FAO, 2009).

Initially honey bees were valued only as honey gatherers, as well as used for beekeeping and production of bee wax and other products. Koelreuter (1761) first identified the value of honeybees as pollinator. Presently, farmers are only managing 11 Apis species out of the total 20,000–30,000 bee species across the globe, with *Apis mellifera* (Italian honeybee), as the vital one amongst them (Parker et al., 1987). Stingless bees are also a major pollinating agent, visiting almost 90 tropical plant species (Heard, 1999). Pollination needs in modern farming systems are mostly fulfilled by honeybees. Numerous adaptations by honeybees have made them an efficient pollinator over the years. Some of the important body modifications of honeybee such as forked body hairs, presence of pollen baskets in hind legs, antennal cleaners in fore legs aids in pollination. They have specially modified mouthparts, honey combs and sacs for storing pollen as well as honey. Honeybees shows social behaviour and function as a unit. The entire bee colony overwinters during the winter and become available in a large number for foraging during the early spring, which coincide with the flowering period of many crops. Bees forage over large distance and communicate each other about the food availability through unique dance, which differentiates them from the other pollinators and makes them an excellent pollinator.

Apart from pollination, honeybees enhance the productivity of the crop with higher fruit set and increased fruit quality. Many earlier works on significant benefits of honeybees on vegetables, pulses, oilseeds, fruit and nuts have been carried out. Some of the beneficial effects of honeybee pollination in increasing quality and yield of different crops reported by researchers earlier are collected and presented in Table 8.2.

Table 8.2 Beneficial effects of honeybee pollination on different crops

S. No.	Crop	Beneficial effects of honeybee pollination	References
1	Apple	Better yield and fruit quality, reduced fruit drop	Dutta and Verma (1987), Partap et al. (2000)
2	Peach	Increased yield, reduced fruit drop	Abrol (1991), Gupta et al. (2000)
3	Plum	Increased yield, reduced fruit drop	Abrol (1991), Gupta et al. (2000)
4	Citrus	Increased yield, reduced fruit drop, increased juice content and sugar percentage	Abrol (1991), Gupta et al. (2000), Partap et al. (2000)
5	Kiwi	Increased yield	Abrol (1991), Gupta et al. (2000)
6	Strawberry	Increased yield, reduced percentage of mishappen fruit	Partap et al. (2000)
7	Cabbage, cauliflower ca, mustard	Improved seed production, quality seeds and fruits and increased yield	Abrol (1991), Partap and Verma (1992, 1994), Verma and Partap (1993, 1994)
8	Rapeseed and mustard	Increased oil quantity and quality	Deodikar and Suryanarayana (1972, 1977), Woyke (1981)
9	Asparagus, onion, carrot, turnip, lettuce	Increased yield and quality	Singh et al. (2000)

Source Abrol (2012)

8.3 Climate Change and Honeybees

In recent years, the population of our major pollinator honey bees are in a decreasing trend, which is an alarming concern for the environmentalists, entomologists as well as agricultural researchers. Intensive use of chemicals in farm lands, fragmentation, mono cropping, habitat loss, introduction of invasive species and climate change are the major factors responsible for reduction in honeybee population (Potts et al., 2010). Climate change is a defining issue of recent times. Industrialization, urbanization, production of greenhouse gases, depletion in ozone layer is contributing to increased global temperature (global warming), which results in the change of rainfall pattern, weather conditions, irregular and prolonged seasons and many more. In short, climate change can be collectively called as global warming and its effects. Honeybees are extremely sensitive to climate change. The declining trend of honeybees in different countries across the globe over many years are depicted in Fig. 8.1. Climate change has direct adverse impacts on the behaviour and physiology of honeybees. However, it alters the floral environment and crop physiology, which affects the honeybee developmental cycle indirectly (Le Conte & Navajas, 2008) which creates disturbances in the synchrony between honeybees and flowers. Pollinators, especially honeybees are major assets of our ecosystem. Declining trend in honeybee population is an alarming sign for severe disturbances in ecological balance, decrease in crop production and quality food; which are also a major threat to our economy.

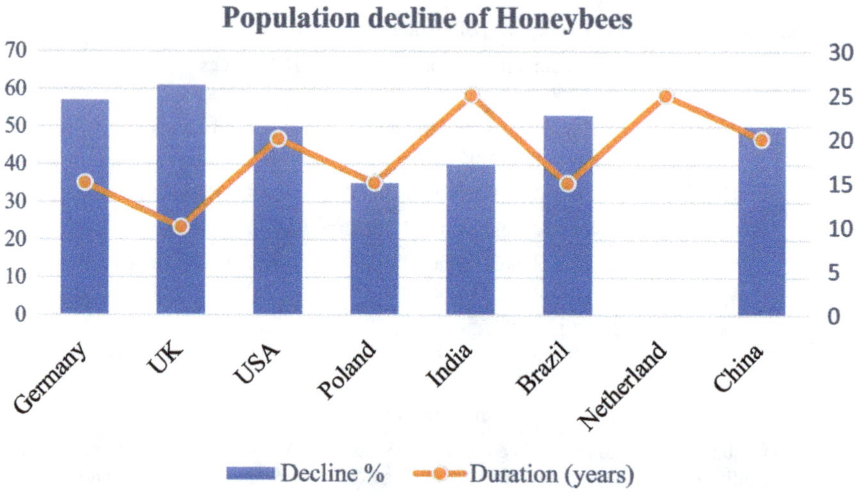

Fig. 8.1 Declining trend of honeybee population: World scenario. *Source* Gallai et al. (2009)

8.3.1 *Effect of Increasing Temperature on Honeybees*

Temperature has a vital role in regulation of developmental cycle and natural balance among honeybees. Intergovernmental panel on climate change (IPCC) suggests a rise in temperature within the range of 1.1–6.4 °C by the end of this decade. The hazardous effects of changing climate on insect depends on their temperature tolerance and flexibility to adopt temperature variations. As, they are ectothermic in nature, the surrounding temperature determines many of their activities and with the increase in temperature, colony temperature rises and creates disturbances in the natural processes like foraging, swarming and absconding. Environmental cues like maximum daily temperature, day length and number of degree days (number of days with mean temperature above certain threshold) controls the phenology of bees (Reddy et al., 2012). Every bee species possesses a particular developmental period at a particular temperature. With the variation in temperature, the cycle gets disturbed, which had measurable consequences on crop pollination. Although bees are capable of thermoregulation, small bees absorb the heat rapidly at high ambient temperature due to their high surface to volume ratio (Reddy et al., 2012). All the bee species having body weight above 35 g (*Apisbombus Megachile* and *Xylocapa*), can perform endo-thermoregulation. But the process of thermoregulation demands foraging. Increased temperature also alters the crop phenology, physiology, flowering time, flowering pattern, pollen and nectar quality, which reduces the efficacy of pollen removal and deposition among pollinators. Continuous high or low temperature for a prolonged period of time, creates adoptive pressure on bee species. With increased temperature, tropical areas may move towards more distinct seasons with prolonged dry spells. In this scenario, honey bees have to modify the honey harvesting techniques to store adequate amount of honey for prolonged dearth period or has to find out escape strategy to migrate towards a new place in accordance with seasons, flowering pattern and disruption. Similarly, Spring season is the ideal time for egg laying by queen bees, development of worker bees and colony expansion. Temperature variation during spring and prolonged winter after spring season will severely affect the activities of worker bees leading to starvation and death among honeybees during dearth period. Hence, with each degree rise in temperature because of global warming, we are losing a good share of honeybees, whose consequences will be very much unlikely. Therefore, there is an alarming concern to study the thermal tolerance of different *Apis* pollinators as well to create synchrony between flowering plants and pollinators.

8.3.2 Effect of Increased Carbon Dioxide (CO_2) Concentration on Honeybees

No direct impact of elevated CO_2 concentration on honeybee has been studied so far. However, increase in the level of carbon dioxide has impact on the plants, which have indirect effects on pollinators and honeybees. Increased CO_2 level change the carbon and nitrogen ratio in plants, which alters the nectar composition (Rusterholz & Erhardt, 1998). Elevated concentration of CO_2 in atmosphere have direct impact on the plant community structure, which alters the number of C3 and C4 plants in a given habitat (Bazzaz, 1998). However, the direct impacts of elevated CO_2 on honeybees are still uncertain. More researches are required in this field to establish a direct relationship between elevated CO_2 and honeybees.

8.3.3 Effect of Changing Climate on Health and Diseases of Honey Bees

A steady rise in the population of certain mites and gut parasites extremely harmful to the honey bees has been recorded due to warming weather conditions. The changing scenario in climate is resulting migration among honeybees to more suitable place for their survival. All the diseases and pathogens affecting honey bees are not distributed worldwide. The pathogen races vary from region to region. Various diseases and mites of honeybee having worldwide distribution are *Varroa destructor* affecting both *Apis mellifera* and *Apiscerana*; bacteria causing American foul brood and European foul brood; *Nosema apis* and *Nosema cerena*; numerous viruses affecting *Apis mellifera* (Le Conte & Navajas, 2008). Whereas, other pathogen like *Tropilaelaps* has very limited distribution. It has been found only in Asia (Sammataro et al., 2000). The virulence of these pathogens varies among different haplotypes. Changing climatic situation leads to transfer of these haplotypes into honeybee population and leads to movement of honey bees from one geographical region to other. This will cause the distribution of new species of honeybee as well as pathogen races to newer places and it may cause encounter of pathogen and honeybees, which have never co-evolved. Similar example was recorded in case of *Varroa destructor* and *Apis mellifera,* where two haplotypes of this parasite has totally invaded into all the *A. mellifera* distributed regions across the world. Such interaction among the pathogens and honeybees were found catastrophic in history. Climate change may influence different pattern of migration among honeybee, which will also increase the risk of pathogen distribution to newer areas having severe detrimental effects against the native bee species. However, the variation in climate will lead to changes in the developmental cycle of honeybees. Therefore, there may be change in the interaction pattern among honeybee and pathogens. *Tropilaelaps* mite is an interesting example to quote here. Generally, the mite does not attack European honeybees; *Apis mellifera* yet, as their developmental cycle includes a period without brood; upon which the pathogens rely for their multiplication (Sammataro et al., 2000). Whereas, warmer winter due to climate change may induce continual brood development in case of *A. mellifera* and may result in the honeybee species more suitable for the pathogen. The colony collapse disorder and higher mortality rate among the honeybees are a clear warning of their sensitiveness to changing climatic scenario. The condition may get worse day by day with the introduction of new pathogen to a particular area. Human intervention and aid to honeybees with appropriate control measures for different diseases are required to save the population of bee species. The impact of climate change on honey bees at different levels is given in Fig. 8.2.

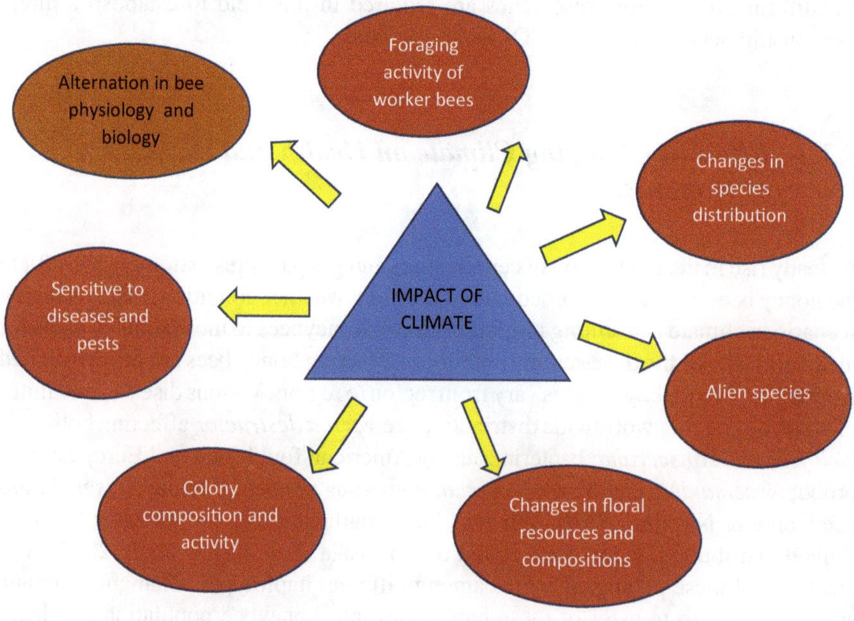

Fig. 8.2 Multilevel impacts of climate change of honeybees. *Source* Reddy et al. (2012)

8.4 Impact of Changing Climate on Crop Production

The prevailing climate in a particular geographical area determines the diversity of fauna and flora in that habitat. Growth and development of crops relay on the climatic situation of that geographical location. Various biological processes i.e., development of flowers, pollen and nectar production, which plays a key role in determining the foraging activity and development of pollinators are climate dependant and requires a specific set of climatic condition. Any deviation will influence the crop production directly and development of pollinators indirectly (Winston, 1987). Heavy rain affects the interaction among flowers and pollinators. When Acacia flowers are washed in rain, their nectar gets diluted and they don't remain attractive to the honeybees. Similarly, an excessive dry climate reduces the nectar of lavender flowers. The honey dew produced by the stinging insect are climate dependant. Growth and development of honeybee broods are dependent on pollen excessively. However, excessive dry climate can cause pollen shortage among the entomophilous crops and their quality may get reduced, making it nutritionally deficient for honeybees. Mean temperature and photoperiod during the growing season of plant determines its growth and development (Nigam et al., 1998). Rising global temperature will make the growing season longer (Rosenzweig et al., 2007). Extreme weather and drought during the crop growing season will affect severely during the anthesis period (Wheeler et al., 1999). Decreased number of flowers and inflorescence in annual garden spice legume *Trigonella was* reported under controlled drought conditions (Akhalkasti & Lösch, 2005). Drought may make flowers less attractive to pollinators with reducing the concentration of attractant, which may ultimately lead to reduction in pollination level and deterioration in quality and quantity of seeds (Philipp & Hansen, 2000; Kudo & Harder, 2005). The impact of changing plant growth pattern on pollination and pollinators are less studied. However, a number of studies conclude that changing climatic trends severely curtail plant growth, reduced fruit set, decrease in seed set and ultimately lower yield. Similarly, poor growth and flower set will ultimately reduce the number of flower and will alter the quality of flower constituents, pollen and nectar quality, which are detrimental to the pollinators. Extremely drought condition, condition due to climate change is lowering the productivity of crops, making the soil dry, unhealthy, saline and infertile. All these deviation in the plant as well as soil will reduce the crop production and yield with passing years, making more no of barren lands and a higher share of global population striving for food and dying in hunger.

Crop production and pollination are directly proportion. Entomophilous crops rely on insect for their pollination. Climate change and global warming is having severe detrimental effects against pollinators and honeybee, which will lead to decrease in the level of pollination. Reduced pollination will decrease the quality and quantity of food produced by the crop and decline in crop yield. On the other hand, reduced quality of pollen and nectar produced by the plant will disrupt the developmental cycle of honeybees and starve them to death. Therefore, pollinators and crops are correlated with each other. Deviation in any will harm the other. More researches are required to study the effects of changing climatic conditions on the pollinators, pollinating crops to develop synchrony among them as well as to restore the crop productivity.

8.5 Expected Mismatch Between Crop and Pollinators

The harmful outcomes of climate change on plant fitness and persistence is becoming more apparent with different researches over the years. The major reasons for crop and pollinator mismatch include spatial and temporal mismatches among crop and pollinator, morphological changes in plants and pollinators and the changes in foraging behavior and level of host attraction. Amongst all the factors cited above, major researches in context to pollinating crops and honeybees has been carried out related to temporal and spatial mismatches as shown in Fig. 8.3.

Crop pollination is a biological process which is synchronous with biological timing of both the crops and pollinators. Changing climatic trends may have significant effects on the crops like mango, litchi and coffee which undergoes mass blooming over a short period. They may shorten or delay the blooming period, which will ultimately stress the developmental cycle of pollinators. Insects and crops respond differently to the changing climatic conditions causing spatial and

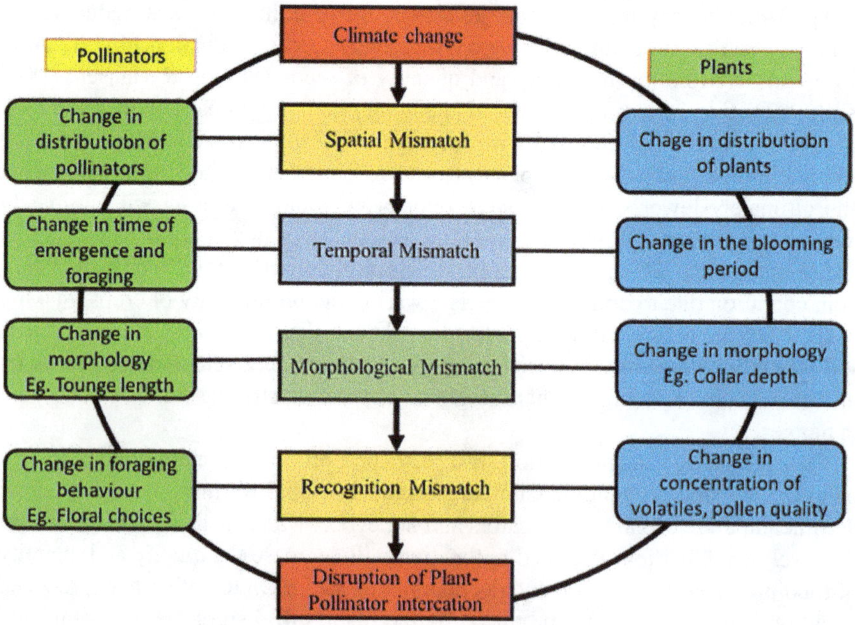

Fig. 8.3 Expected mismatches between pollinators and plants. *Source* Gerard et al. (2020)

temporal mismatches. For example, increased spring temperature postpone the flowering time in crops and don't have any effects on the pollinators. Mismatches among the pollinator and crop will result in less visitation and reduced pollination in the crop. Whereas, pollinators will have reduced food supply. Memmott et al. (2007) reported that the changes in the phenology of floral sources reduces the pollinator services by 17–50%. Reddy et al. (2012) reported a lesser number of pollen collector bees with increase in temperature but increased with increase in relative humidity. Peak foraging activity was recorded during 8:00 AM–11:00 AM. Similarly, reduced foraging activity of *Apisflorae* on mango crop was noticed at a temperature beyond 32 °C.

Honeybees build up sufficient honey store during the spring season to survive the dearth period during winter. Change in climate results in the changes in distribution of flower; a major source of honeybee food (Thuiller et al., 2005). Prolonged dry climate and excessive heat reduces pollen quality and production, which negatively affect the bees present over that region (Stokstad, 2007). Bees usually store sufficient honey during the spring season to survive the dearth period during winter. The developing bee brood requires pollen rich diet for their further growth. However, pollen shortage due to autumn drought will create food shortage for the developing broods, lower immunity and increased susceptibility to pathogens and predators. In the tropic region, global warming will cause prolonged dry period leading them towards a distinct season thorough out the year. In that situation, Asian honeybees need to change their strategy for honey harvesting to store surplus amount of honey to maintain their colony during the dearth period. Otherwise, they could develop migration strategy to move towards a favorable area having suitable climate for growth and abundant supply of food.

8.6 Impact of Climate Change on Pollination, Crop Production and Food Security

One of every three bite of food eaten worldwide relies on services of pollinators especially honeybees. The production of agricultural crops, mostly fruits and vegetables are dependent on insect pollination. The annual economic value of crop pollination has been estimated to be $156 billion across the globe (Gallai et al., 2009). Nearly 75% of crop plant show increased yield due to insect visitation. Pollinators play a vital role in our eco system. They pollinate the crops as well as maintain the diversity in eco system. Amongst the wild plants 80% are insect pollinated. The population of obligate cross-pollinated plants are solely dependent on insect pollination. Thus, wild plants are also at a higher risk due to pollinator loss (Spira, 2001).

With the changing trends of climate and global warming, the pollinators especially honeybees are at a greater risk. With an increase of 1.5–2.5 °C in global temperature, approximately 20–30% of known species are at risk, which may undergo extinction (Rathee and Dalal, 2017). Due to heavy rain and temperature in Mexico, flower buds came in mango resulting in small seedless mangoes. According to International Centre for Integrated Mountain Development, the productivity of apple reduced drastically due to inadequate pollination. Reduced yield in crop production was observed in western countries depending fully on artificially managed honeybee colonies. Staple crops like cereals are mainly wind pollinated and are at a lower risk. But fruits and vegetables contributing major share of proteins, vitamins and minerals are entomophilus in nature and are at a higher risk. Globally, the vulnerability of yield loss due to declining population of honeybees are high for fruits, vegetables, oil crops, nuts, pulses and oils with 23, 12, 16, 31, 4 and 3% respectively. However, the associated risk is almost nil for cereal crops, root crops and tuber crops (Gallai et al., 2009). The values of vulnerability are heterogeneous and vary from region to region. Some regions may possess more risk of pollinator less than others.

The role pollinators specifically honey bees for pollination, crop production and food security are vital. With the growing human population and tremendous pressure of feeding to those millions, the role of pollinators for food security and the possible outcomes of climate change on honeybees and pollinators can't be neglected. Major researches on the diversity of pollinators, their community composition and role of changing climatic parameters on the behavior, physiology and distribution of honeybee is still lacking. Lack of awareness among the farmers and growers regarding the possible impacts of global warming and conservation practices of honeybees are disheartening. Timely measures and appropriate measures should be taken at individual and organizational level to conserve the honeybees for restoring the bio diversity, food quality, nutrition and for food security. The climate smart agriculture has gained attention in recent times among scientific fraternity but more research on a large scale should be carried out to generate a good amount of data as well to formulate suitable conservational strategies. In India, climate change needs to be taken with great concern, since this phenomenon can reduce the crop production to a great extent and India has the 2nd largest population to feed on. Hence, to ensure higher crop production and better nutrition for all, strategies and policies for conservation, restoration and augmentation of honey bees and other pollinators must be developed and implemented for safeguarding them against the disastrous effects of climate change as latter cannot be slowed down but only impacts can be mitigated.

8.7 Global Initiatives for Honeybee Conservation

The western world faced a severe reduction in American and European bee population as a result of climate change during 2007 due to colony collapse disorder. Concerns about global pollinator decline gained momentum after this incident. Globally, the importance of pollinators has been identified by Convention on biological diversity and established an International Initiative, facilitated and coordinated by FAO (otherwise known as the International Pollinators Initiative-IPI), for conserving their population and sustainable use during 2000. This initiative aims at identifying the bee keeping practices and capacity building in the area of pollinator services by formulating projects across seven countries (including India). Various initiative like NAPPC (North American Pollinator Protection Campaign), European Pollinator Initiative (EPI), Brazilian Pollinator Initiative (BPI), Canadian Pollinator Conservation, 2013 were taken in developed counties. However, different initiatives, viz.; African Pollinator Initiative, Pollinator Conservation Action Plan for Sri Lanka were adopted by developing countries like Africa and Srilanka respectively.

8.8 Mitigating Measures Against Climate Change for Honey Bees

Human aids for honeybees are required to mitigate the ill effects of climate change on honeybees. Provision of alternate food sources for the honey bees during dearth period can be done. Promotion of more flowering plants along with the food crops will help to augment the population of honeybees in agro-ecosystem. Conservational method to protect the natural habitat of wild honeybees should be promoted. Appropriate management practices for various honeybee pest and diseases should be taken during the favourable seasons for disease development. Ensuring better harmony of natural habitats with farmlands will allow easy dispersion and aid in range shifts with respect to changing climatic conditions (Reddy et al., 2012).

8.9 Conclusion

The hazardous effects of climatic change to ecosystem are an alarming sign to protect our ecosystem as well to bring balance in biodiversity. Climate change is a global phenomenon. Although the detrimental effects of climate change on pollinators have been realised, scientific literatures pertaining to impact of climate change to pollinators, specifically to honeybees are lacking. Tropical climatic zones believed to have the major share of entomophilous crop but the impact of increasing temperature to those crops and their pollinators are yet to be studied. European honeybee species, *Apis mellifera* is believed to be more adoptable to diverse climatic zones. Therefore, it may use its genetic variability to adopt the changing climatic scenario. Whereas, the Asian honeybees are only restricted to Asia and may face the severe drawbacks of climatic change. Human intervention in the form of bee keeping will be helpful in conserving and augmenting the population of domesticated honeybees. Drastic decline in honeybee population due to colony collapse disorder in recent times realised the impact of climate change o pollinator shortfall. Very few species of honeybee in relation to pollination biology were studied in India. The knowledge about right pollinators for many of the crops are still lacking. Distribution of honeybee across different parts of the world, their relation with different pollinating crops and the phenomenon involved in context to global climatic change are still lacking. It is very difficult to access the magnitude of loss to our pollinators until we identify and understand their biology properly. Therefore, before formulating the management strategies, more number of research works has to be carried out to understand the pollination biology of honeybees and various other pollinators in relation to varied crops under diversified climatic conditions to access the impact of changing climate on them.

References

Abrol, D. P. (1991). Conservation of pollinators for promotion of agricultural production in India. *Journal of Animal Morphology and Physiology, 38*(2), 123–139.

Abrol, D. P. (2009). Plant-pollinator interactions in the context of climate change—An endangered mutualism. *Journal of Palynology, 45*, 1–25.

Abrol, D. P. (2012). Honeybee and crop pollination. In *Pollination biology* (pp. 85–110). Springer.

Aguiar, A. P., Deans, A. R., Engel, M. S., Forshage, M., Huber, J. T., Jennings, J. T., Johnson, N. F., Lelej, A. S., Longino, J. T., Lohrmann, V., Mikó, I., Ohl, M., Rasmussen, C., Taeger, A., & Yu, D. S. K. (2013). Order Hymenoptera. *Zootaxa, 3703*(1), 51–62.

Akhalkatsi, M., & Lösch, R. (2005). Water limitation effect on seed development and germination in *Trigonella coerulea* (Fabaceae). *Flora, 200*, 493–501.

Ascher, J. S., & Pickering, J. (2010). Apoidea species Guide. Available from https://www.discoverlife.org/mp/20q?guideApoideaspeciesguide.

Bazzaz, F. A. (1998). Tropical forests in a future climate: Changes in the biological diversity and impact on the global carbon cycle. *Climatic Change, 39*, 317–336.

Deodikar, G. B., & Suryanarayana, M. C. (1977). Pollination in the service of increasing farm production in India. *Advances in Pollen-Spore Research, 2*, 60–82.

Deodikar, G. B., & Suryanarayana, M. C. (1972). Crop yields and bee pollination. *Indian Bee Journal, 34*, 53–64.

Dutta, P. C., & Verma, L. R. (1987). Role of insect pollinators on yield and quality of apple fruit. *Indian Journal of Horticulture, 44*, 274–279.

FAO. (2009). *Global action on pollination services for sustainable agriculture. Food and Agriculture Organization of the United Nations*. Vialedelle Terme di Caracalla, Rome, Italy.

Gallai, N., Salles, J. M., Settele, J., & Vaissi, B. E. (2009). Economic valuation of the vulnerability of world agriculture confronted with pollinator decline. *Ecological Economics, 68*, 810–821.

Gerard, M., Vanderplanck, M., Wood, T., & Michez, D. (2020). Global warming and plant–pollinator mismatches. *Emerging Topical Life Sciences, 4*(1), 77–86.

Gupta, J. K., Rana, B. S., & Sharma, H. K. (2000). Pollination of kiwifruit in Himachal Pradesh. In M. Matsuka, L. R. Verma, S. Wongsiri, K. K. Shrestha, & U. Partap (Eds.), *Asian bees and beekeeping: Progress of research and development. Proceedings of the fourth international conference*, 23–28 Mar 1998, Kathmandu (p. 274). Oxford and IBH Publishing Co. Pvt. Ltd.

Gupta, R. K. (2010) An annonated catalouge of the bees of the Indian region. beesind.com/beesind2.

Heard, T. A. (1999). The role of stingless bees in crop pollination. *Annual Review of Entomology, 44*, 183–206.

Koelreuter, J. G. (1761). *vorlaufigenachricht von einigen das geschlecht der pflanzenbetreffenden-versuchen und beobachten*. Engelmann, Leipzig, 50.

Kudo, G., & Harder, L. D. (2005). Floral and inflorescence effects on variation in pollen removal and seed production among six legume species. *Functional Ecology, 19*, 245–254.

Le Conte, Y., & Navajas, M. (2008). Climate change: Impact on honey bee populations and diseases. *Rev Sci Technical Official International Epiz, 27*(2), 499–510.

Memmott, J., Craze, P. G., Waser, N. M., & Price, M. V. (2007). Global warming and the disruption of plant-pollinator interactions. *Ecological Letters, 10*, 710–717.

Michener, C. D. (2007). *The bees of the world* (p. 953). The Johns Hopkins University Press.

Nigam, S. N., Rao, R. C. N., & Wynne, J. C. (1998). Effects of temperature and photoperiod on vegetative and reproductive growth of groundnut (*Arachis hypogaea* L.). *Journal of Agronomy and Crop Sciences, 181*, 117–124.

Pannure, A. (2016). Bee pollinators decline: Perspectives from India. *International Research Journal of Natural and Applied Sciences, 3*(5), 2349–4077.

Parker, F. D., Batra, S. W. T., & Tepedino, V. J. (1987). New pollinators for our crops. *Agricultural and Zoological Review, 2*, 279–304.

Partap, U., Shukla, A. N., & Verma, L. R. (2000). Impact of *Apiscerana* pollination on fruit quality and yield in peach and plum in the Kathmandu valley of Nepal. In M. Matsuka, L. R. Verma, S. Wongsiri, K. K. Shrestha, Uma Partap (Eds.), *Asian bees and beekeeping: Progress of research and development. Proceedings of the fourth AAA international conference, Kathmandu, 23–28 Mar 1998* (p. 274). Oxford and IBH Publishing Co. Pvt. Ltd.

Partap, U., & Verma, L. R. (1992). Floral biology and foraging behaviour of *Apiscerana* on lettuce crop and its impact on seed production. *Progress Horticulture, 24*, 42–47.

Partap, U., & Verma, L. R. (1994). Pollination of radish by *Apiscerana. Journal of Apicultural Research, 33*, 237–241.

Philipp, M., & Hansen, T. (2000). The influence of plant and corolla size on pollen deposition and seed set in *Geranium sanguineum* (Geraniaceae). *Nordan Journal of Botany, 20*, 129–140.

Potts, S. G., Jacobus, C. B., Kremen, C., Neumann, P., Schweiger, O., & William, E. K. (2010). Global pollinator declines: Trends, impacts and drivers. *Trends in Ecology and Evolution, 25*(6), 345–353.

Rathee, M., & Dalal, P. K. (2017). Impact of climate change on pollinators, pollination and food security. In *National seminar on agrometeorology for sustainable development, 12–14 Oct* (p. 68). CCS Haryana Agricultural University.

Reddy, P. V., Verghese, A., & Rajan, V. V. (2012). Potential impact of climate change on honeybees (*Apis* spp.) and their pollination services. *Pest Management in Horticultural Ecosystems, 18*(2), 121–127.

Rosenzweig, C., Casassa, G., Karoly, D. J., Imeson, A., Liu, C., Menzel, A., Rawlins, S., Root, T. L., Seguin, B., & Tryjanowski, P. (2007). Assessment of observed changes and responses in natural and managed systems. In M. Parry, O. Canziani, J. Palutikof, & P. van der Linden (Eds.), *Contribution of working group II to the fourth assessment report of the intergovernmental panel on climate change* (pp. 79–131). Cambridge University Press.

Rusterholz, H. P., & Erhardt, A. (1998). Effects of elevated CO2 on flowering phenology and nectar production of nectar plants important for butterflies of calcareous grasslands. *Oecologia, 113*, 341–349.

Saini, M. S., & Rathor, V. S. A. (2012). Species checklist of family Halictidae (Hymenoptera: Apoidea) along with keys to its subfamilies, genera and subgenera from India. *International Journal of Environmental Science and Technology, 3*(1), 134–166.

Sammataro, D., Gerson, U., & Needham, G. (2000). Parasitic mites of honey bees: Life history, implications, and impact. *Annual Review of Entomology, 45*, 519–548.

Singh, B., Gatoria, G. S., & Singh, L. (2000). Effect of priming the queen cell cups on mass queen rearing in *Apis mellifera* Linn. Colonies. *Indian Bee Journal, 63*(3), 68–71.

Spira, T. P. (2001). Plant–pollinator interactions: A threatened mutualism with implications for the ecology and management of rare plants. *Natural Areas Journal, 21*, 78–88.

Stokstad, E. (2007). The case of the empty hives. *Science, 316*(5827), 970–972.

Thuiller, W., Lavorel, S., Araujo, M. B., Sykes, M. T., & Prentice, I. C. (2005). Climate change threats to plant diversity in Europe. *Proceedings of the National Academy of Sciences of the United States of America, 102*(23), 8245–8250.

Verma, L. R., & Partap, U. (1993). *The Asian hive bee, Apiscerana, as a pollinator in vegetable seed production* (p. 52). ICIMOD.

Verma, L. R., & Partap, U. (1994). Foraging behaviour of *Apiscerana* on cabbage and cauliflower and its impact on seed production. *Journal of Apicultural Research, 33*, 231–236.

Wheeler, T. R., Craufurd, P. Q., Ellis, R. H., Porter, J. R. & Prasad, P. V. V. (1999). Temperature variability and the yield of annual crops. In *Conference on food and forestry: Global change and global challenges* (pp. 159–167). Elsevier Science.

Winston, M. L. (1987). *The biology of the honey bee*. Harvard University Press.

Woyke, H. W. (1981). Some aspects of the role of the honeybee in onion seed production in Poland. *Acta Horticulture, 111*, 91–98.

Chapter 9
Opportunities and Challenges to Mitigate the Emerging Fungal Pathogens Exposed to Adaptation Against Climate Change

R. Rajeshwari, V. Devappa, C. G. Sangeetha, and S. E. Navyashree

9.1 Introduction

The change in climate is an important environmental challenge throughout the world. Greenhouse gases viz., carbon dioxide (CO_2), methane (CH_4), water vapour (H_2O), hydro fluorocarbons (HFCs), ozone (O_3) and nitrous oxide (N_2O) entrap reflected radiation in the atmosphere (Mahato, 2014). Due to industrial uprising, there has been drastic changes in earth's climate and is estimated to cause serious problems in the near future. The surface air temperature has increased by $0.74 \pm 0.18°$ on an average over ten decades and by the end of the twenty-first century, it is assumed to rise by 1.0–3.7 °C, due to increased accumulation of greenhouse gases (Huang et al., 2017). This has direct effects on the evolution and development of organisms (Ahanger et al., 2013). In India, global warming was observed along the West coast, Central India, the interior peninsula and Northeast India (Gautam et al., 2013). The earth is experiencing climate change by 30% increase in atmospheric CO_2 and 0.3–0.6 °C rise in temperature every year (Chakraborty et al., 2000).

Climate change is attributed to increased levels of CO_2, higher temperature and rainfall (Pachauri & Reisinger, 2007). This global climate change has effects on the environment, host and pathogen which are the major elements in disease triangle

R. Rajeshwari · S. E. Navyashree
Department of Plant Pathology, College of Horticulture (UHS, Campus), GKVK Post, Bengaluru, Karnataka 560065, India

V. Devappa (✉)
Department of Plant Pathology, College of Horticulture, University of Horticultural Sciences, Sirsi, Sirsi, India
e-mail: devappav@gmail.com

C. G. Sangeetha
Department of Plant Pathology, College of Horticulture, University of Horticultural Sciences, Kolar, Kolar, India

S. Sheraz Mahdi et al. (eds.), *Adapting to Climate Change in Agriculture-Theories and Practices*, https://doi.org/10.1007/978-3-031-28142-6_9

(Legreve & Duveiller, 2010). A vulnerable host when infected by a virulent pathogen, disease develops only if the environmental conditions are conducive. The alteration in CO_2 concentrations, water availability and temperature has unfavorable effects on disease development, because of emergence of disease causing pathogens. Plant resistance pathways such as pattern-triggered immunity and effector-triggered immunity and pathogen virulence mechanisms such as production of toxins and virulence proteins have effects on humidity and temperature (Velasquez et al., 2018).

The relationship between plant diseases, climate change and food security requires international cooperation and integrated solutions, including disease management strategies to meet the food demands of the growing world population. Under changing climate, the integration of different strategies of crop protection and plant health is important to reduce crop losses caused by invasive or rapidly evolving pathogens (Chakraborty & Newton, 2011).

Change in climate offers infinite probability for the emergence of plant pathogens by dispersal of propagules to new locations. Hence, exploitation and adoption of available disease management opportunities/strategies to mitigate the emerging plant pathogens exposed to adaptation against climate change is important.

9.2 Environmental Factors Affecting Plant Disease

The most important variables in environment are temperature, water availability and atmospheric Carbon di oxide concentration. These variables can render the plant to be resistant or susceptible to a pathogen and the pathogen to be highly virulent or weakly pathogenic. The environment-host–pathogen tritrophic interactions are conducive factors for the development of disease.

9.2.1 *Temperature*

Temperature plays an important role in the growth and development of plants as well as pathogens. The increase in temperature affects the resistance and crop physiology to a disease. The change in temperature causes shift in the ecological distribution of plant pathogens (Mina & Sinha, 2008). Disease epidemics is based on several infection processes such as spore germination, survival, virulence, infection, perpetuation, rate of reproduction, spread and development of new pathovars/strains of pathogens in new areas which is interdependent on rise in temperature and duration of its exposure (Albouy et al., 2014; Garrett et al., 2011; Ladanyi & Horvath, 2010). The infection of rust pathogen, *Puccinia substriata* enhances with increase in temperature (Tapsoba & Wilson, 1997). The rust pathogen, *P. striiformis* infections occur when temperature ranges from 18 to 30 °C (Park, 1990). European wheat yellow rust pathogen has shown temperature specific adaptation causing evolution of highly virulent stripe rust isolates (Mboup et al., 2012).

The pathogens having cold, heat or desiccation tolerant surviving structures overwinters for the subsequent crop-growing season with high potential causing severe and frequent epidemics (Ma et al., 2015). The pathogen strains adapt to variable temperatures and emerge as most predominant disease-causing pathogens (Ritchie et al., 2013). New race of rust fungus, *P. striiformis* has been evolved since 2000 that are prevalent worldwide, causing disease at higher temperatures (Milus et al., 2009). The northward shift of leaf spot pathogen, *Cercospora beticola*, in sugar beet was attributable to increase in average temperature by 0.6–1 °C (Richerzhagen et al., 2011). The spread of pathogen, *Erysiphe necator* is dependent on the spore germination at temperatures, 5°–22 °C with relative humidity, 33–90% (Bendek et al., 2007). The temperature determines the diffusion and translocation of systemic fungicides (Coakley et al., 1999).

Disease epidemics are due to transient temperature shifts which are estimated to be prevalent worldwide (Fischer & Knutti, 2015). The accumulation of carbohydrate and proteins are observed in plants grown at increased temperature (Obrêpalska-Stêplowska et al., 2015). The rust resistance genes, Pg4 and Pg3 of oats, Lr217 and Lr210 of wheat, showed susceptible reaction at high temperatures (Das et al., 2016).

9.2.2 Water Availability

Majority of plant diseases are favored by conditions of rain, increased atmosphere humidity and soil moisture. Forecast models for diseases viz., late blight of potato, apple scab and other foliar diseases are based on rainfall, relative humidity and leaf dampness and the pathogens infect plants with increased moisture content (Coakley et al., 1999). The virulence and spread of the disease depends on dispersal mechanisms of the pathogen, suitability of the environment for dispersal, survival between seasons, and modification in host physiology and ecology in the new environment (Chakraborty et al., 1998).

The frequent and extreme rainfall allows moisture to be prevalent as leaf moisture and relative humidity for extended periods resulting in a condition congenial for spread of leaf and soil borne diseases such as powdery mildews, late blights and root rots (Coakley et al., 1999). The virulence of *Sclerotinia sclerotiorum* increases with highest disease development in lettuce as relative humidity surpasses 80% (Clarkson et al., 2014). The leaf wetness for 5 h is critical for the pathogen, *Magnaporthe oryzae* and *P. striiformis* for disease development (Magarey et al., 2005). Moisture in the soil plays an essential role in the interaction between host and pathogens resulting in moist environment (Mcelrone et al., 2005). Under sufficient soil moisture and increased temperature, the evapo-transpiration rate increases resulting in humid microclimate and attracts disease (Coakley et al., 1999). Soil-borne pathogens such as *Rhizoctonia*, and *Sclerotium* are more vulnerable to cause disease under high moisture conditions (Sharma et al., 2010). The dry root rot disease is favored by drought conditions (Gautam et al., 2013).

9.2.3 *Atmospheric CO_2 Concentration*

The carbon di oxide concentration is estimated to increase from 355 to 710 ppm by 2050. Higher CO_2 concentration in the environment increases water use efficiency by plants and results in more biomass production and has stimulatory effect on pathogen growth and development (Manning & Tiedemann, 1995). The changes that occur in plants due to elevated CO_2 are protein metabolism, high accumulation of carbohydrates, waxes and fibre, production of papillae and accumulation of silicon at the sites of appressorial penetration and increased mesophyll cells (Chakraborty et al., 2000). At elevated levels of CO_2, plant shows greater biomass, slower litter decomposition, partitioning of assimilates to roots and increased survival of pathogen on over wintering crop residues which intern increases the initial inoculum levels available to subsequent crops (Coakley et al., 1999).

Increased CO_2 levels leads to increased leaf thickness, leaf area, higher total leaf area/plant, more numbers of leaves, stems and branches with greater diameter (Pritchard et al., 1999). This enhances photosynthesis, water use efficiency and reduced damage from ozone (Von Tiedemann and Firsching, 2000). Dense canopy favours the incidence of rust, powdery mildew, *Alternaria* blight, *Stemphylium* blight and anthracnose diseases (Coakley et al., 1999; Manning & Tiedemann, 1995). Host resistance increases due to physiological changes that occurs under increased CO_2 concentrations (Coakley et al., 1999). The mechanism of altered stomatal opening and leaf chemistry leads to reduced infections of downy mildew (*Perenospora manshurica*), brown spot (*Septoria glycines*) and sudden death syndrome (*Fusarium virguliforme*) that target the stomata (Eastburn et al., 2010; Mcelrone et al., 2005).

9.3 Pathogen Adaptation

Plant pathogens are subjected to adaptation by three important processes: (1) the emergence of new biotypes/races within a pathogen which is endemic to an area (genetic recombination); (2) foreign pathogen races getting introduced in to a new region and (3) the natural population of new pathogen biotype/races emerge from an endemic population because of changes in host and environment (changes in plant cultural practices, fungicide applications etc.). Pathogen emerges due to introduction by spread or dispersal and gets established due to adaptation to the environment. In case of foliar pathogens, leaf wetness, temperature and relative humidity has effects on production and germination of propagules (Huber & Gillespie, 1992). Many pathogens spread into new geographic areas and comes in contact with new potential hosts at high temperatures (Etterson & Shaw, 2001). Pathogens tolerate a broad range of temperatures if they have been evolved to survive at higher latitudes. These pathogens live in cooler climates. Therefore, warming enhances the fitness and disease epidemics (Deutsch et al., 2008).

The biology of the pathogen is directly influenced by environment in changing the plant architecture, and altering the micro-environment. Pathogens that prefer humid conditions grow on plants with high leaf wetness (Huber & Gillespie, 1992). Necrotrophs are to be distinguished between biotrophic pathogens. The infections by necrotrophs are favoured by high temperatures or ozone levels that accelerate tissue death. Factors such as elevated level of CO_2 or increased temperature or drought cause manipulation in the physiology of a host species and further alters the colonization by biotrophic pathogens (Jeger & Xu, 2015).

The fecundity of pathogen increases due to elevated levels of CO_2, leading to evolution under climate change (Chakraborty & Datta, 2003). The atmospheric moisture and temperature govern the speed of reproduction of many pathogens (Caffarra et al., 2012). Climate change influences the sexual reproduction of pathogens and increases the evolutionary potential of pathogens (Legler et al., 2012). Most pathogens have shorter generation period and has greater ability to disperse through wind (Davis et al., 2005). The arrival of new pathogens has been attributed to changing climatic conditions (Bebber et al., 2013), trade movements (Bebber, 2015), host shifts (Santini & Ghelardini, 2015) and lack of host resistance, all contributing towards pathogen harboring crops and ecosystems (Bebber et al., 2014).

9.4 Plant and Pathogen Battle Under Climate Change

Pathogens and plants interact in association. Disease triangle envisages the tritrophic interaction between plant, pathogen and the environment. A virulent pathogen, a susceptible host and a favorable environmental condition favors the disease. Lack of any of these factors does not cause disease (Stevens, 1960). Plant disease outbreaks are due to complex interactions between many factors. The advancement of a virulent strain within a variable population, non availability of appropriate resistance in host to newly developed aggressive strain, plant physiology, cropping system, limited antagonistic activities of nonpathogenic populations and weather conditions play an important and inter-related role (Pangga et al., 2013).

The climate change influences disease phenology in the forefront. Firstly, the development and availability of inoculum enhances under higher temperatures and/or elevated CO_2. This leads to higher levels of penetration, infection and evolution of highly virulent strains (Chakraborty & Datta, 2003). Secondly, the expression of plant resistance traits is influenced by climate change in a positive or negative way. Breeding for disease resistance is lengthy, time consuming and resistant varieties are specifically bred for current agricultural conditions. The expression of quantitative resistance in oilseed rape against *Phoma* dropped drastically as the temperature increased from 19 to 24 °C (Huang et al., 2009). The generation cycles of pathogen was triggered by climate change and lead to selection of more aggressive populations of pathogen (Chakraborty, 2013). This was due to unpredicted or unprecedented, epidemiological outbreaks. Thirdly, the genetic variation of a crop and continued series of cultivation lead to new or adapted strain of a pathogen that became dominant leading to serious effects (Strange & Scott, 2005). Increased thickness of the epicuticular wax layer on leaves was due to cultivation of crops under elevated CO_2 concentration resulting in reduced intake, translocation and metabolism of systemic fungicides. Plants metabolic rate increases at high temperatures with increased uptake of chemicals resulting in high toxicity (Coakley et al., 1999).

9.5 Plant Disease Epidemics (PDE) and Climate Change

Climate change has a major role on the occurrence and disease epidemics (Rosenzweig et al., 2001). It is true that inoculum production and dissemination are vital for epidemic diseases in field crops. The wind can spread fungal spores over long distances, as in case of wheat stem rust pathogen, *Puccinia graminis* in its disease cycle (Meyer et al., 2017). The pathogen, *Cercospora zeae-maydis* in maize sporulate at more than 96% humidity and humid condition, disease is observed at 24–29 °C optimum temperature (Paul & Munkvold, 2005).

From the year, 1970–2012, a model of the connection between climate change and wheat powdery mildew epidemics has been analyzed (Tang et al., 2017). The long-term effect of climate change on the percent acreage (PA) of the disease was evaluated. The multiregression model was constructed using PA and the pathogen's temperature requirements to predict changes in epidemics during the 2030s, 2040s, and 2080s under representative concentration pathways RCP2.6, RCP4.5, and RCP8.5. Mean monthly air temperature increased between, 1970–2012, but hours of sunshine and relative humidity decreased ($P < 0.001$). Year-to-year changes in temperature were negatively related to PA during over-summering and late spring during disease epidemics, whereas positive relationships were noted for other seasons. Year-to year changes in relative humidity were associated with PA changes in the early spring season ($P < 0.001$). These models also predicted that PA increased less under RCP2.6 (13.43%) than under RCP4.5 (13.51%) by the 2030s but would be higher by 2040s and 2080s and would increase least under RCP8.5 (13.37% by the 2030s). Hence, powdery mildew posed greater threat to all crops. The pathogens viz., *Rhizoctonia,* and *Sclerotium* in pulses showed enhanced activity under excess soil moisture conditions (Sharma et al., 2010). Bebber (2019) modelled the influence of climate change on Black Sigatoka disease in banana. It showed that the risk of infection increased by a median of 45.2% across the Latin America and the Caribbean since the 1960s. It was attributed to, increase in the wetness of the canopy and the temperature congenial for the pathogen. This facilitated Black Sigatoka disease establishment and spread making the region conducive for plant disease spread. Disease forecasting models were developed for some of the plant-pathogen systems such as coffee rust (*Hemileia vastatrix*) in Brazil (Ghini et al., 2011), *Cercospora* leaf spot (*Cercospora beticola*) of sugar beet in Germany (Richerzhagen et al., 2011), canker (*L. maculans*) of oilseed rape in the United Kingdom (Evans et al., 2008), *Phoma* stem, powdery mildew

(*Erysiphe necator*) (Caffarra et al., 2012) of grapevine in Northern Italy and banana black sigatoka (*Mycosphaerella fijiensis*) in Brazil (Ghini et al., 2007).

9.6 Plant Disease Management Under Climate Change

The effectiveness of physical, biological and chemical control methods under changing climatic conditions were assessed considering research in different areas and strategies were devised to the change in the climate. Several agronomic practices, such as tillage, crop rotation, irrigation, fertilization, use of resistant/tolerant varieties, selection of the production site and sanitation to reduce the amount of overwintering inoculum, can be used to prevent the disease risks linked with climate change (Juroszek & Von Tiedemann, 2011). Fungicides are an effective way of managing plant diseases and continuously used as normal disease suppression agents besides cultural and biological control methods. Weather parameters greatly define the persistence of chemical pesticides in the phyllosphere. Changes in amount and rate of precipitation and duration of exposure influence the efficiency of the pesticides. Temperature affects the degradation of chemicals and alters the plant phenology and physiology. Hence, indirectly affects the penetration, translocation, persistence and mode of action of several systemic fungicides (Coakley et al., 1999).

9.7 Opportunities and Challenges to Mitigate Climate Change

Novel and eco-friendly approaches are to be used in plant disease control for epidemic diseases under climate change to achieve sustainable crop production. Inoculum monitoring and rapid diagnostics are the recent approaches for crop management practices along with weather-based disease monitoring. Biopesticides and organic amendments aid in the mitigation of climate change due to reduction of nitrous oxide emission (Pathak et al., 2010). Rosenzweig et al. suggested that farmers have to adjust the time of planting and harvesting to avoid heat stress, follow crop rotation, alter the cropping pattern, substitute crops with cultivars that require longer time to mature, land management and change the agronomic operations that suit to new climatic conditions. The mitigation strategies to follow under climate change are alteration in the date of sowing to prevent disease epidemics, efficient crop cultivation practices, growing of resistant genotypes at elevate temperatures, new and promising forecasting models for disease prediction, use of bio agents having high adaptability to temperature, application of newer molecules having high efficacy at varied temperatures for disease management. Plant quarantine and disease exclusion are the two important strategies for disease management in plants where pathogens fail to establish at altered temperatures. The threat posed by the emerging pathogens under climate change depends on the use of Geographical Information Systems and climate matching tools.

Exploitation of biotechnological tools such as sequence data mining, sequencing of the genome, and new gene finding would be a novel approach to combat the effect of climate change. Marker-assisted selection along with plant breeding protocols following selection for disease resistance and quality seed production can be achieved within a short period of time. The marker-aided selection has enhanced the understanding of multiplicity of agriculturally important crops at variety and species levels for successful cross-breeding and accelerating the pace of genetic improvement (Chittaranjan et al., 2015).

Identification of stress that manipulate the growth and productivity of crops under certain climate change conditions might be one of the better approaches for new cultivar development. The accessions from a particular stress-prone environment can be recognized with the help of geographic information systems for making passport data. Selection of robust accessions are through DNA fingerprinting and mapping of genes. Genetic engineering can be the better approaches for enhancing crop yields under the climate change. With the help of this technique, integration of important gene in the target crop is possible to obtain a phenotype with desired characteristics (Varshney et al., 2006).

Invasion biology and biodiversity dispersal is very important for native plant communities (Mordecai, 2011). The decreased biodiversity frequently increases disease spread in plants and anchorage a greater pool of pathogens (Keesing et al., 2010). In common, conservation of diverse plant communities has less risk of disease incidence (Turnbull et al., 2010). Plant breeding for climate-related traits such as drought resistance (Khan et al., 2010) should also consider disease resistance. The adaptation strategies developed and mitigation related to plant disease control becomes a key factor (Fitt et al., 2010). Another way is to use planting mixtures of varieties of crops that cover an array of climatic conditions and pathogen specificity. This approach approximately enhanced the yields of rice fields infected by *Magnaporthe oryzae* (Zhu et al., 2000). Likewise, native landraces of potato in the Andes protects the farmer from the irregularity of climate and pathogen infection (Zimmerer, 1998).

Wild varieties of cultivated crops having traits for combined biotic and abiotic resistance need to be used in further breeding programs for development of varieties that favor plant resistance to pathogen infection. These characters might be introduced into cultivated plant varieties through marker-assisted selection, whole genome-wide association mapping (GWAS) analysis, gene pyramiding etc. Efforts have been made to introduce abiotic and biotic resistance into cultivated plant species through introgression of chromosomal segments of entire genome of wild grass species, *Festuca pratensis* into forage grasses (King et al., 2013).

A major concern of genetic manipulation of plant character is the unwillingness of the public to allow transgenic crops. To illustrate, many countries have major restrictions on the use of genetically modified organisms (GMO). With the advent of the CRISPR technology, such discrepancies may be abolished. Novel technologies will also permit us to accurately forecast upcoming pathogen epidemics by using nano-sensors to identify pathogen populations in the field (Kwak et al., 2017). Additionally, the plants resistant to both disease and environmental stresses have to be identified by high-throughput phenotyping technology (Fahlgren et al., 2015).

9.8 Conclusion

Challenges and opportunities for sustainable agriculture programs are based on integrated disease management under climate change. Broader collaborative approaches have to be made to develop preventive adaptive strategies resulting in climate resilient and robust cropping patterns to tackle diseases and pests under climate change. Global networking of researchers and stakeholders at all levels of crop production in different locations are important. Due to increasing pressures on crop protection, it is better to utilize the resources available to address the challenges faced in twenty-first century.

Basic studies on effect of temperature, humidity and other abiotic stresses on plant immune system need to be undertaken as it is more resilient to environmental fluctuations. Expression of immunity related genes in plants offers balance between plant growth and disease resistance under varied climatic conditions (Xu et al., 2017).

Research carried out on the impact of climate change on plant diseases under field situation is inadequate. Although, some assessments for few seasons, crops and specific food safety pathogens are available, focus must be on impact evaluation to designing plants and option for adaptation and mitigation. Firstly, the effectiveness of present chemical, physical, and biological control strategies together with resistant cultivars for disease should be estimated in the perspective of change in climate and secondly, future climate scenarios must be included in all investigations which meant for developing new instruments and tactics. Disease risk analysis should be carried out on the basis of host pathogen interactions and host response and adaptation studies must be carried out to know how plant diseases could be affected by eminent climate change.

References

Ahanger, R. A., Bhat, H. A., Bhat, T. A., Ganie, S. A., Lone, A. A., Wani, I. A., Ganai, S. A., Haq, S., Khan, O. A., Junaid, J. M., & Bhat, T. A. (2013). Impact of climate change on plant diseases. *International Journal of Modern Plant and Animal Sciences, 1*(3), 105–115.

Albouy, C., Velez, L., Coll, M., Colloca, F., Loc'h, F. (2014). From projected species distribution to food web structure under climate change. *Global Change Biology, 20*, 730–741.

Bebber, D. P. (2015). Range-expanding pests and pathogens in a warming world. *Annual Review of Phytopathology, 53*, 335–356.

Bebber, D. P. (2019). Climate change effects on black sigatoka disease of banana. *Philosophical Transactions of the Royal Society, 374*, 20180269.

Bebber, D. P., Holmes, T., Smith, D., & Gurr, S. J. (2014). Economic and physical determinants of the global distributions of crop pests and pathogens. *New Phytologist, 202*(3), 901–910.

Bebber, D. P., Ramotowski, M. A. T., & Gurr, S. J. (2013). Crop pests and pathogens move polewards in a warming world. *Nature Climate Change, 3*, 985–988.

Bendek, C. E., Campbell, P. A., Torres, R., Donoso, A., & Latorre, B. A. (2007). The risk assessment index in grape powdery mildew control decisions and the effect of temperature and humidity on conidial germination of *Erysiphe necator*. *Spanish Journal of Agricultural Research, 5*(4), 522–532.

Caffarra, A., Rinaldi, M., Eccel, E., Rossi, V., & Pertot, I. (2012). Modeling the impact of climate change on the interaction between grapevine and its pests and pathogens: European grapevine moth and powdery mildew. *Agriculture, Ecosystems and Environment, 148*, 89–101.

Chakraborty, S. (2013). Migrate or evolve: Options for plant pathogens under climate change. *Global Change Biology, 19*(7), 1985–2000.

Chakraborty, S., & Datta, S. (2003). How will plant pathogens adapt to host plant resistance at elevated CO_2 under a changing climate? *New Phytologist, 159*, 733–742.

Chakraborty, S., Murray, G. M., Magarey, P. A., Yonow, T., O'Brien, R. G., & Croft, B. J. (1998). Potential impact of climate change on plant diseases of economic significance to Australia. *Australasian Plant Pathology, 27*, 15–35.

Chakraborty, S., & Newton, A. C. (2011). Climate change, plant diseases and food security: An overview. *Plant Pathology, 60*, 2–14.

Chakraborty, S., Von Tiedemann, A., & Teng, P. S. (2000). Climate change: Potential impact on plant diseases. *Environmental Pollution, 108*, 317–326.

Chittaranjan, K., Muthamilarasan, M., & Henry, R. J. (2015). Application of genomics-assisted breeding for generation of climate resilient crops: Progress and prospects. *Frontiers in Plant Science, 6*, 563.

Clarkson, J. P., Fawcett, L., Anthony, S. G., & Young, C. (2014). A model for *Sclerotinia sclerotiorum* infection and disease development in lettuce, based on the effects of temperature, relative humidity and ascospore density. *PLoS ONE, 9*(4), e94049.

Coakley, S. M., Scherm, H., & Chakraborty, S. (1999). Climate change and plant disease management. *Annual Review of Phytopathology, 37*, 399–426.

Das, T., Hajong, M., Majumdar, D., Tombisana Devi, R. K., & Rajesh, T. (2016). Climate change impacts on plant diseases. *SAARC Journal of Agriculture, 14*(2), 200–209.

Davis, M. B., Shaw, R. G., & Etterson, J. R. (2005). Evolutionary responses to changing climate. *Ecology, 86*, 1704–1714.

Deutsch, C. A., Tewksbury, J. J., Huey, R. B., Sheldon, K. S., Ghalambor, C. K., Haak, D. C., & Martin, P. R. (2008). Impacts of climate warming on terrestrial ectotherms across latitude. *Proceedings of the National Academy of Sciences of the United States of America, 105*, 6668–6672.

Eastburn, D. M., Degennaro, M. M., Delucial, E. H., Demody, O., & McElrone, A. J. (2010). Elevated atmospheric carbon dioxide and ozone alter soybean diseases at SoyFACE. *Global Change Biology, 16*, 320–330.

Etterson, J. R., & Shaw, R. G. (2001). Constraint to adaptive evolution in response to global warming. *Science, 294*, 151–154.

Evans, N., Baierl, A., Semenov, M. A., Gladders, P., & Fitt, B. D. L. (2008). Range and severity of a plant disease increased by global warming. *Journal Royal Society Interface, 5*, 525–531.

Fahlgren, N., Gehan, M. A., & Baxter, I. (2015). Lights, camera, action: High-throughput plant phenotyping is ready for a close-up. *Current Opinion Plant Biology, 24*, 93–99.

Fischer, E. M., & Knutti, R. (2015). Anthropogenic contribution to global occurrence of heavy-precipitation and high-temperature extremes. *National Climate Change, 5*, 560–564.

Fitt, B. D. L., Evans, N., Gladders, P., Hughes, D. J., Madwick, J. W., Jeger, M. J., Townsend, J. A., Turner, J. A., & West, J. S. (2010). Climate change and arable crop disease control; mitigation and adaptation. In *Proceedings of the 16th international Reinhardsbrunn symposium, April 25–29* (in press).

Garrett, K. A., Forbes, G. A., Savary, S., Skelsey, P., & Sparks, A. H. (2011). Complexity in climate change impacts: An analytical framework for effects mediated by plant disease. *Plant Pathology, 60*, 15–30.

Gautam, H. R., Bhardwaj, M. L., & Kumar, R. (2013). Climate change and its impact on plant diseases. *Current Science, 105*(12), 1685–1691.

Ghini, R., Hamada, E., Goncalves, R. R. V., Gasparotto, L., & Pereira, J. C. R. (2007). Risk analysis of climatic change on black sigatoka on banana in Brazil. *Fitopatologia Brasileira, 32*(3), 197–204.

Ghini, R., Hamada, E., Pedro Junior, M. J., & Gonçalves, R. R. V. (2011). Incubation period of *Hemileia vastatrix* in coffee plants in Brazil simulated under climate change. *Summa Phytopathologica, 37*(2), 85–93.

Huang, J. P., Yu, H. P., Dai, A. G., Wei, Y., & Kang, L. T. (2017). Drylands face potential threat under 2 °C global warming target. *National Climate Change, 7*, 417–422.

Huang, Y. J., Pirie, E. J., & Evans, N. (2009). Quantitative resistance to symptomless growth of *Leptosphaeria maculans* (phoma stem canker) in *Brassica napus* (oilseed rape). *Plant Pathology, 58*, 314–323.

Huber, L., & Gillespie, T. J. (1992). Modeling leaf wetness in relation to plant disease epidemiology. *Annual Review of Phytopathology, 30*, 553–577.

Jeger, M. J., & Xu, X. M. (2015). Modelling the dynamics of a plant pathogen and a biological control agent in relation to flowering pattern and populations present on leaves. *Ecological Modelling, 313*, 13–28.

Juroszek, P., & von Tiedemann, A. (2011). Potential strategies and future requirements for plant disease management under a changing climate. *Plant Pathology, 60*(1), 100–112.

Keesing, F., Belden, L., Daszak, K., Dobson, P. A., Harvell, C. D., Holt, R. D., Hudson, P., Jolles, A., Jones, K. E., Mitchell, C. E., Myers, S. S., Bogich, T., & Ostfeld, R. S. (2010). Impacts of biodiversity on the emergence and transmission of infectious diseases. *Nature, 468*, 647–652.

Khan, H. R., Paull, J. G., Siddique, K. H. M., & Stoddard, F. L. (2010). Faba bean breeding for drought-affected environments: A physiological and agronomic perspective. *Field Crops Research, 115*(3), 279–286.

King, J., Armstead, I., Harper, J., Ramsey, L., Snape, J., Waugh, R., James, C., Thomas, A., Gasior, D., & Kelly, R. (2013). Exploitation of interspecific diversity for monocot crop improvement. *Heredity, 110*, 475–483.

Kwak, S. Y., Wong, M. H., Lew, T. T. S., Bisker, G., Lee, M. A., Kaplan, A., Dong, J., Liu, A. T., Koman, V. B., & Sinclair, R. (2017). Nanosensor technology applied to living plant systems. *Annual Review of Analytical Chemistry, 10*, 113–140.

Ladanyi, M., & Horvath, L. (2010). A review of the potential climate change impact on insect populations general and agricultural aspects. *Applied Ecology and Environmental Research, 8*(2), 143–152.

Legler, S. E., Caffi, T., & Rossi, V. (2012). A nonlinear model for temperature dependent development of Erysiphe necator chasmothecia on grapevine leaves. *Plant Pathology, 61*(1), 96–105.

Legreve, A., & Duveiller, E. (2010). Preventing potential diseases and pest epidemics under a changing climate. In M. P. Reynolds (Ed.), *Climate change and crop production* (pp. 50–70). CABI.

Ma, L., Qiao, J., Kong, X., Zou, Y., Xu, X., Chen, X., & Hu, X. (2015). Effect of low temperature and wheat winter hardiness on survival of *Puccinia striiformis* f. sp *tritici* under controlled conditions. *PLOS One, 10*, e0130691.

Magarey, R. D., Sutton, T. B., & Thayer, C. L. (2005). A simple generic infection model for foliar fungal plant pathogens. *Phytopathology, 95*, 92–100.

Mahato, A. (2014). Climate change and its impact on agriculture. *International Journal of Scientific and Research Publications, 4*(4). www.ijsrp.org

Manning, W. J., & Tiedemann, A. V. (1995). Climate change: Potential effects of increased atmospheric carbon dioxide (CO_2), ozone (O_3), and ultraviolet-B (UVB) radiation on plant diseases. *Environmental Pollution, 88*(2), 219–245.

Mboup, M., Bahri, B., Leconte, M., De Vallavieille-Pope, C., Kaltz, O., & Enjalbert, J. (2012). Genetic structure and local adaptation of European wheat yellow rust populations: The role of temperature-specific adaptation. *Evolutionary Applications, 5*(4), 341–352.

Mcelrone, A. J., Reid, C. D., Hoye, K. A., Hart, E., & Jackson, R. B. (2005). Elevated CO_2 reduces disease incidence and severity of a red maple fungal pathogen via changes in host physiology and leaf chemistry. *Global Change Biology, 11*(10), 1828–1836.

Meyer, M., Cox, J. A., Hitchings, M. D. T., Burgin, L., Hort, M. C., Hodson, D. P., & Gilligan, C. A. (2017). Quantifying airborne dispersal routes of pathogens over continents to safeguard global wheat supply. *Nature Plants, 3*(10), 780–786.

Milus, E. A., Kristensen, K., & Hovmøller, M. S. (2009). Evidence for increased aggressiveness in a recent widespread strain of *Puccinia striiformis* f. sp. *tritici* causing stripe rust of wheat. *Phytopathology, 99*(1), 89–94.

Mina, U., & Sinha, P. (2008). Effects of climate change on plant pathogens. *Environews, 14*(4), 6–10.

Mordecai, E. (2011). *Pathogen impacts on plant communities: Unifying theory, concepts, and empirical work*. Ecological Monographs. https://doi.org/10.1890/10-2241.1 (in press)

Obrêpalska-Stêplowska, A., Renaut, J., Planchon, S., Przybylska, A., Wieczorek, P., Barylski, J., & Palukaitis, P. (2015). Effect of temperature on the pathogenesis, accumulation of viral and satellite RNAs and on plant proteome in peanut stunt virus and satellite RNA-infected plants. *Frontiers Plant Science, 6*, 903.

Pachauri, R. K., & Reisinger, A. (2007). Climate change synthesis report. In *Contribution of working groups I, II and III to the fourth assessment report of the inter governmental panel on climate change*. IPCC.

Pangga, I., Hanan, J., & Chakraborty, S. (2013). Climate change impacts on plant canopyarchitecture: Implications for pest and pathogen management. *European Journal of Plant Pathology, 135*, 595–610.

Park, R. F. (1990). The role of temperature and rainfall in the epidemiology of *Puccinia striiformis* f. sp *tritici* in the summer rainfall area of eastern Australia. *Plant Pathology, 39*(3), 416–423.

Pathak, H., Bhatia, A., Jain, N., & Aggarwal, P. K. (2010). Green house gas emission and mitigation in Indian agriculture-A review. In *ING bulletins on regional assessment of reactive nitrogen*. Indian Nitrogen Group Society for Conservation of Nature.

Paul, P. A., & Munkvold, G. P. (2005). Influence of temperature and relative humidity on sporulation of *Cercospora zeae-maydis* and expansion of gray leaf spot lesions on maize leaves. *Plant Disease, 89*(6), 624–630.

Pritchard, S. G., Rogers, H. H., Prior, S. A., & Peterson, C. M. (1999). Elevated CO_2 and plant structure: A review. *Global Change Biology, 5*(7), 807–837.

Richerzhagen, D., Racca, P., Zeuner, T., Kuhn, C., Falke, K., Kleinhenz, B., & Hau, B. (2011). Impact of climate change on the temporal and regional occurrence of Cercospora leaf spot in Lower Saxony. *Journal of Plant Diseases and Protection, 118*(5), 168–177.

Ritchie, F., Bain, R., & Mcquilken, M. (2013). Survival of sclerotia of *Rhizoctonia solani* AG3PT and effect of soil-borne inoculum density on disease development on potato. *Journal of Phytopathology, 161*, 180–189.

Rosenzweig, C., Iglesias, A., Yang, X. B., Epstein, P. R., & Chivian, E. (2001). Climate change and extreme weather events; implications for food production, plant diseases, and pests. *Global Change and Human Health, 2*(2), 90–104.

Santini, A., & Ghelardini, L. (2015). Plant pathogen evolution and climate change. *CABI Reviews, 10*, 035.

Sharma, M., Mangala, U. N., Krishnamurthy, M., Vadez, V., & Pande, S. (2010). Drought and dry root of chickpea. In *Proceedings of the 5th international food legumes research conference (IFLRC V) and 7th European conference on grain legumes (AEP VII), Antalya, Turkey*.

Stevens, R. B. (1960). Cultural practices in disease control. In J. G. Horsfall (Ed.), *Plant pathology: An advanced treatise, volume III: The diseases population epidemics and control* (pp. 357–429). Academic Press Inc.

Strange, R. N., & Scott, P. R. (2005). Plant disease: A threat to global food security. *Annual Review of Phytopathology, 43*, 83–116.

Tang, X., Cao, X., Xu, X., Jiang, Y., Luo, Y., Ma, Z., Fan, J., & Zhou, Y. (2017). Effects of climate change on epidemics of powdery mildew in winter wheat in China. *Plant Disease, 101*(10), 1753–1760.

Tapsoba, H., & Wilson, J. P. (1997). Effects of temperature and light on germination of uredin-iospores of the pearl millet rust pathogen, *Puccinia substriata* var. *indica*. *Plant Disease, 81*(9), 1049–1052.

Turnbull, L. A., Levine, J. M., Fergus, A. J. F., & Petermann, J. S. (2010). Species diversity reduces invasion success in pathogen-regulated communities. *Oikos, 119*(6), 1040–1046.

Varshney, R. V., Hoisington, D. A., & Tyagi, A. K. (2006). Advances in cereal genomics and applications in crop breeding. *Trends Biotechnology, 24*(11), 490–499.

Velasquez, A. C., Christian, D. M. C., & He, S. Y. (2018). Plant pathogen warfare under changing climate conditions. *Current Biology, 28*(10), R619–R634.

Von Tiedemann, A., & Firsching, K. H. (2000). Interactive effects of elevated ozone and carbon-dioxide on growth and yield of leaf rust-infected versus non-infected wheat. *Environmental Pollution, 108*, 357–363.

Xu, G., Yuan, M., Ai, C., Liu, L., Zhuang, E., Karapetyan, S., Wang, S., & Dong, X. (2017). uORF-mediated translation allows engineered plant disease resistance without fitness costs. *Nature, 545*, 491–494 [PubMed: 28514448].

Zhu, Y., Chen, H., Fan, J., Wang, Y., Li, Y., Chen, J., Fan, J., Yang, S., Hu, L., & Leung, H. (2000). Genetic diversity and disease control in rice. *Nature, 406*, 718–722. [PubMed: 10963595]

Zimmerer, K. S. (1998). The ecogeography of Andean potatoes. Versatility in farm regions and fields can aid sustainable development. *BioScience, 48*, 445–454.

Chapter 10
Development Prospective and Challenges of Nanotechnology in Sustainable Agriculture

Shalini Singh Visen, Dinesh Jinger, Manoj Parihar, Gopal Tiwari, Rajendra Prasad Meena, Manoj Kumar Chitara, and Surendra Singh Jatav

10.1 Introduction

Agriculture is the most novel occupation and way of life since old age and had been practiced with nature's law. Significant increment in production was recorded after green revolution due to efficient use of fertilizers, irrigation systems, mechanization and prominent crop variety (Lowry et al., 2019). Globally, each year more than 3000 million metric tonnes of food grain are producing (USDA, 2018), for that huge amount agriculture inputs are needed (Zhang, 2018) which enhanced the cost of cultivation. However, under present situation, availability of resources (water, fertilizers, pesticides, herbicides) for intensive agriculture is limited, which poses serious threat to environmental integrity (De Oliveira, 2014). Agrochemicals cause cascading

S. Singh Visen
Amity Food and Agriculture Foundation, Amity University, Lucknow, Uttar Pradesh 226010, India

D. Jinger (✉)
ICAR-Indian Institute of Soil and Water Conservation, RC-Vasad, Anand, Gujarat 388306, India
e-mail: dineshjinger28@gmail.com

M. Parihar · R. P. Meena
ICAR-ICAR-Vivekananda Parvatiya Krishi Anusandhan Sansthan, Almora, Uttarakhand 263601, India

G. Tiwari · M. K. Chitara
ICAR-National Bureau of Soil Science and Land Use Planning, Nagpur, Maharashtra 440033, India

Govind Ballabh Pant University of Agriculture and Technology, Pantnagar, Uttarakhand 263153, India

S. S. Jatav
Banaras Hindu University, Varanasi, Uttar Pradesh 221005, India

S. Sheraz Mahdi et al. (eds.), *Adapting to Climate Change in Agriculture-Theories and Practices*, https://doi.org/10.1007/978-3-031-28142-6_10

environmental repercussions (Willett et al., 2019) including pesticides and fertilizer contamination in natural water bodies (Diaz & Rosenberg, 2008) which raises serious health concern. Moreover, agriculture also contributes in climate change as 24% of total annual greenhouse gases emission of the glob especially N_2O associated with this sector (Smith et al., 2014). In addition, monocropping with poor input use efficiency, globalization of diseases, pest and weeds turns agriculture system toward low resilient and unproductive. Thus, we can conclude that existing agricultural activities are intrinsically inefficient and unsustainable and are prone to environmental degradation (Willett et al., 2019). Challenges associated with existing agriculture system are sustainable use of natural resources, ecosystems services, food and nutritional security. To address these challenges, use of novel and efficient technologies would be indispensable for future and sustainable agriculture (Willett et al., 2019). These technological interventions are required to set new paradigm in agriculture to ensure the sustainable nutritional and food security (Glenn & Florescu, 2016). Among such development, use of nanotechnology in agriculture may contribute significantly for addressing the problem of existing agriculture sustainability.

Nanotechnology as a multidisciplinary approach made remarkable progress in various sector. However, its application in agriculture is still at infancy stage (Mullen, 2019). Therefore, it is an urgent need to explore their potential areas in agriculture field. The nanotechnology can be used in various agricultural fields such as crop cultivation, animal husbandry, food preservation etc. With the help of nanoproducts such as NFs, NPs, NHs we can achieve higher efficiency of inputs and reduce associated environmental problems. Use of nano-fertilizers offers a way for agronomic biofortification of cereals to address the poor dietary intake of micronutrients. It would be true win–win situation, with better environmental safety, reducing the cost of cultivation and more farm profitability (Mullen, 2019). In this chapter, we will be discussing about the development and application of NT in agriculture sector as well as assessment of associated risk and future roadmap to present this technology as a sustainable way to achieve food, nutritional as well as livelihood security.

10.2 Application of NT in Agriculture

The usefulness of NT in agricultural domain has developed an immense scope, as it can increase crop yield with minimum input. NT provides numerous products in agriculture sector that include NFs, NPs and NHs etc. (Grillo et al., 2016). Use of NT in agriculture especially for crop production and crop protection has been published and patented (Mishra & Singh, 2015). Agricultural sector is being benefited through NT by various ways (Fig. 10.1).

For instances: encapsulated NPs for their controlled and sustained release; better seed germination of rainfed crops through carbon nanotube (CNT); NFs for enhanced crop production; NPs for plant protection and NHs for weed management (Servin &

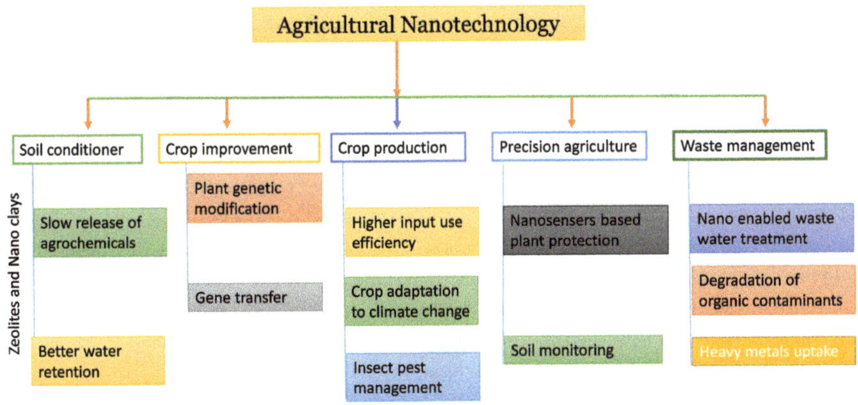

Fig. 10.1 Representing the potential applications of nanotechnology in agriculture

White, 2016). Nanotechnology has also potential role in remediating the environmental contaminants like heavy metals in soil and water (Ion et al., 2010). The potential roles of NT are described in following sections.

10.3 Role of Nanotechnology in Plant Nutrition

The NT enables the use of NFs for agriculture production. Nano-fertilizers have greater stability than ordinary fertilizers. NFs may contain nano zinc (Zn), silica (Si), iron (Fe) and titanium dioxide (TiO_2), Zinc Cadmium Selenium/Zinc sulfide (ZnCdSe/ZnS) core shell quantum dots (QDs), Indium phosphide/ZnS (InP/ZnS) core shell QDs, manganese/zinc selenium (Mn/ZnSe) QDs, etc. which impart sustained release and enhance its quality (Zhang et al., 2016). The strength of the coated NFs decreases the rate of dissipation of the fertilizer, enables slow and sustained release of nutrients which leads to efficient absorption by plant root system. Currently, innovative products like slow-release fertilizers (SRFs) are being used to mitigate environmental pollution and to save fertilizer consumption (Wu & Liu, 2008). Fertilizers having Sulphur (S) nanocoating ($\leq$ 100 nm layer) are more useful than SRFs since S is advantageous especially in S deficient soils (Brady & Weil, 1999). Urea and phosphate coated with nanomaterials (NMs) have potential to release nutrients in controlled manner, hence, synchronize the soil and crop demands as most of the soils are deficient in nitrogen (N). Various kind of natural and synthetic polymers have been explored for controlled delivery of nutrients. Biodegradable polymeric chitosan nanoparticles have reported dramatic outcome for the slow and steady release phenomenon of fertilizers (Corradini et al., 2010). Kaolin and polymeric biocompatible nanoparticles also impart in the slow-release function of fertilizers (Wilson et al., 2008). NFs synchronize the release of N and phosphorus (P) with the absorption by the plant root system, thereby checking the nutrient losses

(Emadian, 2017). Nutrients absorption from soil can be improved by application of NFs in crops. Nanotechnology has also potential role in enabling the availability of micronutrients to plants. Nano-formulations of micronutrients are applied either on plants or in soil for uptake by roots and to improve the soil health (Peteu et al., 2010). Foliar application of iron nanoparticles at 500 ppm to black-eyed peas significantly increased the number of pods (28%), seed index (4%), Fe content (45%), and chlorophyll content (12%) as compared to regular iron salt (Delfani et al., 2014). Application of Fe nanoparticles significantly increased the promising effect of magnesium (Mg) nanoparticles used as NFs on black-eyed peas. In the recent decade, NFs are easily available in the market (Table 10.1). Nanotechnology has potential to bring agriculture scenario into sustainable agriculture as NFs have promised to meet the burgeoning food demand of the world. Nano-fertilizers could be a best alternative to alleviate problem of nutrient deficiencies and eutrophication (Shukla et al., 2019). Conventional nitrogen fertilizers are prone to leaching, volatilization, degradation which eventually decreases the efficiency of fertilizers (Yang et al., 2016). Porous NMs, such as zeolites, clay or chitosan not only regulate the demand-based release and uptake of nutrient by plant but also minimizes the nutrient losses (Panpatte et al., 2016). NFs are mostly developed by the encapsulation of nutrients with NMs. Peteu et al. reported that supplementation of nano entrapped micronutrients has also potential for sustained release of nutrients which would lead to efficient uptake nutrients and ultimately better crop productivity and soil quality. Kale and Gawade (2016) also reported that fertilization of ZnO-nanoparticles in Zn deficient soil not only improved NUE but also tremendously increased the crop yield by 91% over control.

10.4 Role of Nanotechnology in Plant Growth and Quality

Nanotechnology involves the development of innovative application approaches for improving different phases of crop plants viz. seed germination, growth, and development of the plant. Application of NMs has significantly positive impacts on germination, and crop yield (Table 10.2). For instance, multi walled carbon nanotubes (MWCNTs) have beneficial effect on seed germination of barley, garlic, maize, peanut, soybean, tomato, and wheat (Joshi et al., 2018). Similarly, application of nano silicon dioxide (SiO_2), TiO_2 and zeolite enhanced seed germination potential and seed germination index in crop plants (Manjaiah et al., 2019). Application of nano-ZnO in zinc deficient soil increased the length of the root and shoot, turgid and dry weight and photosynthesis in various crops over control (Ali et al., 2019). Nanomaterials of Fe/SiO_2 have significantly improved seed germination potential in maize and barley (Disfani et al., 2017). Nanomaterials have capability to penetrate the seed coat and improve absorption capacity of water, which stimulates enzymatic system and eventually improves germination and seedling growth (Banerjee & Kole, 2016). NMs, such as ZnO, TiO_2, MWCNTs, ferrous oxide (FeO), and hydroxy fullerenes have been reported to improve crop growth as well as quality in groundnut, green

Table 10.1 Commercially available nano fertilizers, nano insecticides and nano herbicides

Nano product	Composition	Application/ action mode	Company	Country
Zintrac™ 700	ZnO suspension concentrate	Foliar application, improve uptake and long term feeding ability	Yara Fertilizers India Pvt. Ltd.	India
Primo Maxx	Trinexapac-ethyl	Growth regulator	Syngenta, Greensboro	USA
Tropical agro tag Nano phos granules	Proteino-lacto-gluconate formulation of phosphorus fertilizer	Foliar and granular application, improves root growth and photosynthesis, oil percentage and pod filling	Tropical Agro System India Private Ltd.	India
Nano zinc	Chelated Zn 12%	Alleviate Zn deficiency	Alert Biotech	India
Nano-5	P and Ca	Improve soil nutrients	Natural organic fertilizer	Taiwan
Nano bor	Boron 20%	Beneficial in production of high-quality fruits	Alert Biotech	India
Nano guard	Organic nano fungicide	Improve plant growth	Vision Mark Biotech	India
Mega harvest liquid fertilizer	Ca and Zn + sticker	Improve plant growth	Biomedica Marketing	Philippines
Nanoboost	Proprietary blend of elemental compounds and derivatives (8%) and Linear Ethoxylated Compound (1.5%)	For weed control	Monty's Plant Food Company	Louisville, Kentucky
Nanomax-Cal	Ca, Mg, S and all secondary nutrient	Foliar and granular application for all crops, improve overall growth and yield	JU Pvt. Ltd.	India
Zeolites and nano-clays	–	Improve water and nutrients absorption	Geohumus-Frankfurt, DE	Germany
Nanogrip	Metsulfuron methyl 20%WP	For weed control	Bhavya industries	India

(continued)

Table 10.1 (continued)

Nano product	Composition	Application/ action mode	Company	Country
Nano virus biopesticide	Virus	Insect control	Futures Industries	India
Nanoemulsion	Neem oil (*Azadirachta indica*)	Insect control	VIT University	India

Source Khan and Rizvi (2017)

gram, mustard, onion, potato, spinach, soybean, tomato, and wheat (Dubey & Mailapalli, 2016). Application of fullerol as seed dressing in bitter melon (*Momordica charantia*) enhanced the fruit weight, fruit size, and yield by 128%. Further, it has also improved the content of cucurbitacin-B, lycopene, charantin and inulin (Kole et al., 2013). Application of FeNFs improved agronomic parameters as well as essential oil contents of *Dracocephalum moldavica* (Yousefzadeh & Sabaghnia, 2016).

Application of Znand BNFs as foliar spray increased total soluble solids by 4.4–7.6%, maturity index by 20.6–46% juice pH by 0.28–0.62 pH unit and decreased titratable acidity by 9.5–29% in pomegranate (Davarpanah et al., 2016). The shoot length of barley and maize seedlings were enhanced by 8.25% and 20.8%, respectively by the application of nano Fe/SiO_2 at 15 ppm (Disfani et al., 2017). The total chlorophyll, total carbohydrate and essential oil content of *Ocimum basilicum* enhanced by the foliar application of ferrous ferric oxide (Fe_3O_4). Further, it also increased the Fe content, plant height, number of branches and leaves, turgid and dry weight (El-Feky et al., 2013). Nanoparticles of TiO_2 increased growth of spinach by improving *Rubisco activase* activity, light absorbance and N metabolism (Yang et al., 2007). CNTs and nanoparticles of gold (Au), SiO_2, ZnO, and TiO_2 can improve vegetative as well as reproductive growth of plants by enhancing nutrient uptake (Khot et al., 2012). However, the positive impacts of NMs depend on their formulation, concentration, size, surface charge and physio-chemical properties (Lambreva et al., 2015). Application of ZnO nanoparticles at lowerdose enhanced the growth of chickpea (Mahajan et al., 2011). Foliar application of Fenanoparticles at 500 ppm significantly increased the number of pods, seed index, Fe and chlorophyll content as compared to controls in black eyed peas (Delfani et al., 2014).

10.5 Role of Nanotechnology in Plant Protection

Although, synthetic pesticides have higher efficacy towards the targeted pests but, they have cause deleterious effects on environment. Hence, there is a need to come towards NT in pest management. Nanoparticles have more potential to suppress the insect-pests of modern agriculture. NPs with nano-encapsulation impart greater absorption of active ingredient (AI) into the plant system and release the AI in controlled manner which remains for longer period of time with in the plant (Scrinis &

Table 10.2 Impact of nanoparticles on crop growth, yield and insect-pest suppression

Nanoparticle	Crop	Experimental condition	Results	References
Ag	White Radish (*Raphanus sativus*)	White radish grown in soil with presence of snails in lab experiment. Application of Ag nanoparticles were done in soil matrix	Reduction in viability of the snail as well as the frequency of fungi	Ali et al. (2015)
Ag	Wheat (*Triticum aestivum*)	A cavity slide experiment to determine the combative strength of Ag-nanoparticles against *B. sorokiniana*	Complete inhibition of conidial germination of the fungus	Mishra et al. (2014)
Ag	Fenugreek (*Trigonella foenum-graecum*)	Various doses of Ag-nanoparticles (0, 10, 20, 30 and 40 μg mL^{-1}) were tested	Tremendously improve seed germination, seed vigor index, seedling fresh and dry weight	Hojjat & Hojjat (2015)
FeS_2	Beetroot, carrot, fenugreek, alfalfa, mustard and sesamum	Seeds were soaked for 10–12 h with nano-FeS_2 solution before sowing	Yield enhancement	Das et al. (2016)
TiO_2	Spinach (*Spinacia oleracea*)	Seeds were soaked in 0.25% nano-TiO_2 and up to 2 days at 10 °C under light	Improve spinach growth	Yang et al. (2007)
CuO	Lettuce (*Lactuaca sativa*)	Application of nano-CuO to lettuce crop	Improve transpiration rate with higher stomatal conductance	Wang et al. (2019)
MgO	Tomato (*Lycopersicum esculantum*)	Drenching of tomato roots in solution of MgO nanoparticles	No disease incidence	Imada et al. (2016)

Source Shang et al. (2019)

Lyons, 2007). The slow release, solubility, specificity, permeability and stability are some of the major properties of NP shaving nano-encapsulation (Bhattacharyya et al., 2016). Nano-encapsulated NPs are always applied at very low dose which led to reduction in the number of application and quantity of applied pesticide which is environmentally benign (Nuruzzaman et al., 2016). Applications of nanoparticles like silver (Ag) nanoparticles, aluminum oxide (Al_2O_3), ZnO and TiO_2 for the management of rice weevil (*Sitophilus oryzae* L.) have been studied (Goswami et al.,

2010). Application of nano Al_2O_3in wheat crop reported to significant destruction of *Sitophilus oryzae* L. and *Rhyzopertha dominica* within 3 days. Shyla et al. (2014) reported that soil application of NPs of ZnO, Ag and TiO_2in pulse and oilseed crops significantly reduced the incidence of *Macrophomina phaseolina.* They have also observed that Ag nanoparticles have higher efficacy against this pathogen even at lower concentrations than nanoparticles of ZnO and TiO_2. The application of NMs to agricultural crops has played a vital role in suppression of fungal disease and noxious weed. Further, many nanoparticles, viz. CaO, MgO, MnO ZnO, SiO_2, TiO_2, Cu, and Ag nanoparticles have potential to control bacterial diseases (Servin et al., 2015). Application of nanoparticles of ZnO effectively controlled the growth of *Penicillium expansum, Fusarium oxysporum, F. graminearum, Alternaria alternate, Rhizopus stolonifer, Mucor plumbeus and Aspergillus flavus* (Dwivedi et al., 2016). Tomato leaf blight caused by *Phytophthora infestans* has been effectively controlled by application of nano Cuformulation over non-nano Cu formulation (Giannousi et al., 2013). It inhibits the growthof conidiophores and eventually leads to the demise of fungal hyphae. Karate® ZEON a nano encapsulated broad-spectrum pesticide has been developed by Syngenta to control insect pests of rice, cotton, soybeans and peanuts. The lambda-cyhalothrin, AI of this product is released when it comes in contact with leaves. Similarly, "gutbuster" a nano-insecticide releases its AI it comes in contact with pH more than 7.5 such as insect's stomach (Prasad et al., 2014). NMs containing herbicides offer control of weeds in ecological benign way. Sharifi-Rad et al. (2016) reported that application of NPs of SiO_2 drastically reduced the germination, growth, photosynthetic pigments and total protein of weeds. Pectin NPs amalgamated with Metsulfuron methyl applied at low dose of AI are more cytotoxic to *Chenopodium album*over the commercial herbicide (Kumar et al., 2017). Usually, commercial herbicides cause detrimental effect on weeds by inhibiting the growth of leaves, stem and flowers but do not affect the under above ground portion like roots, rhizomes and tubers. Thus enable the weeds to persist for longer period of time in the field. However, NHs prevents the reoccurrence of weeds (Dwivedi et al., 2016).

10.6 Role of Nanotechnology in Remediation of Contaminants and Soil Quality Improvement

Nanoparticles also play a great role in improving the soil health and soil quality by degrading the contaminants. Sekhon (2014) has revealed that hydrogels, nano clays, and nano zeolites enhanced the moisture retention capacity of soil by releasing water in a sustained manner, thus, mitigating the water scarcity during crop growing periods. Therefore, waste lands or fallow-uncultivated agriculture lands can be restored by the use of such nanoparticles. Nano-metals, nano-metal-oxides, organic polymer and CNT have good capacity to engross all the contaminants (Khin et al., 2012) and improve soil restoration ability. Camargo et al. (2009) have observed that application of nano-composites has potential to get rid from pollutants or contaminants from the environment. Anandan et al. (2006) have revealed that application of ZnO/Hβ containing 5% ZnO caused drastic degradation of monocrotophos. Similarly, application of nano-composites of Fe/zeolite and TiO_2/zeolite resulted inabsolute degradation of monocrotophos, methomyl and dichlorvos (Tomasevic et al., 2010). Nano-bio-composites are the modern composite materials where nano fillers are amalgamated into bio-polymer matrix by various methods. Bio-polymers have hydrophilic properties which enable them to adsorb pollutants from aqueous environment (Averous & Pollet, 2012). Removal of atrazine and organophosphate pesticide by Ag/chitosan nano-bio-composite has been reported by Saifuddin et al. (2011) and Dehaghi et al. (2014) respectively. Therefore, nano-bio-composites conglomerated with bio-polymers like, CuO-montmorillonite-chitosan and CuO-montmorillonite-gumghatti has the excellent potential for degradation of dichlorvos pesticide.

10.7 Role of Nanotechnology in Climate Change Mitigation

The long span changes in the statistical distribution of weather parameters over periods of times. Climate change brings changes in the pattern of rainfall (water), temperatures, pH and environment involving contamination of pernicious metals. Hence, to mitigate these environmental stresses researchers should focus on to improve adaptation capacity of plants without affecting the ecosystem (Vermeulen et al., 2012). To achieve this task a multifaceted approach is needed, such as stimulation of enzymes of plant system, modulation of hormones, activation of stress gene, inhibition of pernicious metal uptake and keep the plant away from moisture stress by shortening the ontogeny of plants. Scientists are developing technologies for mitigating ill impacts of environmental stresses to achieve the goal of sustainable agricultural (Dubey & Mailapalli, 2016). Nano-fertilizers have potential for sustainable agricultural production system even when the environmental conditions are not congenial. Soil salinity decreases production of food crops on about 23% of the agricultural soils across the world (Onaga & Kerstin, 2016). In contrast, Haghighi et al. (2012) have observed that nano SiO_2 applied to tomato and squash grown under salinity stress resulted in improvement in seed germination, greenness index and proline content. Similarly, Torabian et al. (2017) reported that application of nanoparticles of $FeSO_4$ to sunflower grown under salinity stress improved the leaf area index, net CO_2 assimilation rate, sub-stomatal CO_2 concentration, greenness index and photosynthetic efficiency. It has also been explored that application Si-nanoparticles in wheat crop significantly diminished the UV-B stress (Tripathi et al., 2017). Nano-zeolite application to crop plants improve the soil fertility by enhancing the availability and mobility of nutrients which leads to better germination and growth of the crop plants (Manjaiah et al., 2019). Ontogeny of wheat is reduced by 24% due to the application of NFs over traditional fertilizer (Abdel-Aziz et al., 2016). Nanomaterials have found effective in reclamation of toxic pollutants such as heavy metals. Cd toxicity reduced by the application of nano-Si at 2.5 mM concentration by decreasing its uptake and accumulation by rice plant (Wang et al., 2015). Nano-Si fertilizers are safer from human and environment point of view because they have potential to diminish the accumulation of pernicious heavy metals over conventional fertilizers (Wang et al., 2016). The positive effects of NMs on crop growth under stress conditions might be due to improved stimulation of stress enzymes within the plant system (Shojaei et al., 2019). Nano-SiO_2 improved the plant tolerance to extreme environmental stress condition by increasing the content of stress amino acids like proline and reaction of antioxidant enzymes like superoxide dismutase, catalase and peroxidase (Shalaby et al., 2016). Microarray analysis of Arabidopsis revealed that Ag-nanoparticles control the expression of stress gene by increasing or decreasing their sensitivity. The genes with increased sensitivity are linked with reaction towards heavy metal stress and the genes with decreased sensitivity are linked with reaction towards pathogens and hormone activation (Banerjee & Kole 2016).

10.8 Challenges and Risk Assessment

The diverse application of nanotechnology in the area of biomedicine, informatics, energy resource, astronautics, cosmetics, clothing, sporting goods oceanography and national defense etc., become most efficient and sustainable technology of the world in recent time. Within a short time, span NT capture the large area of domestic as well as international market and, are still emerging. Nanotechnology has the enormous applications in agriculture sectors like NFs, NPs, NHs and precise shipping systems for sustained release of applied nano products (Grillo et al., 2016). The physical, chemical and biological characteristics of the soil affect the distribution, fate and movement of the nanoparticles when applied in the soil and plant system. Numerous studies have reported that application of nanoparticles causes ill effects on soil micro fauna. The application of TiO_2-nanoparticles and ZnO-nano-particles cause adverse effect on soil bacteria (Ge et al., 2012). Similarly, application Ag-nanoparticles affects the growth of *Rhizobium* and *Azotobacter* by reducing nitrate reductase (Shahrokh et al., 2014). ZnO-nanoparticles have eco-toxicological effects of on soil microbes (Shen et al., 2015); phyto-toxicity by Ag-NPs in plant (Dietz & Herth, 2011); TiO_2-nanoparticles (> 0.5 g L^{-1}) caused significant damage to the duckweed (*Lemna minor*) (Song et al., 2012) and denitrifying bacterial community is highly susceptible to nanoparticles toxicity (Vandevoort & Arai, 2012).

Every technology has mixed blessing. The potential and advantages of the NT are escorted by some risks, and greater skills for supervising and controlling the risks efficiently. Today therefore constitutes an appropriate platform, where brainstorming session in respect of challenge and risk assessment of NTs involving scientists, consumers, workers, industrialists, and other stakeholders in keeping view of its effect on public health, and safety at the workplace, and the surrounding environment. Risk evaluation techniques are required to facilitate administrators to hastily evaluate the risks inflicted by the use of NTs. The main objective of the risk assessment methods

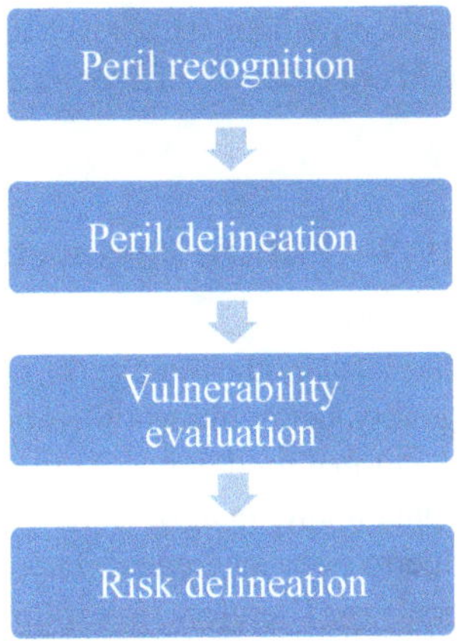

Fig. 10.2 Steps for assessment of risk/steps of risk assessment. *Source* Haimes (2004)

is delineation of a range of probable outcomes which mitigate parametric uncertainty of the mathematical models.

Most common obstacles that NT face while risk evaluation is mentioned below.

1. Scarcity of information related to environment, health, and safety NMs.
2. Lacking of index to measure the toxicity of NMs.
3. Nomenclature of NT is under in development, with no shared understanding.
4. Rising chances exposure due to ready environmental transport of NMs.
5. Difficulty in real time monitoring of NMs.

Four phases of the risk evaluation namely; Peril recognition, Peril delineation, Vulnerability evaluation, Risk delineation (Fig. 10.2).

Peril recognition stage consists of risk filtration, classification, and management (Haimes, 2004). Identifying risks at early stage is one of the techniques for implementing adjustable control in such management systems. Incertitude could be managed by the early peril recognition stage. The purpose of peril recognition is to recognize the probable danger and to determine the probability and enormity of exposure. Hazards are also known as peril that may be defined as those substance or activity which impart pernicious effects to the human beings and the atmosphere or that have noxious effects, like outbreak of fire. The characterization or delineation of peril involves valuation of toxicity by the different laboratory test and evaluation of particular biological reaction. The probability and intensity of exposure of NMs at every phase of its life span is evaluated during vulnerability or exposure evaluation. Characterization of the NMs prerequisites for assessing human health risks, and it is known that a large number of nanoparticle characteristics may influence the overall hazard (Chaudhry, 2012; Hankin, 2012). A robust method for risk assessment of nanomaterials is case by case study and this should be performed in order to take the unique properties of specific NMs into consideration (Kobe, 2012). This may the only way in which one can obtain a uniform risk assessment of a given nanoparticle and its specific properties. Nevertheless, there are scopes to think various other perceptive like what level and how existing test guidelines should be adapted for NMs in order to differentiate their effects to those of bulk materials.

10.9 Future Perspective

Use of nanotechnology in precision agriculture, enhancement of food texture and quality, better inputs utilization through NFs, NPs, NHs, packaging etc. may address the global food demand in sustainable manner. Despite numerous benefits of NT, its applicability in agriculture for sustainable production is still limited, which might be due to poor awareness, lack of legislative and regulatory framework and biosafety concern. Therefore, to exploit this technology at larger scale our future research must be focused on following points:

1. Development of suitable carrier material for unloading the various inputs i.e., nutrients, pesticides and fertilizers to achieve the higher input use efficiency.
2. Comparative study of newer nano-product with existing products is required to present the realistic results.
3. Detailed evaluation and assessment of life cycle and risk associated with nano-product on secondary or non-target species are needed to predict its behavior in environment.
4. Multidisciplinary or collaborative approach is required to develop the suitable analytical methodology for quantifying the concentration of engineered NMs in the environment.
5. In order to define the non-toxic dose of nanoparticles in environment, the permissible limits of nanoparticles concentration within safe category need to be validated carefully.
6. Future research must be emphasized on green or biosynthesized based nano-products to enhance its practical utility with minimum negative impact on environment.
7. Debate and discussion with various stake holders should be opened to familiarize this technique for higher acceptance among consumers.

10.10 Conclusion

In summary, under present climatic scenario and burgeoning food demand, introduction of NT in agriculture may potentially address the sustainability issue. The use of nano compound for the delivery of costly agriculture inputs may increase their effectiveness or reduce their amount without compromising the crop productivity. In addition, NT can be used in waste management and its recycling along with energy efficient precision agriculture. However, before citing the benefits of this emerging technology, side by side assessment of risk and cost involved in terms of economic and environmental perspective is much needed. For full exploitation of this technology, high-end analytical methods are required for characterization, enumeration and validation of nanoparticles in environment and their effect on human health. Overall, development of conceptual framework based on comprehensive database along with clearly defined regulatory guidelines is needed to encourage this technology among various stakeholders.

References

Abdel-Aziz, H. M. M., Hasaneen, M. N. A., & Omer, A. M. (2016). Nano chitosan-NPK fertilizer enhances the growth and productivity of wheat plants grown in sandy soil. *Spanish Journal of Agricultural Research, 14*(1), 1–2.

Ali, S. M., Yousef, N. M. H., & Nafady, N. A. (2015). Application of biosynthesized silver nanoparticles for the control of land snail *Eobania vermiculata* and some plant pathogenic fungi. *Journal of Nanomaterials, 1*, 1–10.

Ali, S., Rizwan, M., Noureen, S., Anwar, S., Ali, B., Naveed, M., AbdAllah, E. F., Alqarawi, A. A., & Ahmad, P. (2019). Combined use of biochar and zinc oxide nanoparticle foliar spray improved the plant growth and decreased the cadmium accumulation in rice (*Oryza sativa* L.) plant. *Environmental Science and Pollution Research, 26*, 11288–11299.

Anandan, S., Vinu, A., Venkatachalam, N., Arabindoo, B., & Murugesan, V. (2006). Photocatalytic activity of ZnO impregnated Hβ and mechanical mix of ZnO/Hβ in the degradation of monocrotophos in aqueous solution. *Journal of Molecular Catalysis, 256*, 312–320.

Averous, L., & Pollet, E. (2012). Green nano-biocomposites. In L. Avérous, & E. Pollet (Eds.), *Environmental silicate nano-biocomposites: Green energy and technology*. Springer.

Banerjee, J., & Kole, C. (2016). Plant nanotechnology: An overview on concepts, strategies, and tools. In C. Kole, D. Kumar, & M. Khodakovskaya (Eds.), *Plant nanotechnology* (pp. 1–14). Springer.

Bhattacharyya, A., Prabha, D., Marimuthu, G., Abdul, A. B., & Ram, P. (2016). Nano-biofungicides: Emerging trend in insect pest control. *Advances and applications through fungal nano biotechnology* (pp. 307–319). Springer.

Brady, N. R., & Weil, R. R. (1999). *The nature and properties of soils* (pp. 415–473). Prentice Hall.

Camargo, P. H. C., Satyanarayana, K. G., & Wypych, F. (2009). Nanocomposites: Synthesis, structure, properties and new application opportunities. *Materials Research, 12*, 1–39.

Chaudhry, Q. (2012). Current and projected applications of nanomaterials. *WHO workshop on nanotechnology and human health: Scientific evidence and risk governance*. Bonn, Germany, 10–11 December 2012.

Corradini, E., Moura, M. R., & Mattoso, L. H. C. (2010). A preliminary study of the incorporation of NPK fertilizer into chitosan nanoparticles. *Food Policy, 37*(2), 191–198.

Das, C. K., Srivastava, G., Dubey, A., Roy, M., Jain, S., Sethy, N. K., & Singh, S. K. (2016). Nano-iron-pyrite seed dressing: A sustainable intervention to reduce fertilizer consumption in vegetable (beetroot, carrot), spice (fenugreek), fodder (alfalfa), and oilseed (mustard, sesamum) crops. *Nanotechnology for Environmental Engineering, 1*(1), 2. https://doi.org/10.1007/s41204-016-0002-7

Davarpanah, S., Tehranifar, A., Davarynejad, G., Abadia, J., & Khorasani, R. (2016). Effects of foliar applications of zinc and boron nano-fertilizers on pomegranate (Punicagranatum cv. Ardestani) fruit yield and quality. *Scientia Horticulturae, 210*, 57–64.

De Oliveira, J. L., Campos, E. V., Bakshi, M., Abhilash, P. C., & Fraceto, L. F. (2014). Application of nanotechnology for the encapsulation of botanical insecticides for sustainable agriculture: Prospects and promises. *Biotechnology Advances, 32*(8), 1550–1561.

Dehaghi, S. M., Rahmanifar, B., Moradi, A. M., & Azar, P. A. (2014). Removal of permethrin pesticide from water by chitosan-zinc oxide nanoparticles composite as an adsorbent. *Journal of Saudi Chemical Society, 18*, 348–355.

Delfani, M., Firouzabadi, M. B., Farrokhi, N., & Makarian, H. (2014). Some physiological responses of black-eyed pea to iron and magnesium nanofertilizers. *Communication in Soil Science and Plant Analysis, 45*, 530–540.

Diaz, R. J., & Rosenberg, R. (2008). Spreading dead zones and consequences for marine ecosystems. *Science, 321*, 926–929.

Dietz, K. J., & Herth, S. (2011). Plant nanotoxicology. *Trends in Plant Science, 16*, 582–589.

Disfani, M. N., Mikhak, A., Kassaee, M. Z., & Maghari, A. (2017). Effects of nano Fe/SiO_2 fertilizers on germination and growth of barley and maize. *Archives of Agronomy and Soil Science, 63*, 817–826.

Dubey, A., & Mailapalli, D. R. (2016). Nano-fertilizers, nanopesticides, nanosensors of pest and nanotoxicity in agriculture. In E. Lichtfouse (Ed.), *Sustainable agriculture reviews* (Vol. 19, pp. 307–330). Springer.

Dwivedi, S., Saquib, Q., Al-Khedhairy, A. A., & Musarrat, J. (2016). Understanding the role of nanomaterials in agriculture. In D. P. Singh, H. B. Singh, & R. Prabha (Eds.), *Microbial inoculants in sustainable agricultural productivity* (pp. 271–288). Springer.

El-Feky, S. A., Mohammed, M. A., Khater, M. S., Osman, Y. A., & Elsherbini, E. (2013). Effect of magnetite nano-fertilizer on growth and yield of *Ocimum basilicum* L. *International Journal of Indigenous Medicinal Plants, 46*, 1286–1293.

Emadian, S. E. (2017). *Physiological responses of loblolly pine (Finustaeda L.) to silicon and water stress* (pp. 27–37). Texas A & M University (Ph.D. thesis).

Ge, Y., Schimel, J. P., & Holden, P. A. (2012). Identification of soil bacteria susceptible to TiO_2 and ZnO nanoparticles. *Applied and Environmental Microbiology, 78*, 6749–6758.

Giannousi, K., Avramidis, I., & Dendrinou-Samara, C. (2013). Synthesis, characterization and evaluation of copper based nanoparticles as agrochemicals against Phytophthora infestans. *RCS Advances, 3*, 21743–21752.

Glenn, J. C., & Florescu, E. (2016). Millennium project team. 2015–16 state of the future. *Journal of Socialomics, 5*(3), 1–6.

Goswami, A., Roy, I., Sengupta, S., & Debnath, N. (2010). Novel applications of solid and liquid formulations of nanoparticles against insect pests and pathogens. *Thin Solid Films, 519*, 1252–1257.

Grillo, R., Abhilash, P. C., & Fraceto, L. F. (2016). Nanotechnology applied to bio-encapsulation of pesticides. *Journal of Nanoscience and Nanotechnology, 16*, 1231–1234.

Haghighi, M., Afifipou, Z., & Mozafariyan, M. (2012). The effect of N-Si on tomato seed germination under salinity levels. *Journal of Biological and Environmental Sciences, 6*, 87–90.

Haimes, Y. Y. (2004). *Risk modeling, assessment and management* (2nd ed.). Wiley.

Hankin, S. (2012). REACH implementation projects on nanomaterials: Outcomes and implementation. In *WHO workshop on nanotechnology and human health: Scientific evidence and risk governance*, Bonn, Germany, 10–11 December 2012.

Hojjat, S. S., & Hojjat, H. (2015). Effect of nano silver on seed germination and seedling growth in fenugreek seed. *International Journal of Food Engineering, 1*(2), 106–110.

Imada, K., Sakai, S., Kajihara, H., Tanaka, S., & Ito, S. (2016). Magnesium oxide nanoparticles induce systemic resistance in tomato against bacterial wilt disease. *Plant Pathology, 65*(4), 551–560.

Ion, A. C., Ion, I., & Culetu, A. (2010). Carbon-based nanomaterials: Environmental applications. *University of Politehnica of Bucharest, 38*, 129–132.

Joshi, A., Kaur, S., Dharamvir, K., Nayyar, H., & Verma, G. (2018). Multi-walled carbon nanotubes applied through seed-priming influence early germination, root hair, growth and yield of bread wheat (*Triticum aestivum* L.). *Journal of the Science of Food and Agriculture, 98*, 3148–3160.

Kale, A. P., & Gawade, S. N. (2016). Studies on nanoparticle induced nutrient use efficiency of fertilizer and crop productivity. *Green Chemistry and Technology Letters, 2*, 88–92.

Khan, M. R., & Rizvi, T. F. (2017). Application of nanofertilizer and nanopesticides for improvements in crop production and protection. In *Nanoscience and plant-soil systems* (pp. 405–427). Springer.

Khin, M. M., Nair, A. S., Babu, V. J., Murugan, R., & Ramakrishna, S. (2012). A review on nanomaterials for environmental remediation. *Energy and Environmental Science, 5*, 8075–8109.

Khot, L. R., Sankaran, S., Maja, J. M., Ehsani, R., & Schuster, E. W. (2012). Applications of nanomaterials in agricultural production and crop protection: A review. *Crop Protection, 35*, 64–70.

Kobe, A. (2012). Nanomaterials in European Union regulation. In *WHO workshop on nanotechnology and human health: Scientific evidence and risk governance*, Bonn, Germany, 10–11 December 2012.

Kole, C., Kole, P., Randunu, K. M., Choudhary, P., Podila, R., Ke, P. C., Rao, A. M., & Marcus, R. K. (2013). Nanobiotechnology can boost crop production and quality: First evidence from increased plant biomass, fruit yield and phytomedicine content in bitter melon (*Momordica charantia*). *BMC Biotechnology, 13*, 37.

Kumar, S., Bhanjana, G., Sharma, A., Dilbaghi, N., Sidhu, M. C., & Kim, K. H. (2017). Development of nanoformulation approaches for the control of weeds. *Science of the Total Environment, 586*, 1272–1278.

Lambreva, M. D., Lavecchia, T., Tyystjarvi, E., Antal, T. K., Orlanducci, S., & Margonelli, A. (2015). Potential of carbon nanotubes in algal biotechnology. *Photosynthesis Research, 125*, 451–471.

Lowry, G. V., Avellan, A., & Gilbertson, L. M. (2019). Opportunities and challenges for nanotechnology in the agri-tech revolution. *Nature Nanotechnology, 14*, 517–522.

Mahajan, P., Dhoke, S. K., & Khanna, A. S. (2011). Effect of nano-ZnO particle suspension on growth of mung (*Vigna radiata*) and gram (*Cicer arietinum*) seedlings using plant agar method. *Journal of Nanotechnology, 7*, 1–7.

Manjaiah, K. M., Mukhopadhyay, R., Paul, R., Datta, S. C., Kumararaja, P., & Sarkar, B. (2019). Clay minerals and zeolites for environmentally sustainable agriculture. In M. Mariano, S. Binoy, & L. Alessio (Eds.), *Modified clay and zeolite nanocomposite materials* (pp. 309–329). Elsevier.

Mishra, S., Singh, B. R., Singh, A., Keswani, C., Naqvi, A. H., & Singh, H. B. (2014). Biofabricated silver nanoparticles act as a strong fungicide against *Bipolaris sorokiniana* causing spot blotch disease in wheat. *PLoS ONE, 9*(5), e97881.

Mishra, S., & Singh, H. B. (2015). Biosynthesized silver nanoparticles as a nanoweapon against phytopathogens: Exploring their scope and potential in agriculture. *Applied Microbiology and Biotechnology, 99*, 1097–1107.

Mullen, A. (2019). Expectations from nano in agriculture. *Nature Nanotechnology, 14*, 515–516.

Nuruzzaman, M., Rahman, M. M., Liu, Y., & Naidu, R. (2016). Nanoencapsulation, nano-guard for pesticides: A new window for safe application. *Journal of Agricultural and Food Chemistry, 64*, 1447–1483.

Onaga, G., & Kerstin, W. (2016). Advances in plant tolerance to biotic stresses. In I. Y. Abdurakhmonov (Ed.), *Plant genomics* (pp. 167–228). Tech.

Panpatte, D. G., Jhala, Y. K., Shelat, H. N., & Vyas, R. V. (2016). Nanoparticles: The next generation technology for sustainable agriculture. In *Microbial inoculants in sustainable agricultural productivity* (pp. 289–300). Springer.

Peteu, S. F., Oancea, F., Sicuia, O. A., Constantinescu, F., & Dinu, S. (2010). Responsive polymers for crop protection. *Polymers, 2*, 229–251.

Prasad, R., Kumar, V., & Prasad, K. S. (2014). Nanotechnology in sustainable agriculture: Present concerns and future aspects. *African Journal of Biotechnology, 13*(6), 705–713.

Saifuddin, N., Nian, C. Y., Zhan, L. W., & Ning, K. X. (2011). Chitosan-silver nanoparticles composite as point of use drinking water filtration system for household to remove pesticides in water. *Asian Journal of Biochemistry, 6*, 142–159.

Scrinis, G., & Lyons, K. (2007). The emerging nano-corporate paradigm nanotechnology and the transformation of nature, food and agrifood systems. *International Journal of Sociology of Agriculture and Food, 15*, 22–44.

Sekhon, B. S. (2014). Nanotechnology in agri-food production: An overview. *Nanotechnology, Science and Applications, 7*, 31–53.

Servin, A. D., & White, J. C. (2016). Nanotechnology in agriculture: Next steps for understanding engineered nanoparticle exposure and risk. *Nano Impact, 1*, 9–12.

Servin, A., Elmer, W., Mukherjee, A., De la Torre-Roche, R., Hamdi, H., White, J. C., Bindraban, P., & Dimkpa, C. A. (2015). Review of the use of engineered nanomaterials to suppress plant disease and enhance crop yield. *Journal of Nanoparticle Research, 17*, 92.

Shahrokh, S., Hosseinkhani, B., & Emtiazi, G. (2014). The impact of nanosilver on bacterial aerobic nitrate reductase. *Journal of Bioprocessing and Biotechnologies, 4*, 162.

Shalaby, T. A., Bayoumi, Y., Abdalla, N., Taha, H., Alshaal, T., Shehata, S., Amer, M., Domokos-Szabolcsy, E., & El-Ramady, H. (2016). Nanoparticles, soils, plants and sustainable agriculture. In R. Shivendu, D. Nandita, & L. Eric (Eds.), *Nanoscience in food and agriculture* (pp. 283–312). Springer.

Shang, Y., Hasan, K., Ahammed, G. J., Li, M., Yin, H., & Zhou, J. (2019). Applications of nanotechnology in plant growth and crop protection: A Review. *Molecules, 24*, 2558.

Sharifi-Rad, J., Sharifi-Rad, M., & da Silva, J. (2016). Morphological, physiological and biochemical responses of crops (*Zea mays* L., *Phaseolus vulgaris* L.), medicinal plants (*Hyssopuso cinalis* L., *Nigella sativa* L.), and weeds (*Amaranthus retroflexus* L., *Taraxacumocinale* F. H. Wigg) exposed to SiO_2 nanoparticles. *Journal of Agriculture, Science and Technology, 18*, 1027–1040.

Shen, Z., Chen, Z., Hou, Z., Li, T., & Lu, X. (2015). Ecotoxicological effect of zinc oxide nanoparticles on soil microorganisms. *Frontiers of Environmental Science and Engineering, 9*, 912–918.

Shojaei, T. R., Salleh, M. A. M., Tabatabaei, M., Mobli, H., Aghbashlo, M., Rashid, S. A., & Tan, T. (2019). Applications of nanotechnology and carbon nanoparticles in agriculture. In A. R. Suraya, R. Niro, & Z. H. Mohd (Eds.), *Synthesis, technology and applications of carbon nanomaterials* (pp. 247–277). Elsevier.

Shukla, P., Chaurasia, P., Younis, K., Qadri, O. S., Faridi, S. A., & Srivastava, G. (2019). Nanotechnology in sustainable agriculture: Studies from seed priming to post-harvest management. *Nanotechnology for Environmental Engineering, 4*, 11.

Shyla, K. K., Natarajan, N., & Nakkeeran, S. (2014). Antifungal activity of zinc oxide, silver and titanium dioxide nanoparticles against *Macrophomina phaseolina. Journal of Mycology and Plant Pathology, 44*, 268–273.

Smith, P., Bustamante, M., Ahammad, H., Clark, H., Dong, H., Elsiddig, E. A., Haberl, H., Harper, R., House, J., Jafari, M., Masera, O., Mbow, C., Ravindranath, N. H., Rice, C. W., Robledo, A. C., Romanovskaya, A., Sperling, F., & Tubiello, T. (2014). Agriculture, forestry and other land use (AFOLU). In O. Edenhofer, R. Pichs-Madruga, Y. Sokona, E. Farahani, S. Kadner, K. Seyboth, A. Adler, I. Baum, S. Brunner, P. Eickemeier, B. Kriemann, J. Savolainen, S. Schlomer, C. Von Stechow, T. Zwickel, & J. C. Minx (Eds.), *Mitigation of climate change. Contribution of working group III to the fifth assessment report of the intergovernmental panel on climate change* (pp. 811–922). Cambridge University Press.

Song, G., Gao, Y., Wu, H., Hou, W., Zhang, C., & Ma, H. (2012). Physiological effect of anatase TiO_2 nanoparticles on Lemna minor. *Environmental Toxicology and Chemistry, 31*, 2147–2152.

Tomasevic, A., Kiss, E., Petovic, S., & Mijin, D. (2010). Study on the photocatalytic degradation of insecticide methomyl in water. *Desalination, 262*, 228–234.

Torabian, S., Zahedi, M., & Khoshgoftar, A. H. (2017). Effects of foliar spray of nano-particles of $FeSO_4$ on the growth and ion content of sunflower under saline condition. *Journal of Plant Nutrition, 40*, 615–623.

Tripathi, D. K., Singh, S., Singh, V. P., Prasad, S. M., Dubey, N. K., & Chauhan, D. K. (2017). Silicon nanoparticles more effectively alleviated UV-B stress than silicon in wheat (*Triticum aestivum*) seedlings. *Plant Physiology and Biochemistry, 110*, 70–81.

USDA. (2018). *World agricultural production circular series WAP 4–18.*

Vandevoort, A. R., & Arai, Y. (2012). Effect of silver nanoparticles on soil denitrification kinetics. *Industrial Biotechnology, 8*, 358–364.

Vermeulen, S. J., Aggarwal, P. K., Ainslie, A., Angelone, C., Campbell, B. M., Challinor, A. J., Hansen, J. W., Ingram, J. S. I., Jarvis, A., & Kristjanson, P. (2012). Options for support to agriculture and food security under climate change. *Environmental Science and Policy, 15*, 136–144.

Wang, S. H., Wang, F. Y., & Gao, S. C. (2015). Foliar application with nano-silicon alleviates Cd toxicity in rice seedlings. *Environmental Science and Pollution Research, 22*, 2837–2845.

Wang, S. H., Wang, F. Y., Gao, S. C., & Wang, X. G. (2016). Heavy metal accumulation in different rice cultivars as influenced by foliar application of nano-silicon. *Water, Air and Soil Pollution, 227*(7), 228.

Wang, Y., Lin, Y., Xu, Y., Yin, Y., Guo, H., & Du, W. (2019). Divergence in response of lettuce (var. ramosa Hort.) to copper oxide nanoparticles/microparticles as potential agricultural fertilizer. *Environmental Pollutants and Bioavailability, 31*(1), 80–84.

Willett, W., Rockstrom, J., Loken, B., Springmann, M., Lang, T., Vermeulen, S., & Jonell, M. (2019). Food in the anthropocene: The EAT–Lancet Commission on healthy diets from sustainable food systems. *The Lancet, 393*(10170), 447–492.

Wilson, M. A., Tran, N. H., Milev, A. S., Kannangara, G. S. K., Volk, H., & Lu, G. Q. M. (2008). Nanomaterials in soils. *Geoderma, 146*, 291–302.

Wu, L., & Liu, M. (2008). Preparation and properties of chitosan coated NPK compound fertilizer with controlled release and water-retention. *Carbohydrate Polymers, 72*, 240–247.

Yang, F., Liu, C., Gao, F., Su, M., & Wu, X. (2007). The improvement of spinach growth by nano-anatase TiO_2 treatment is related to nitrogen photoreduction. *Biological Trace Element Research, 119*(1), 77–88.

Yang, H., Xu, M., Koide, R. T., Liu, Q., Dai, Y., Liu, L., & Bian, X. E. (2016). Effects of ditch-buried straw return on water percolation, nitrogen leaching and crop yields in a rice-wheat rotation system. *Journal of the Science of Food and Agriculture, 96*, 1141–1149.

Yousefzadeh, S., & Sabaghnia, N. (2016). Nano-iron fertilizer effects on some plant traits of dragonhead (*Dracocephalum moldavica* L.) under different sowing densities. *Acta Agriculturae Slovenica, 107*, 429–437.

Zhang, Q., Han, L., Jing, H., Blom, D. A., Lin, Y., & Xin, H. L. (2016). Facet control of gold nanorods. *ACS Nano, 10*, 2960–2974.

Zhang, W. (2018). Global pesticide use: Profile, trend, cost/benefit and more. *Proceedings of the International Academy of Ecology and Environmental Sciences, 8*, 1–27.

Chapter 11
Climate Resilient Development for Discourse the Disastrous Confront

Neelam Yadava

11.1 Climate Resilience

There are innumerable definitions about the climate resilience which start from the origin which is about the climate change. The evidence of science is overwhelming in that the climate is altering and manmade action is the primary factor of changing the climate as well as to pace the acceleration since last several centuries. The capability to formulate strategies and plans for convalesce from and implement to these impressions is called climate resilience. The resilience become a kind of term which becomes more binocular when the extreme weather conditions shows that resilience is the important component which is about the climate action program because climate change globally and on a local level also. There are lot of evidences which reflects that the risks because of the climate change are going to address and which is not on the protecting the people and property but which can generate the economic value also that helps to create domestic jobs and prosperity.

According to the Oxford English dictionary the original orientation to resilience given by Francis Bacon in seventeenth century and it describes that the physical features of an eco and how it rebounds back of a wall. But there are many people they have define the term resilience in their own context there is something which is bounced back from that is what is aliens is or not absolute term it is a relative term. In certain countries like Western US and Perry acute in Cape Town South Africa there are the problems where individual's necessity is to be resilient to not enough water the planning of the places is essential to rebound back from impact of Climate Change climate resilience, that require climate risk be further wholly tacit. The National Academy of Sciences define resilience as the capability to formulate and strategy for engross convalesce from and additional effectively adopt to antagonistic actions.

N. Yadava (✉)
Tata Institute of Social Sciences, School of Rural Development, Tuljapur Campus, Osmanabad, India
e-mail: neelamunique@tiss.edu

S. Sheraz Mahdi et al. (eds.), *Adapting to Climate Change in Agriculture-Theories and Practices*, https://doi.org/10.1007/978-3-031-28142-6_11

11.2 Climate Resilience and Risk

The relationship of risk with resilience When there are certain infrastructural facilities in that case the community become more resilient to face and stress, that require to understand the concept of risk as well as to management of the risk. It can be defined as risk is the involvement of some related thoughts these are threats and vulnerability. These terms are often used interchangeably or synonymously, but it all have precisely identical diverse meaning. The activities can be negatively impact on a strength or structural vulnerability is the grade of probable impairment to the strength and risk is the purpose of threat as vulnerability. The mechanisms of vulnerability like compassion acquaintance and adaptive capacity these are the important and perilous factor to assess resilience.

Risk = Vulnerability (Sensitivity exposure and adaptive capacity) × Threat

Vulnerability is determined by sensitivity, exposure, and adaptive capacity of an individual or system.

Source: C2ES (Centre for Climate and Energy Solutions, April 2019).

11.3 The Outcome of Resilience

In the past ten years there were tropical storm, cyclone, flood, rainfall for the aftermath of the extreme climatic conditions, which can be shape back better or tougher than before it all has done under the name of the resilience. The argument was centred about the reappearance on the speculation of resilient building that is what it can be called as benefit cost ratio.

Climate resilient as the capacity for social economic system which includes to engross pressures and preserve functions in appearance of exterior pressures that executed upon it by climate change. It may include to adapt recognise and evolve into some desirable configurations such as to improve the sustainability of the system which can make a prepare for a future climate change impact. Definition of climate resilience is still debatable because it is having in both the terms in practical as well as in conceptual terms. In case of climate resilience there are multiple things in world specially like how climate resilience related to climate change adaptation, it is degree to which it must in compare active grounded, versus system grounded tactics to improve steadiness, it is related with the equilibrium of nature theory for homeostasis equipoise with the understanding of ecological structure. In the current situation there are a lot of focus are given on the climate resilience and activities taken to preserve the prevailing system and structure.

11.4 Additional Resources and Tools

There are additional resources and toolkits for the discovering of climate changes, its hazards and developing solutions, which can provide library of tools for various levels ranging from individual to cities. There is package-based information tailored according to the needs of the areas of interest. There are mapping tools which allows to expression at forecasts for rainfall and its impact in various ranges of season over the period of time, such as dry, humid, wet etc. there are the World wide fund, which looked at the publicly available screening tools, policies, international financial institutions for assessing the sustainability and climate resilience.

- U.S. Resilience toolkit: https://toolkit.climate.gov/
- Climate Change Adaptation Resource Centre (ARC-X) EPA'S Adaptation Resource Centre: (ARC-X) https://www.epa.gov/arc-x
- Climate Resilience Evaluation And Awareness Tool (CREAT) Climate Scenarios Projection: https://epa.maps.arcgis.com/apps/MapSeries/index.html?appid=3805293158d54846a29f750d63c6890e
- World Wildlife Fund Tool Screening: https://www.worldwildlife.org/publications/review-of-screening-tools-final-report-sep-2017

11.5 Application of Resilience Framework to Address the Vulnerability

Because of the climate resilience it creates an idea of multi stable social economic situation where the modern interpretations of resilience mention that it has been established as a social logical system which can actually stabilizer roundup multiple possible state. Along with the same climate resilience state an important role and emphasize in the importance on preventive actions when it accesses the effect of climate change. Climate change perspective can encourage a great across scale connectedness of systems where it can be argued solely rely on concepts of adaptation is a preventive factor since fundamentally the perception does not necessity it as abundant full system consistency as a resilience viewpoint.

As per the **International Panel on Climate Change**, vulnerability can be demarcated as using three features:

- Adaptive capacity
- Sensitivity
- Exposure to the effects of climate change.

Adaptive capacity means communities capacity where it can create resilience infrastructure compassion and acquaintance, these are tied to economic and Geographic essentials that are very in different societies.

Vulnerability can be broadly divided into two categories, economic variable vulnerability which is on the basis of social economic conditions as well as the Geographic vulnerability.

- At a very elementary level the community, which is economically vulnerable that is not properly prepared for the consequences of effects of the climate change since it doesn't have the initial resources formulating. Climate resilient culture requires lot of investment of infrastructure of city planning, engineering, justifiable energy resources and prepared arrangements. At a global level it is found that it more likely the people living in below scarcity line they are more sufferer, because of the climatic change and it becomes more vulnerable because they don't have much amount of resources to invest in the resilient infrastructure. It has very less resource for clear-out efforts that's why there are very many chances for occurrence of climate change related disasters.
- In case of geographical mobility, it related to the geography where are the location to climate change that are affecting more side by side it is having a natural threat like rising sea levels or histrionic changes in ecosystem amenities which can include food accessibility. Island countries are considered more vulnerable because the community which resides in a rely heavily on the sustenance lifestyle up to a larger risk. Abaco islands is the example of low elevation island society, which is more affected by raising the sea level due to climatic conditions.

11.6 Climate Justice

The national and international carbon market that approach and proposed which try to resolve the concerns of using market fluctuations to make carbon used less reasonable. But the vulnerable host societies that are for getting the benefits, but it does not receive any benefits or very minimal. The major issues which notified from the carbon market is that the intrinsic struggle of concentration which a body between developed and sustain in space societies in that case developed Nations have frequently priorities development of their own Gross National Product over the executing change, which can confront the issues of the climate change which concerned by the taxing carbon and it could effect on GDP.

11.7 Environmental Stress

Environmental stress defines the Physical, chemical and biological limitations on production of species and other expansion of ecosystem. Environmental stresses and please or decrease in intensity ecological response resolves stresses is research can be normal environmental features or may be from the actions of human. Environmental stresses can be grouped under its following categories such as Physical stresses, Wildfire, contamination, Thermal stress, Radiation stress, Climatic stress and biological stress.

- Physical stress test about intense exposure to kinetic energy which is the ecological trouble due to its effect, severe episodic nature like winds storms and explosions.
- Wildlife is a kind of disturbance which can be because of abundant of the Biomass in the ecosystem is combustion and foremost species can be killed.
- Pollution happens when chemical is existing in concentration large adequate to affect the organism that is what there are ecological changes there are some toxic gases hey such as Sulphur dioxide, ozone, lead, Mercury and pesticides. These all inputs of nutrient can cause various types of pollution which is called as eutrophication.
- Thermal stress is about the heat influence ecosystem which occurs in the locality of natural hot water outlets on the ocean ground or a nearby the industrial expulsions of impassioned water.
- Radiation stress is the extreme masses of ionizing energy which can occurs in the foothill top where the concentrated exposure of ultraviolet radiation or radioactive material.
- Climatic stress is the extreme for inadequate regions of temperature and moisture solar radiation or combination of all these such as in tundra and desserts.
- Biological stress is about diverse interaction which occurs among organism of the same or different species, which consequently have competition herbivory, predation or disease.

11.8 Political Economy Perspective on Climate Change

There are certain researches which focus and approaches towards the political economy. Brandt and Svendsen (2003) the introduced the term political economy Framework which is grounded on the political sustenance function model by Hillman which analyse the optimal of instrument to regulatory climate change in the European Union policy to instrument its Kyoto Protocol target label. Another report prepared by Bank for Reconstruction and Development in 2011 that considered the political economic tactic to elucidate why some countries adopted climate change strategies although some other countries do not? Precisely, the nations which is in the evolution area. The analysis of the research was the changed political economy aspect of the features of Climate Change strategies which is to comprehend the probable factor lashing climate change justification outcome in many changeover nations the main conclusion of the report was given. The label of egalitarianism is not the foremost driver of Climate Change strategy it means the expectation of involvement to global climate change extenuation may not essentially restricted by the political administration of the country.

- The familiarity of public which formed by numerous features counting threat of Climate Change in a specific nation the education level and existence of unrestricted media perilous component in climate change strategy approval are expressively more probable to implement climate change policies. Therefore, the emphasis of that was to promote community awareness on the crucial threat of Climate Change and prevent evidence irregularities in the many transition nations.
- The carbon concentrated commerce is the main different to adopt the climate change policies sometimes it stimulus government choice construction on climate change strategy that demands for change of inducements supposed by these trades and a changeover of them to a low-carbon construction outline it means that comprise the energy price restructuring and it introduces the international carbon interchange mechanism.
- In the competitive age it gains nationwide economy is in the evolution area in the global economy while it increases international pressure to place to diminish emissions that augment their political Administration internal legitimacy that could support to discourse the intrinsic economic businesses with underline like of economic divergence in global economic crisis.

As per the study of Tanner and Allouche (2011) to develop innovative intangible and methodological Outline for analysing the political economy of Climate Change in the work that could focus on climate change procedures and products in relations of idea authority and capitals. New political economic approach it could expect to go beyond the dominant political economic to switch articulated by international developed countries to analyse the climate change initiative which is ignore the way the idea and ideologies determine the policy outcome (Table 11.1).

Table 11.1 Assessment among the innovative and outmoded political economy investigation of climate change ingenuities

Subject	Foremost approach	Innovative political economy
Policy procedure	Linear, knowledgeable by evidence	Complex, knowledgeable by ideology, actors and power relations
Dominant scale	Global and inter-state	Transformation of international to national and sub-national level
Climate change science and research	Role of objective science in informing policy	Social structure of science and driving narratives
Scarcity and poverty	Distributional outcomes	Political procedures interceding contending entitlements for resources
Decision-making	Collective achievement, balanced choice and payment looking for	Ideological drivers and inducements, power relations

Source Tanner and Allouche (2011)

11.9 Measuring Climate Resilience

There are various agencies they are spending lot of time as well as investment to sustenance resilience building involvements. Measurement of resilience helps to make a lot of contribution in administrative resource allocation towards resilience construction which can include targeted documentation of vulnerability Hotspots, well considerate of the drivers of resilience and tools of in for the influence and efficiency of resilience structure interferences in the current time. There are a amount of Resilience measurement tools also has been appeared which is providing a way to pathway and degree resilience at our range of scales. There are various challenges faced in number in measuring climate resilience such as the description of resilience profoundly questioned it becomes problematic to choose suitable features and pointers, to pathway another difficulty is the resilience or families are societies cannot be restrained by means of a solitary observable metric. There are range of procedures and features which can be imperceptible and problematic to perceive such as social capital that is what the person will get should have a large list of proxy indicators.

In the current time there are a lot of creativities to measure resilience in the rural development context which can have certain shortcoming such as Complexity and high cost. In the practical sense the tools of measuring climate resilience can be appropriate across project of diverse scales which can have equivalent access for a slighter creativity to comprehend the level and opportunity of resilience in a public. The tools of resilience measurement can be classified into subjective and objective methods which may include the two core traits these are how resilience is defined and how it is measured?

- Regarding the definition of resilience who decide what resilience is and the features which make a house resilient and in terms of how to measure the resilience restrained by means of exterior remark or self-assessed verdicts.

Global Framework for climate resilience services. There is an intergovernmental procedure which recognized a task force of high-level autonomous advisory to make recommendations to make significances and propose events of the Global forum of climate services. In 2012 there is an astonishing station of World Meteorological Congress has published the intergovernmental board on climate service as the governing body of GFCS. In 2019 as a form to Reform the WM governance World metrological Congress decided to liquify the IDC's and accept a new climatic direction board as the oversight and execution mechanism for GFCS. Guiding principles includes to priorities capacity expansion in the developing countries which is vulnerable to climate in packed it promotes permitted and exposed exchange of climate relevant data whenever and wherever it would be possible which provide climate amenities principally for the public goods and simplifying the strength and prevailing actions and not replicating. Reflects to shape partnership relating all the stakeholders which is important for evolving, distributing and using climate amenities for generating a marketplace of organizations carrying capitalizing in in using service. There are numerous examples of GSCS success at regional level, national level.

As per the UN Millennium Development Goals, there are eight Millennium Development Goals (MDGs)—which includes from thrilling poverty to controlling spread of HIV/AIDS, providing universal education, were all agreed by all the all the world's developing institutions. The MDGs were adopted in September 2000, with the aims to eliminate extreme poverty and starvation, universal primary education, endorse gender equality and permit women, reduce child mortality, progress mental health etc., like wise 8 goals are there.

11.10 Confronting Disastrous Challenges of Climate Change

Confrontation of disaster challenge on the climate change as an individual it could be the responsibility to take a small action where it can help to reduce the nation with of climate changes. Regarding the large-scale options, government need to take certain actions with the fundamentals to protect the Welfare of the people such as they are high rate of greenhouse gases in Asian, which is paired with environmental degradation as well as over exploitation of natural resources. Almost 97% of the scientist are in believe that all these climate destructions have happened because the human activities, so when we flop to break global warming, the change will be catastrophic. **United Nations Climate Conference** at as every year the Global leaders encounter to deliberate schedules which can be taken to avoid the climatic disaster risk conditions. There is some confrontation mechanism to address the disastrous issues because of the climate changes, These are

- It can protect and restore the key ecosystem
- by supporting small agricultural producers
- by promoting green energy and
- combat short lived climate pollutants as well as
- focus on adaptation rather than just mitigation only.

These are some of the mechanisms is in order to mitigate the disaster situations because of climate change.

11.10.1. In case of protection and Restoration of ecosystem government can protect ecosystem by including the rivers wetlands oceans forest and mangroves engross large numbers of carbon decelerating global warming. Mangroves assist as a barricade in contradiction of the Tropical Storms and wetlands captivate excess water from the floods both extreme weather actions can make by the climate change.

11.10.2. The small agricultural producers can also be supported conferring to the meat Industry that is accountable for 15–18% of altogether greenhouse gas emissions which exceeds from those of the transportation. About 80% of the agricultural production goes for the feeding animal rather than feeding to the people, that is what the expansion of land for livestock as well as crop to feed them become a significant cause of deforestation in the Amazon.

11.10.3. By supporting small local producers' government can make some difference which is not like the large factories' farmers which employee sustainable practices care about land registration in making animal and crop more resilient to climate change.

Another way to combat the climate change disaster situation is to promote green energy in this case 35% of all the Global innovation comes from the energy production but in certain countries where there is a more development take and also consider the more energy production (Fig. 11.1).

11.10.4. Longest back the thermoelectric and hydroelectric energy is considered as the cheapest option because of the technological expansions that permitted us to discovery better cheaper and in a more effective manner. In terms of making extensive Planning Commission can avoid cold climate aggravated energy sources because hydropower it is not a dream source it can adopt for small wind solar geothermal oceanic and other schemes.

11.10.5. Carbon dioxide is the greatest in all Greenhouse gas that leftovers in the atmosphere for centuries even though if we stop to emit source today only

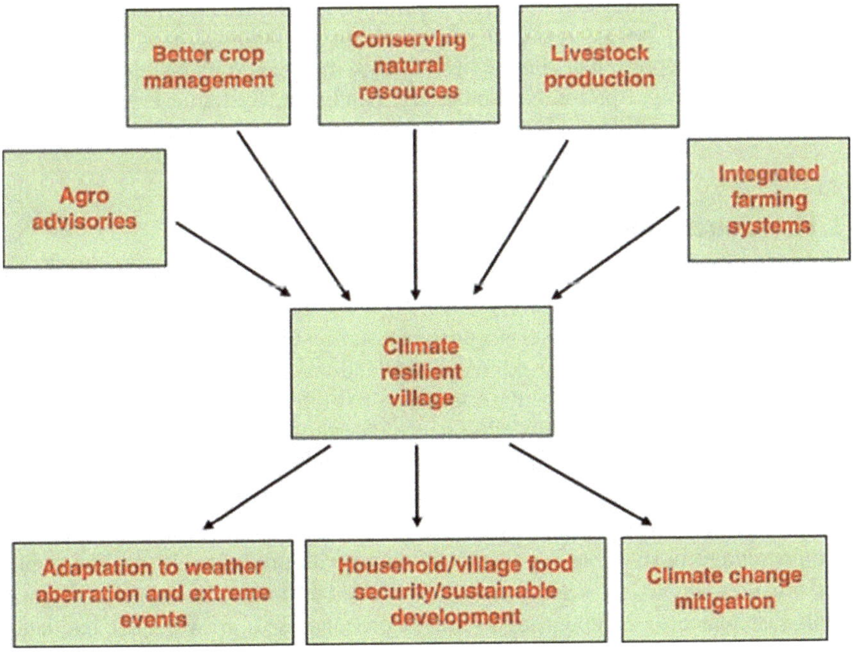

Fig. 11.1 Climate resilient villages for sustainable food security in tropical India: concept, process, technologies, institutions, and impact

the effect of Climate Change can remain last. The good part is that the contamination happen that subsidize to climate change and only in the last few days or years it is in the atmosphere that is called as a short-lived climate pollutant which is accountable for 30–45% of emissions of global warming. The contaminants which include black carbon Methane Ozone hydrofluorocarbon which found in refrigerants taken effective control can be done through the national level of regulations and policies which can quicken the competition in contradiction of climate change in the short term.

11.10.6. In the fight of making change in the climatic conditions are the disaster situations which aim to reduce emissions stop their effects and future consequences that is called is mitigation but this is also important that approximately communities are previously having experience of catastrophic significances due to change in the climate so it needs to talk about the preventing the scanners trophies by increasing resilience and reduce vulnerabilities that is called as adaptation. It needs it reflects that rather than talking about the mitigation and adaptation part is also important because there are a number of projects to mitigate emissions are more attractive initially but it could be design for adaptation which are usually attentive on the maximum vulnerable communities.

11.10.7. Regarding the landscape of climate change, the stakeholders which can discourse the risk modelled by climate change only if it can comprehend it clearly and see it properly so it becomes so complex to confront. Possibilities of physical climate risk which is having several characteristics like Increasing, Spatial, Nonstationary, Nonlinear, Systemic Progressive and Unprepared (Fig. 11.2).

11.11 Conclusion

According to WBG (World Bank groups) by 2025 the aim of delivering fund are disaster risk management, water security, coastal resilience, human development, financial protection and forest and integrated land management. Where in Disaster risk management, is will enlarge the admittance to high quality hydrometeorological data as well as early cautionary system, which can add 250 million more public in at least 30 developing nations and will sustenance 100 interventions for flood forecasting system, hydrological and meteorological system. For water security it would support for 100 river basins with climate informed management tactics as well as improvement in river basin governance along with providing 15 million public with enhanced flood and scarcity risk management structure. In coastal resilience, it will sustenance to 20 countries to accept procedures to growth their resilience for any climate-based shock and stresses in the coastal areas. In human development it would support to 20 climate hotspot nations along with human development engagement. Which will include all the developmental aspect like education,

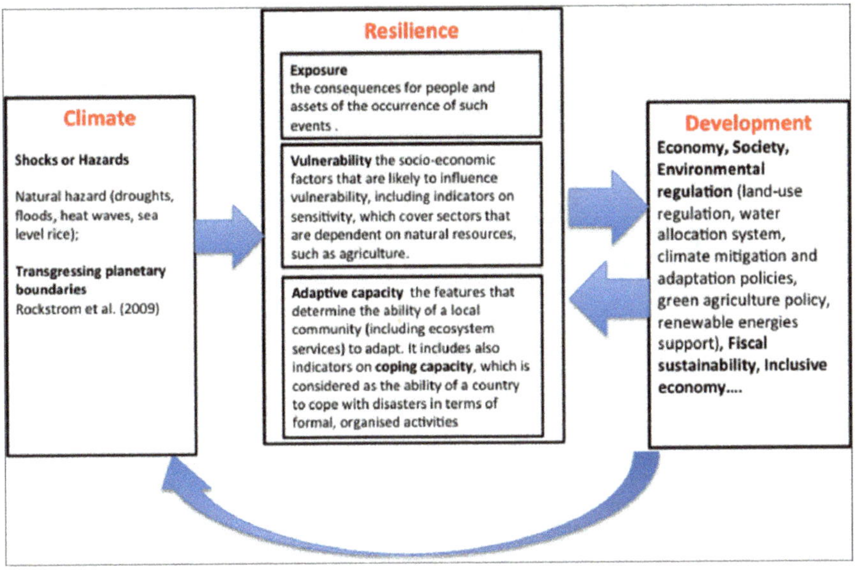

Fig. 11.2 Climate resilient index. *Source* A. Miola (April 2015)

health, nutrition, social protection, and livelihood. For financial protection, it would support for 20 countries to recover from the climatic and disastrous shock quickly and faster, through monetary sector supervisory reforms. Regarding forest and integrated landscape management, it would sustenance through some interventions like avoid deforestation, promote landscape restoration and sustained forest management for 120 million hectares of forest in 50 nations.

References

Brandt, U. S., & Svendsen, G. T. (2003) *The political economy of climate change policy in the EU: Auction and grandfathering*. IME Working Papers No. 51/03.

Tanner, T., & Allouche, J. (2011). Towards a new political economy of climate change and development. *IDS Bulletin Special Issue: Political Economy of Climate Change, 42*(3), 1–14.

Chapter 12
Cropping Systems for Sustainable Millet Production

T. S. Sukanya, Ajay Kumar, A. L. Narayanan, K. Sathya, Kaushal Kishore, Manisha Shyam, Narendra Kumar Nag, D. Krishna Murthy, and Sheela Barla

T. S. Sukanya (✉)
Project Coordinating Unit, ICAR-AICRP on Small Millets, GKVK, Bangalore, Karnataka 560065, India
e-mail: tssukanya@gmail.com

A. Kumar
AICRP on Small Millets, Ranichauri V.C.S.G., Uttarkhand University of Horticulture and Forestry, Bharsar, Uttarkhand 249199, India

A. L. Narayanan
Pandit Jawaharlal, Nehru College of Agriculture and Research Institute, Government of Puducherry, Karaikal 609603, India

K. Sathya
ICAR-AICRP on Small Millets, Centre of Excellence in Millets, Athiyandal, Tamil Nadu 606603, India

K. Kishore
Rajendra Prasad Central Agriculture University, Pusa, Bihar 848125, India

M. Shyam
ICAR-AICRP on Small Millets, Dindori, Mandhya Pradesh 48188, India

N. K. Nag
ICAR-AICRP on Small Millets, Indira Gandhi Agricultural University, Raipur, Chhattisgarh 493211, India

D. K. Murthy
Agriculture Research Station, University of Agricultural Sciences, Hagari, Raichur 583111, India

S. Barla
ICAR-AICRP on Small Millets, Birsa Agricultural University, Ranchi, Jharkhand 834006, India

S. Sheraz Mahdi et al. (eds.), *Adapting to Climate Change in Agriculture-Theories and Practices*, https://doi.org/10.1007/978-3-031-28142-6_12

12.1 Millet—Climate Resilient Crop

The area of irrigated millet cultivation is around 10% indicating their mere suitability under rainfed agriculture. Millet's cultivation is gaining importance in the present era and there are many reasons for the shift towards millet cultivation. They are climate-resilient and can be grown in arid regions, drought-prone areas efficiently as compared to another crop. Their water requirement is very less, 80% less water is required compared to other crops such as rice, wheat, or sugarcane. Another, often positive side of growing millets is that they fit excellent for soil conservation because of their root network. Millets, similar to grasses which have a fibrous root network that helps to maintain soil integrity. A suitable package of practices for the cultivation of different small millets such as time of sowing/planting, choice of varieties, time, and method of fertilizer application for different agro-ecological regions have been developed. Early, mid, and late-season drought mitigation practices have been developed in response to aberrant weather conditions. Profitable cropping systems involving different pulse/cereal/oilseed crops in millet for different ecological regions have been evolved.

12.2 Cropping System and Their Importance

Farmers prefer mixed cropping, especially under dryland conditions, to minimize the risk of crop failure. The basic concept of cropping systems is as old as agriculture. Farmers preferred mixed cropping, especially under dryland conditions, to minimize the risk of total crop failure. Mono-cropping, inter-cropping, and multiple cropping are common among the farming community in Indian agriculture (Chatterjee & Maite, 1982).

Indian agriculture still lacks intensive planning because India has a diversified agro-climatic zone, which is unfortunately not giving sufficient production (IPCC, 1992). If the Indian farming system relies on modern cropping patterns and cropping system, then food grain crops will be predominant, then Indian farming will also be inclined towards commercial crops. In India, the cropping pattern is determined mainly by climate, temperature, rainfall, type of soil.

The important factors that influence Cropping Pattern in India are:

- **Natural Factors**: Rainfall, Soil, Climate, etc.
- **Historical Factors**: Land Type. Type of ownership, Land Tenure System.
- **Social Factors**: Customs, Traditions, Social Environment.
- **Economic Factors**: Prices, Incomes, Input Prices, Size of holdings.
- **Government Policies**: Taxes Subsidies MSP, Export Policy.

The cropping system comprises of on-farm cropping patterns followed and their interaction with other farm resources, farm enterprises, and available technology which determine their makeup (Panda, 2016; Shankar et al., 2013). Cropping systems are designed and managed to achieve human goals so they are purposeful systems.

12.3 Basic Principles of Cropping System

- Choose the crops that complement to each other
- Choose crops and a cropping rotation which utilize available resources efficiently
- Choose crops and a cropping rotation that can restore, retain and enhance soil fertility
- Choose crops that have diversity of growth cycles
- Choose diverse species of crops
- Keep the soil covered
- Strategically plan and modify cropping system as needed

Cropping systems are decided mainly by soil moisture or rainfall received, soil type and length of growing period. Rainfall being the important determinant which influences the choice and success of crop through deciding the length of growing period (Table 12.1). Soil that stores moisture also decides the cropping systems again through deciding the length of growing period (Tables 12.2 and 12.3).

Millets can survive in areas with as little as 350 mm of seasonal rainfall. The minimum water requirement is 400 mm for sorghum and 500–600 mm for maize (Table 12.2).

Table 12.1 Choice of cropping system based on length of growing period

LGP (days)	Cropping system
< 75	Perennial vegetation, mono-cropping with short duration pulses/millets
75–140	Mono-cropping
140–180	Intercropping
> 180	Double cropping

Table 12.2 Choice of cropping system based on rainfall received

Rainfall (mm)	Soil moisture storage (mm)	Cropping system
350–625	100	Single crop (*kharif*)
650–750	100	Intercropping (can be attempted)
780–900	150	Sequential cropping is possible
> 900	200	Sequential cropping is assured

Table 12.3 Potential cropping systems in relation to rainfall and soil type

Rainfall (mm)	Soil type	LGP (weeks)	Cropping system
350–650	Alfisol and shallow vertisol	20	Single crop (*kharif*)
	Deep aridsols, entisols	20	Single crop (*kharif* or *rabi*)
	Deep entisols	20	Single crop (*rabi*)
650–750	Alfisol, entisol, vertisol	20–30	Intercropping
750–900	Deep alfisols, deep vertisols, entisols, inceptisols	30	Double cropping with monitoring
> 900		> 30	Sequential and double cropping assured

12.4 Multiple Cropping

Intensification of cropping in time and space dimensions. Growing two or more crops in the same field in a year.

12.4.1 Forms of Multiple Cropping

Intercropping

Cultivation of two or more than two crops simultaneously in the same field. Crop intensification is in both time and space dimensions.

(a) Mixed intercropping: Growing two or more than two crops simultaneously without distinct row proportion. Also referred to as mixed cropping. Ex: Sorghum, pearl millet and cowpea are mixed and broadcasted in rainfed conditions, finger millet and mustard in Southern Karnataka.
(b) Row intercropping: Raising of two or more than two crops simultaneously where these crops are planted in rows often referred to as intercropping (Table 12.4). Ex: Finger millet + Red gram (8:2).
(c) Strip intercropping: Cultivation of two or more crops simultaneously in strips wide enough to allow their independent cultivation but narrow enough for the crops to interact agronomically. Ex: Ground nut and redgram in 6:4.
(d) Relay intercropping: Cultivation of two or more crops simultaneously in which only a part of the life cycle is overlapped. Second crop is sown only after the first crop reached its reproductive stage of growth, but, before its physiological maturity. Often simply referred to as relay cropping. Ex: Rice fallow pulse/ sorghum.

Advantages of intercropping

(i) Efficient use of growth resources including light, water and nutrients (ii) Weed suppression (iii) yield stability; assured income can be expected from one crop even if

Table 12.4 Prominent millet intercropping systems of millets observed

Cropping system	States
Sorghum + pigeon pea (2:1)	Telangana, Marathwada, Vidarbha, Malwa Plateau, Parts of Gujarat
Pearl millet + pigeon pea (2:1)	Telangana, Karnataka, Maharashtra, MP, Rajasthan, Haryana, Tamil Nadu
Finger millet + pigeon pea (2:1) 8:2)	Karnataka, Odisha
Cotton + sorghum + pigeon pea	Maharashtra
Pearl millet + green gram	Rajasthan, Haryana
Finger millet + soybean (4:1)	Karnataka, Jharkhand
Finger millet + field bean (8:1)	Karnataka
Castor + finger millet (1:2)	Karnataka
Cowpea + finger millet	Karnataka
Rabi season	
Black gram + sorghum	Karnataka, Maharashtra, Madhya Pradesh

Source Anonymous (2014a, 2014b)

another crop fails to aberrant weather conditions (iv) successful intercropping results in higher equivalent yields (yield of base crop + yield of intercrop), higher cropping intensity (v) reduction in pest and disease incidences and (vi) improvement of soil health and agro-eco system.

Sequential cropping: Cultivation of two or more crops in sequence on the same piece of land in a farming year. The succeeding crop can be taken after the harvesting of preceding crop. Crop intensification is only in time dimension (Table 12.5).

(a) Double, triple and quadruple cropping: Growing two, three and four crops, respectively, on the same piece of land in a year in sequence.
(b) Ratoon cropping: The cultivation of crop re-growth after harvest, although not necessarily for grain. The various terms defined above bring out essentially two underlying principles, that of growing crops simultaneously in mixture, i.e., intercropping; and of growing individual crops in sequence, i.e., sequential cropping. The cropping system for a region or farm may comprise either or both of these two principles.

12.4.2 Millet Based Cropping Systems

Under dryland conditions finger millet is often sown by broadcasting the seeds at the rate of 25 kg/ha. Thinning of plants needs to be done after two weeks when they are about 2–3 cm tall to reduce the plant density. The plants which are thinned can be used for animal feed. When sown in rows, optimum spacing recommended for best yields. Generally, in southern India, finger millet is intercropped with crops like

Table 12.5 Millet sequence cropping in different parts of country

Sequence cropping	State
Sorghum–chickpea	Karnataka, Maharashtra, Madhya Pradesh, Rajasthan
Sorghum–wheat	Punjab
Sorghum–safflower	Karnataka
Black gram–sorghum	Karnataka, Maharashtra
Ground nut–sorghum	Karnataka
Pearl millet–wheat	Haryana
Pearl millet–chickpea	Haryana, Uttar Pradesh
Pearl millet–cluster bean	Haryana
Pearl millet–cluster bean/moth bean/mung bean	Rajasthan
Pearl millet–barley/black gram	Jammu
Pearl millet–mustard	Rajasthan, Haryana
Groundnut–potato–pearl millet	Gujarat
Paddy–finger millet	Odisha, Madhya Pradesh
Foxtail millet–safflower	Karnataka

Source Anonymous (2014a, 2014b)

fodder sorghum, field beans, niger, castor and pigeon peas and in some places even with pearl millet (bajra). Finger millet is also rotated with other dry land crops like groundnut, horse gram, other millets, cotton, tobacco and oil seed crops like sesame.

In some of the finger millet growing regions, monocropping is followed. It may be by a fallow or mixed with other crops in the same season. In most situations, however, it is rotated with sorghum, millet, cotton and tobacco. In other parts of India, the dry crop is sometimes rotated with groundnut, cotton, sorghum or millets (Table 12.6). In the better-fertilized areas with supplementary irrigation, finger millet is grown either before or after crops of gingely, onion, sweet potato, chillies, tobacco, wheat, gram or cotton.

Intercropping with other crops is also common. It is frequently grown in association with sorghum, pigeonpea, cotton or gram. Proportion of component crops in intercropping depends on the smothering effect of the crops. Results of experiments at Bangalore (Karnataka) have shown the finger millet and soybean drilled in alternate rows 22.5 cm apart did not depress the finger millet yield (2.5 t ha^{-1}) with a bonus yield (200 kg ha^{-1}) of soybean (Anonymous, 2014a, 2014b). Staggered planting of pigeonpea 3.3 m apart in May followed by planting finger millet in July in the interspaces between sowing of the two crops for improving the yield advantage. To minimize the weed problem in the interspaces, cowpea can be planted and ploughed back as green manure after 45–50 days of vegetative growth. Generally, in southern India, finger millet is intercropped with crops like fodder sorghum (jowar), field beans, Niger, castor and pigeon peas and in some places even with pearl millet

Table 12.6 Finger millet-based cropping systems observed in country

Inter/mixed cropping	Ratio	States
Finger millet + pigeon pea	8–10:2	Karnataka and Tamil Nadu
	6:2	Bihar
Tansplanting of pigeon pea as intercrops in Finger millet	2:8	Chhattisgarh
Finger millet + field bean	8:1	Karnataka and Tamil Nadu
	6:2	Bihar
Finger + soybean	4:1	Karnataka
Finger millet + moth bean/black gram	4:1	Maharashtra
Finger millet + soybean (mixed cropping system)	90:10	Uttarakhand
Finger millet + ground nut strip cropping	6:9	Karnataka
Sequential cropping		
Finger millet + soybean (Kharif)-oat	Northern hilly region	
Potato-paddy-finger millet	Northern Bihar	
Finger millet-potato-maize Finger millet-onion-finger millet Cowpea/sesamum/green gram/black gram-finger millet Finger millet-horse gram	Southern Karnataka or Deccan plateau	
Finger millet—mustard Finger millet—barley Finger millet—linseed Finger millet—tobacco and Finger millet—gram	North India	
Finger millet—potato—maize Finger millet—potato—ragi Finger millet—groundnut Finger millet—sugarcane Finger millet—tobacco	South India	

(bajra). Finger millet is also rotated with other dry land crops like groundnut, horse gram, other millets, cotton, tobacco and oil seed crops like sesame (Thimmegowda et al., 2018).

Finger millet intercropping systems most commonly seen in South India are with pulses, Finger millet/dolichos, finger millet/pigeonpea, finger millet/black gram, finger millet/castor and with cereals: Finger millet/maize, finger millet/foxtail millet, finger millet/jowar, finger millet/little millet and with other species: Finger millet/brassicas, finger millet/mustard (Chapke et al., 2018; Sukanya et al., 2022).

Bajra

Agronomic research for low-input arid regions has focused mainly on cropping system with legumes and on the moisture conservation techniques as pearl millet is traditionally grown in drier areas in mixture or in rotation with pulses and legumes to obtain stability in production and soil fertility. Suitable cropping systems have been formulated for diverse regions. Moisture conservation techniques forms an important recommendation in dry regions. These techniques include wide spaced crop and use of mulch either through manipulating topsoil or by organic means and to grow short duration cultivars to avoid moisture stress situation. Though, adoptions of agronomic recommendations are mainly confined to better rainfall/assured moisture available areas but need to adopt improved agronomic practices is more in drier pearl millet growing areas to sustain the productivity (Moharana et al., 2012).

The majority of farmers in Rajasthan prefer mixed cropping to sole cropping, mixed cropping of pearl millet, mung bean, sesame and moth bean. The proportion of pearl millet in the mixture is reduced if sowings are delayed beyond the first fortnight of July Cluster bean is usually sole- cropped and seldom put in mixtures. Table 12.7 shows the most commonly practiced cropping patterns in this zone.

Mixed systems: Varying proportions (2–25%) of crops like sorghum, pigeon pea, green gram, cowpea, field bean, horse gram and sesame with pearl millet (75–90%) are practiced in mixed cropping systems during *kharif* season. If the component crops are of longer duration, pearl millet is harvested early.

Intercropping: Replacement series of intercropping with crops like pigeon pea, groundnut, cowpea is adopted in many bajra growing regions during kharif season (row proportions; 8:2, 10:2; 10:4; 2:1). Different crops are recommended for intercropping with pearl millet in various parts of the country to ensure higher production, high land equivalent ratio and monitory returns from pearl millet under aberrant weather conditions (Table 12.8).

Sequential systems: Because of its shorter duration and its suitability to all three seasons, it fits into many rotational systems with crops like rice, cotton, sorghum, groundnut under irrigation. More common irrigated system is Pearl millet- wheat. Under rainfed conditions, it is rotated in two-year rotation with crops like cotton (Rajasthan), Chick pea (Punjab) and Tobacco (Gujarat).

Table 12.7 Existing cropping pattern in agroclimatic zone in western Rajasthan

First year		Second year	
Rainy season	Post rainy season	Rainy season	Post rainy season
Pearl millet	Fallow	Fallow	Barley
Moth bean	Fallow	Pearl milet	Fallow
Pearl millet	Chickpea	Fallow	Fallow
Cluster bean	Fallow	Pearl millet	Fallow
Sesame	Fallow	Fallow	Fallow
Pearl millet + grain legumes	Fallow	Sesame	Fallow
Fallow	Fallow	Grain legume	Wheat
Fallow	Mustard	Cluster bean	Fallow

Table 12.8 Suitable pearl millet-based intercropping in different states

Rajasthan	Pearl millet + cluster bean/moth bean/sesame
Haryana	Pearl millet + green gram/sesame
Gujarat	Pearl millet + green gram/sesame
U.P.	Pearl millet + green gram/sesame
M.P.	Pearl millet + black gram/soybean
Delhi	Pearl millet + pigeon pea/groundnut/castor
Maharashtra	Pearl millet + moth bean/pigeon pea
Karnataka	Pearl millet + pigeon pea
Tamil Nadu	Pearl millet + cowpea/sunflower

Sorghum

- Sorghum is grown in sequence with crops like wheat, pea, Bengal gram, potato (in north India) and cotton, tobacco, or ragi (in south India). In rabi dominant area, systems like groundnut- sorghum, pulses—sorghum are practiced.
- Sorghum fits into three crop sequences also. Some three crop sequences are Sorghum-wheat-moong, sorghum-wheat-lobia (in north India) as well as sorghum-ragi-groundnut (south India).
- Sorghum is also grown with soybean, pigeonpea, moong and blackgram in intercropping systems. Some hybrids like CSH-6 are considered more suitable for intercropping systems.
- Rabi sorghum is generally intercropped with Bengal gram, safflower or even sunflower.

Kodo millet

Kodo millet + pigeon pea in 2:1 or Kodo millet + green gram/black gram in 2:1 is profitable in Madhya Pradesh. The other alternative inter cropping systemsare kodo millet + soybean in 2:1 and kodo millet + horse gram in 4:2 proportion. Kodo–soybean-kodo or kodo-niger-kodo crop rotation was found to be sustainable system.

Foxtail millet

Results of experiments indicated yield advantage due to intercropping foxtail millet with pigeon pea in 6:1 proportion. Foxtail millet + pigeon pea in 4:1 or cotton in 5:1 in heavy soils and Foxtail millet + castor in 7:1 ratio is an ideal and remunerative cropping systems. If monsoon is early and favourable in Andhra Pradesh, the practice of sowing foxtail millet at 45 cm row spacing and introducing rabi jowar as a relay crop when foxtail millet is nearing maturity is practiced. In medium deep black soils of Andhra Pradesh, growing 2 crops—foxtail millet-mustard or foxtail millet-green gram—in sequence is profitable than taking one crop of foxtail millet.

Among the *rabi* crops tested after *kharif* foxtail millet growing Jowar found to bring higher returns. The next best was noticed to be Foxtail millet-Bengal gram cropping system which gave higher foxtail millet grain equivalent yield as against fallow-Bengal gram.

Foxtail millet as irrigated crop in garden lands during summer is usually preceded by rice during both kharif and rabi or by rice in *kharif* and groundnut in *rabi*. Summer irrigated foxtail millet can fit into many intensive cropping systems due to its shorter duration.

Proso millet

Intercropping of proso millet with black gram or greengram in 2:1 proportion is recommended for Bihar and Uttar Pradesh. Potato-proso millet rotation is profitable in western Bihar. Many sequential systems include proso millet either before or after kharif or rabi crop. Some such systems are proso millet—wheat/barley; proso millet-chick pea, maize- potato- proso millet or Maize- wheat-proso millet. In India, it is rarely grown as intercrop. But in USA, it is successfully grown as inter crop with crops in high intensity cropping systems.

Barnyard millet

Barnyard millet + ricebean/niger in 4:1 proportion is recommended for Uttaranchal. Mixed cropping with cotton, pigeon pea or short duration pulse crops is also practiced. Crop mixtures of barnyard millet 90% and soybean 10% was found feasible system. The next best is found to be the mixed cropping of barnyard millet with amaranths (90:10 by weight).

Little millet

Little millet + black gram in 2:1 proportion in Orissa; little millet + sesamum or soybean or Pigeonpeain 2:1 proportion in Madhya Pradesh; little millet + pigeonpea in 2:1 proportion in South Bihar is profitable. Little millet and pigeon pea in 8:2 ratioand opening a conservation furrow between paired row of pigeon pea is found to be a better practice to be followed. Little millet–Niger and little millet-toria are found to be promising sequence cropping.

Millets in alternate land use system

Beneath alternate land use system, Custard + finger millet gave maximum custard apple equivalent yield (1511 kg/ha) compared to sole custard apple (620 kg/ha) whereas intercropping of Amla + finger millet resulted higher amla equivalent yield and B: C ratio (1810 kg ha^{-1} and 2.71, respectively) compared to sole amla crop (667 kg ha^{-1}, 2.48, respectively) as reported from AICRP on Dryland Agriculture from Bangalore centre in 2018. Among different intercrops like finger millet, fodder maize, field bean, grain Amaranth, cowpea, horse gram in amla based intercropping, higher crop equivalent yield, net returns and more rain water use efficiency were observed with Amla + Finger millet, Amla + Field bean and Amla + cowpea (Korwar, 1992; Anonymous, 2014a, 2014b).

Millets in fallow system

The area under rice fallows has good scope for increase in area and crop intensification. Integrating millets in existing agricultural practices: Rice and wheat are an integral part of the staple Indian population's diet and finding an equal position with these two is difficult and we cannot expect all the farmers to start cultivating only millets. Millets, on the other hand, are an excellent crop to grow during the fallow periods or the time between harvesting one crop and planting the next. This is because millets have a greater propensity to form a symbiotic association with mycorrhizal fungi a study says. The fungi colonize the root system of the plant are colonized by the fungi and they help in increased water and nutrient absorption capabilities. Their productive utilization can help to tackle many social and economic issues, such as unemployment, low income, and labor migration. After the harvest of *Kharif* rice, climatic conditions of rice fallow lands are suitable for growing cool and warm-season pulses profitably in many areas. The study of Chapke et al. (2011) showed that agro-economic variables such as landholding, an area under sorghum crop cultivation, fertilizers cost, total input cost, labor cost, cost of irrigation and pesticides used and grain yield had a significant correlation with profits obtained from sorghum cultivation. Higher returns were reaped by the resourceful farmers who utilized maximum inputs from the sorghum cultivation. Hence, for a decade, farmers in Andhra Pradesh are cultivating sorghum after the harvest of rice on the residual soil moisture under zero tillage. The cropping patterns of a region are closely influenced by geographical, climatic, historical, socioeconomic, and political factors.

12.5 Conclusion

Agriculture still lacks intensive planning because the country has diversified agro-climatic zone, which is unfortunately not giving sufficient production. The cropping system should ensure sufficient food for family needs, fodder for cattle, and generate enough cash income for domestic and cultivation expenses, the millet cropping system serves the purpose to meet the above socio-economic conditions. Cropping pattern plays a key role in determining the level of agricultural production, which would, in turn, reflect on the agricultural economy of an area. Climate-resilient Nutri-cereals have the future of food and farm security. Systematic efforts to address socio-economic, political, and genetic improvements and technological advancements facilitating the adoption of these crops which ensures food and nutritional security with reference to climate change. So far as the agronomic crop and cropping pattern is concerned, it varies from situation to situation. Conserving water and soil resources and maintaining long-term soil productivity depends largely on the management of cropping systems, which influence the magnitude of soil organic matter dynamics and soil erosion.

References

Anonymous. (2014a). *Annual progress reports of All India Coordinated Research Project on Small Millets*. ICAR, GKVK.

Anonymous. (2014b). *Annual progress report of All India Coordinated Research Project on Dryland Agriculture*. ICAR, GKVK.

Chapke, R. R., Prabhakar, Shyam Prasad, G., Das, I. K., & Tonapi, V. (2018). *Improved millet production technologies and their impact. AICRP on small millets*. University of Agricultural Science, Bengaluru. ISBN: 69-81.

Chapke, R. R., Rakshit, S., Mishra, J. S., & Patil, J. V. (2011). Factors associated with sorghum cultivation under rice fallows. *Indian Research Journal of Extension Education, 11*(3), 67–70.

Chatterjee, B. N., & Maite, S. (1982). *Cropping system (theory and practice)*. Oxford & IBH Publishing C.

IPCC. (1992). Climate change. The supplementary report to the IPCC scientific assessment (J. T. Houghton, B. A. Callandar, & S. K. Varney (Eds.)). Cambridge University Press.

Korwar, G. R., et al. (1992). Alternate land use systems. In L. L. Somani (Ed.), *Dryland agriculture in India State of Agricultural Research in India* (pp. 143–168). Scientific Publishers.

Moharana, P. C., Sharma, B. M., Biswas, D. R., Dvidevi, B. S., & Singh, R. V. (2012). Long term effect of nutrient management on soil fertility and soil organic carbon pools under 6yrs old pearl millet-wheat cropping system in an inceptisols of subtropical India. *Field Crop Research, 136*, 32–41.

Panda, S. C. (2016). Cropping and farming system. Agrobios, ISBN: 978-81-7754-205-9.

Shankar, M. A., Ramachandrappa, B. K., & Thimmegowda, M. N. (2013). Conservation of soil and water Vis-à-vis production systems in Karnataka. In *FFCSWR 2013 held from 14th to 16th, March 2013 at UAS, Bangalore* (pp. 7–22).

Sukanya, T. S., Prabhakar, Krishne Gowda, K. T., Swarna, R., Hariprasanna, K., & Tonapi, V. A. (2022). *Good agronomic practices for higher yield in small millets*. ICAR-Indian Institute of Millets Research.

Thimmegowda, M. N., Mudalagiriyappa, Puneetha, K. M., Savitha, M. S., Vasanthi, B. G., Devaraja, K., & Jayashree, H. T. (2018). Evaluation of different integrated nutrient management approaches for sustainable productivity of finger millet + Pigeonpea cropping system under rainfed condition. *International Journal of Pure Applied Bioscience, 6*(6), 1039–1041.

Chapter 13
Conventional and Advance Breeding Approaches for Developing Abiotic Stress Tolerant Maize

Shyam Bir Singh, Bhupender Kumar, Alla Singh, and Sushil Kumar

13.1 Introduction

Maize (*Zea mays* L.) is a multipurpose cereal crop of the world with wider adaptability to different agro-climatic regions. It is the most versatile food crop being grown on 193.7 million ha area in more than 170 countries across the globe with 1147.6 million ton of production and 5.9 tonnes/ha productivity (Anonymous, 2020a). Maize is the principal staple food crop in large parts of Africa and many parts of developing countries. Globally, maize is predicted to become the crop with greatest production by 2025 (Cairns et al., 2012). In India, during 2019–20, maize is cultivated on area of around 9.0 million ha with production and productivity of 28.07 million tonnes and 3.11 t/ha, respectively (Anonymous, 2020b). The most important use and demand driver of maize in India is poultry feed, which accounts for 47% of total maize consumption. Further, it is an important industrial raw material where more than 3000 products are being made from it providing wide opportunity for value addition. After the wheat and rice, it is the third most important food grain crop in India. Being an international crop, its demand is increasing day by day. However, sustaining maize

S. B. Singh (✉) · S. Kumar
Regional Maize Research and Seed Production Centre (ICAR-IIMR), Begusarai, Bihar 851129, India
e-mail: singhsb1971@rediffmail.com

S. Kumar
e-mail: sushilccsu999@gmail.com

B. Kumar
ICAR-Indian Institute of Maize Research, Unit Office, Pusa Campus, New Delhi, India

S. B. Singh · A. Singh · S. Kumar
ICAR-Indian Institute of Maize Research, Ludhiana 141012, Punjab, India
e-mail: allasingh.iimr@gmail.com

S. Sheraz Mahdi et al. (eds.), *Adapting to Climate Change in Agriculture-Theories and Practices*, https://doi.org/10.1007/978-3-031-28142-6_13

yield stability over the year and locations is the big challenge in the era of climate change.

There are many biotic and abiotic factors which are hindering maize production and productivity worldwide. Different forms of abiotic stresses like drought, cold, heat, water logging, salinity and climate change are some of the major factors that reduce the agriculture crop production considered major threats to global crop production and this is also at the time when supply of basic food required to be enhanced at a significantly higher level (Meena et al., 2017a, 2017b; Zorb et al., 2019). These abiotic stresses (drought, heat, cold waterlogging and salinity) drastically reduce plant biomass, yields and survival of major food crops up to 70% (Meena et al., 2017a, 2017b). Drought stress caused nearly 52% grain yield reduction (Kumar et al., 2016). These stresses threaten the food security worldwide because the crops productivity is not increasing as much as the food demand rising up. The abiotic stresses disturb the plant growth and development and therefore prevent the genotype(s) from realizing their full genetic potential (Waqas et al., 2019). Now, challenges are to reduce the crop and yield losses caused by different abiotic stresses and to meet the increasing food demand. Grain yield in general is a trait which have low heritability under selection leading to slow progress in developing varieties/cultivars with higher yield. Relief from abiotic stresses is possible either by changing/avoiding the stressful environment or by changing the genetic constitution of the plant itself. Several varieties and hybrids in maize were developed and recommended for cultivation using conventional breeding approaches. The recent advent of molecular tools, have revolutionized the genetic analysis of traits for crop plants and provided tremendous opportunities not only to the plant breeders but also to physiologists, agronomists, and biochemist to identify traits of importance in improving tolerance to abiotic stresses. To get more success in achieving target production of maize, conventional breeding requires integration of advance technological tools. To design and implement advanced breeding strategies for developing abiotic stress tolerant maize genotypes, new tools and technologies like doubled haploids (DH), whole genome sequencing, high-density genotyping, high-throughput phenotyping, genomics-assisted breeding (rapid-cycle genomic selection, marker-assisted recurrent selection), and crop modelling play a greater role (Prasanna et al., 2013). During last two decades, the development of new genomics and breeding tools changed the methodology of doing plant breeding. Now new tools and techniques like New breeding techniques like genomic selection and rapid generation advancement, DNA/RNA sequencing, high-throughput genotyping/phenotyping, trait mapping, functional characterization etc. are now available to speed up the breeding process (Nepolean et al., 2018). Thus, the aim of this chapter is to report the important conventional and advance breeding approaches for abiotic stress tolerance with special emphasis on drought and water logging as they are the major ones and affect almost 80% of the total cultivated maize in India.

13.2 Abiotic Stresses in Maize

Abiotic stresses are considered in the form of adversarial effects on developmental phases of plant due to the poor climatic or soil conditions which results in loss of millions of dollars annually (Joshi & Karan, 2013). Abiotic stresses are severely affect plant development and productivity either individually or in combination and finally reduce the grain yield. Some abiotic stresses are interrelated like drought and heat, excess moisture and salinity. The abiotic stresses cause changes in physiological, morphological, molecular and biochemical processes. Major abiotic stresses like drought, heat, cold, waterlogging and salinity induce damage to plant cellular parts of the plant, including maize. There is always a strong need of mitigating abiotic stresses by adopting various scientific strategies including either management practices or breeding climate resilient genotypes. The lateral one is more preferable and sustainable as it provides in-built capacity in genotypes for surviving under various abiotic stresses. Following are the different types of abiotic stresses affecting maize crop.

13.2.1 Drought

Drought stress is the major abiotic stress followed by waterlogging casing remarkable yield reduction during rainfed crop. Maize production as mainly dependent on rainfall, its yield fluctuates more widely due to drought as compared to rice and wheat (Prasanna, 2016) that is why drought is considered as the major constraint and destabilizing factor in maize grain production. Physiological responses to drought are complex and are unpredictable as drought stress affects the crop in differential way at different growth levels of the plant. Moisture stress during vegetative stage reduced plant height, dry stover yield, grain yield and its components traits. However, the major effect of drought in maize is delayed silking which results in widening of ASI which is the major reason of reduced kernel setting in maize ears. The Highest reduction was recorded in kernels per ear from 11.26 to 54.59% (Singh et al., 2009). The probable mechanisms in maize in response to drought include redirecting of growth and accumulation of dry matter in roots (Hsiao & Xu, 2000) and osmotic adjustments. At the time of moisture stress, the water potentials and turgor are reduced enough which affect the normal functioning and well-being of the plants (Zhu, 2002). The excessive desiccation of water from the plant systems lead to blockage of various important metabolic pathways and hence may lead to plant death. Maize being principally grown as rainfed crop is more prone to both low as well as high moisture stresses (Kumar et al., 2016). The effect of these abiotic stresses in maize can be very easily seen in the form of variable yield level in different genotypes. In nature, there are several in-built mechanisms which results in combating of all these stress effects and therefore help in plant survival. These mechanisms and traits are needs to be explored through scientific breeding strategies.

13.2.2 Waterlogging

Waterlogging is a condition when water accumulates in excess of the plant requirement for long period and hinders the different process of plant. Waterlogging occurred due to inadequate rainfall distribution either enhancing number of rainy days or high intensity of daily rains. Waterlogging is the natural phenomena in Southeast Asia including eastern India. In these areas waterlogging and floods affects nearly 18% maize growing areas which reduce nearly 25–30% annual production (Cairns et al., 2012). The intensity of losses caused by flooding or waterlogging on plant differ from one stage to other. (Mano et al., 2002). Water-logging causes depletion of oxygen as rate of oxygen diffusion is slower in water than in air which subsequently reduces the oxygen availability in plant roots due to the imbalance between slow diffusion and rapid consumption (Erdmann et al., 1986). Oxidative stress induced by waterlogging is encountered by various mechanisms including enhanced availability of soluble sugars, greater activity of glycolytic pathway, aerenchyma formation, and anti-oxidant defence mechanism. During deficiency of oxygen in plant ethylene reported to play an important role in changing the plants mechanism (Hossain & Uddin, 2011). One of the water-logging adaptive traits in maize is the development of brace roots in tolerant genotypes that provide anchorage the maize plant.

13.2.3 Heat Stress

During growing period, increasing temperature disrupt the crop production system in two ways, firstly, by affecting the pollination resulted in reduced grain development and secondly by enhancing the growth rate thereby reducing the grain filling duration in maize (White & Reynolds, 2003). The optimum temperature for maximum grain filling in maize is 25–32 °C above which germination capacity of pollen is reduced on silk resulting in fewer grain development (Basra, 2000) and production of starch is also reduced affecting the kernel development (Singletary et al., 1994). Reproductive stage in maize is most vulnerable to elevated temperatures and when supplemented with drought conditions lead to significant yield losses (Cairns et al., 2013). The impact of heat becomes more severe when it occurs with drought stress. Metabolic activities at cellular level are altered by high temperature stress. Heat stress affects the biomass accumulation, net photosynthesis, leaf area, and seed test weight (Cheikh & Jones, 2006).

13.2.4 Cold Stress

The maize crop is very sensitive to cold, therefore, its cultivation in temperate region is limited. The temperatures between 18 and 27 °C during the day and around 14 °C during the night is most suitable for growth of maize plant. The temperature below 10 °C reduce the growth of the plant and further reduction in temperature below 5 °C cause chilling injury to leaf tissues. Prolong duration of low or chilling temperature slow down the plant metabolism and affects various processes. In maize low temperatures for long duration slow the cell cycle and reduce the cell division in susceptible genotypes (Rymen et al., 2007). Cold stress also affects the reproductive development of maize plant. It mitigates the initiation of floral development and reduce the rate of tassel branch meristem development and also reduce the branches and spikelets of tassel (Bechoux et al., 2000). In maize exposure to chilling stress reduce the rate of development and cell division and leading to which decreased leaf area and plant growth. A reduction of 20% in leaf size was reported in cold stress affected maize (Rymen et al., 2007). During the reproductive developmental stage cold stress induce abnormalities in reproductive system, which lead to failure of fertilization. Cold stress also leads to low pollen formation and extreme cold for longer duration during floral development leads to complete failure of pollen development. The low pollen availability for fertilization or complete pollen non availability results partial seed setting or complete failure of kernel setting in maize ears.

13.2.5 Salinity Stress

Salinity is the excess quantity of dissolvable salts present in the soil which affects various growth and developmental process of plant. The low rainfall in a area, poor soil and water management and high rate of evapotranspiration are some of the causes for salinity of soil (Munns, 2002). During salinity stress external water potential is low which results high osmotic stress. Salinity stress also caused ion toxicity sodium and/or chloride. Due to salinity plant face imbalance of essential nutrients as the salinity caused the interference in absorption and transport of the nutrients. Salinity leads to stunted maize plants growth causing wilting of plants at a salinity level 0.25 M NaCl or more (Menezes-Benavente et al., 2004). The germination, seedling and early plant developmental stages are more sensitive to salinity. High amount of sodium and chloride reduce the absorption of potassium, nitrogen, magnesium, calcium, and iron. The deficiency of Phosphorus enhanced the salt tolerance in maize plants as it reduces the Sodium (Na^+) accumulation and selective absorption of K^+ over Na^+ (Tang et al., 2019). To enhance and sustain the maize production in salinity prone areas development of salinity tolerant genotypes are essentially required (Farooq et al., 2015).

Most of the abiotic stresses are polygenic in nature and show large amount of G × E interaction. Due to higher interaction and effect of environments they show low heritability and small number of potentials gains during selection in crop improvement programme. Screening is required to be done at multiple stages and locations which are very critical and sensitive for particular stress. Further, screening must be coinciding with high and uniform level of stresses across the field which is it is unpredictable and therefore have low success rate.

13.3 Current Understanding of Abiotic Stresses Responses in Maize

Various abiotic stresses like low moisture, waterlogging, high salt and high temperature, influence the yield of maize crop. While the maize plant has developed strategies to cope with each, certain common set of genes have been deciphered that respond to all stresses. Li et al. (2017) studied genes expressed differentially upon treatment of maize seedlings (third leaf) with four types of stresses, viz., low moisture, high salt, low and high temperatures. The authors reported 167 genes that responded to all the four stresses. These common genes also included transcription factors (TFs), ten of them were always up-regulated, while the two of them were always down-regulated in response to the stresses. Apart from transcription factors, the common set of genes includes those which regulated abscisic acid (ABA) biosynthesis, and very-long-chain fatty acid and lipid signaling. While the differentially expressed genes belonged to various TF families, viz., ERF, MYB, bZIP, bHLH, WRKY, and NAC; Ethylene Responsive Factor (ERF) family TFs, constituted the largest TF family that responded to the stresses. Most of the TFs, belonging to bZIP (basic leucine zipper-type) family were implicated in abiotic stress in maize. These constitute an important reservoir of molecular factors, which can be potentially used in genome edited/cis- or trans-genic plants, for imparting resistance to multiple stresses in maize.

Hussain et al. (2018) implicated the physiological and biochemical processes that are common between drought and cold stress tolerance in maize. Among the processes that maize shares when responding to low moisture and low temperature stresses, are stomatal closure, production of reactive oxygen species (ROS), regulation of dehydrins, calcium signaling, loss of turgor etc. There are certain responses that are unique to each stress and not expressed in the other. Adee et al. (2016) reported that the incorporation of traits that collectively constitute drought tolerance (DT) in maize does not result in yield penalty in low stress/high yield environments. DT hybrids were able to conserve water for the critical early and late stages of reproduction. The authors found that in environments with low evapotranspiration or those under full irrigation regardless of the evapotranspiration potential, non-DT hybrids show significant improved performance. However, in poor yield environments, DT hybrids show better yield than non-DT counterparts. Cooper et al. (2014) reported differential water usage by DT maize hybrids as compared to non-DT hybrids. Hence, it appears that differential usage of resources leads to stress tolerance in one genotype, while resulting in susceptibility to stress in a different genotype. The physiological and biochemical responses to stress often entail consequent mechanisms that have the potential to reduce yield, while protecting the plant from stress condition. For example, stomatal closure can lead to reduced photosynthesis. ROS generation can lead to oxidative damage. Loss of turgor can lead to poor growth. All these may translate to altered plant-nutrient relationships, impaired flowering and pollination, and overall yield penalty. The genotypes that are able to balance intensity and duration of molecular phenomena leading to stress tolerance, preventing the corrective mechanisms from cascading to events that reduce yield, would constitute stress tolerant material. Adee et al. (2016) report that the DT hybrids showed less phenotypic plasticity than non-DT hybrids. This means that the expressions of individual characteristics of a DT maize genotype change to a lesser extent by drought environment, when compared to a non-DT genotype. Therefore, the ability to respond, but also to balance the response, preventing initiated molecular process to translate to yield reduction, through different physiological and biochemical mechanisms, would be helpful in maintaining plant yield under stressed environments.

The cross-talk of different stresses and the elucidation of common genes that respond to multiple stresses, raises the possibility of utilizing advanced tools of genetic manipulation for stress management. Determination of key regulators at the hierarchical top of stress-responsive metabolic pathways would lead to identification of molecular factors that could be deployed to impart stress tolerance. In addition to genotype-specific molecular mechanisms, another physiological phenomenon, stress memory, can be used to impart stress tolerance. In some cases, when a plant faces a stress, it becomes primed to handle another episode of that or some other stress. The plant develops a stress memory and keeps a repertoire of molecules in the form of epigenetic marks, lingering proteins, stalled RNA polymerase, which respond to a future episode of stress in a faster and effective way (Ding et al., 2012; Herman & Sultan, 2011; Schneider et al., 2019). Stress memory can also be utilized for stress management in maize. On the basis of understanding of abiotic stress effective breeding strategies to be formulated to develop stress tolerance maize genotype. The steps involved in stress tolerance breeding and various breeding approached have been shown in Fig. 13.1.

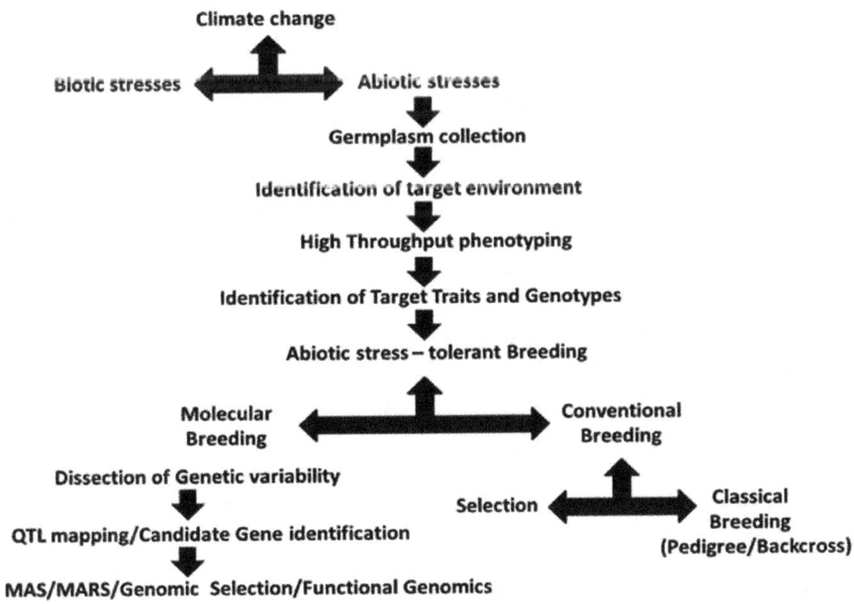

Fig. 13.1 A schematic representation of stress tolerance breeding approaches in maize

13.4 Conventional Breeding Approaches for Abiotic Stress Tolerance in Maize

A combined approach of cultural practices and genetic improvement should be followed to mitigate the negative effect of abiotic stresses on maize productivity. Selection for stress-adaptive traits for waterlogging tolerance, like early brace root development found helpful. Recurrent selection for several cycles results in the development of stress tolerant inbred lines in maize. Genetic improvement though conventional methods involve development of cultivars tolerant to abiotic stresses using various population improvement methods to enhance yield. Adjustment and modification in agronomic practices such as change in planting date, plant density and soil and water management minimize the stress effects but development of plant genotype with improved genetic make-up is the best substitute to fight and beat the abiotic stress without sacrificing yield. Physiological and genetic studies indicate that most stress traits are governed by polygenes which are highly influenced by the environment. Under these circumstances recurrent selection of their modifications can be used to develop stress tolerance population. The simple steps which breeders follow to develop abiotic stress tolerant cultivars include:

(a) Creation of variability: by intercrossing parents possessing desirable characteristics breeders produce new gene combinations and useful variability among genotypes or it is introduced by bringing new germplasm lines from other breeding programs.
(b) Selection: genotypes performing best in the target environment are selected and used in the further breeding programs.

13.5 Selection Approaches

Breeding varieties or hybrids having tolerant to drought or water logging tolerance is different from normal breeding methods in the sense that additional features are required along with the yield which contributes in survival under aberrant situation. The breeding strategies required are dependent on types of cultivar required, the nature of inheritance and type of gene action prevails for the traits conferring abiotic stress tolerance. Conventional breeding has made significant contribution in the development and deployment of stress resilient genotypes since long back. Around 60–65% of the total maize area in India is under improved maize hybrids, off this around 30% is specifically occupied by the single cross hybrids. Generally maize hybrids are having better roots system, strong stalk, high yield, uniform middle cobs placement, better brace roots, low ASI, uniform plant stand, better germination, stay green traits, better photosynthetic ability and high nutrient use efficiency etc. Therefore, they may be having better adaptability under drought as well as water logging stresses conditions. However, further there is huge amount of genetic variability available, which may be exploited by developing and selecting diverse inbred lines and or hybrids having in-built tolerance to drought and water logging stresses. This will be helpful in increasing the efficiency as well as effectiveness of stress resilient breeding programme. Direct selection for abiotic stress tolerance under field conditions sometimes may not be effective due to quantitative inheritance and involvement of environmental variance. Therefore, the more appropriate technique to identify abiotic stress tolerant plants is to evaluate the breeding materials in managed stress environment and select the genotypes with higher yield potential. Inbred lines can be developed by selfing desirable plants in a source population continuously for five to six generations by following ear to row selection. The tolerance of inbred lines for abiotic stress can be better judge by evaluating their test crosses performance developed using a susceptible tester under managed abiotic stress conditions. The hybrids developed from tolerant inbred lines will increase the chance of getting stress resilient hybrids, however its tolerance need to be established by evaluating in suitable experimental trials. Sometimes, if we have good yielding hybrids can directly be use for screening under drought and water logging trials. The suitable hybrids may be selected on the basis of their yield potential under the trials and best one can be recommended for further testing and cultivation in stress prone ecologies. Wherever there is no access of hybrids seeds, the improved synthetic/composites may be developed and evaluate under stress conditions. The stress tolerant superior combinations can also be identified in the basis of selection indices reported that Some selection indices such as Geometric Mean Productivity, Stress Tolerance Index, Harmonic Mean, Stress Susceptibility Index, and Yield Index were found more accurate criteria for selection of waterlogging-tolerant genotypes (Singh et al., 2017). The frequency of favorable alleles for abiotic stresses tolerance in the cultivars can be increases by adopting different ways of population improvements like recurrent selection methods. The conventional breeding approaches followed to develop stress tolerant genotypes include:

(a) Use of recurrent selection method and evaluation of segregating lines under managed stress and multi-location environment.

The most critical step in developing stress tolerant cultivar is the choice of selection strategy. Selection for yield under non-stress/optimum environment and horizontal evaluation of those selections under multiple locations trial with variable stress level is the most widely used selection strategy (Magorokosho & Tongoona, 2003). Again, the rate of improvement or genetic gain is dependent on the choice of testing environment(s). Ideally, the screening environment must have same environmental conditions as target environment with respect to climatic conditions like physical properties and nature of soils, rainfall distribution, temperature etc. (Ribaut et al., 2009).

(b) Selection of secondary traits under stress environment

Besides yield there are certain traits which give information of the performance of the plant are considered as secondary traits (Lafitte et al., 2003). The ideal secondary trait should have high association with grain yield under selection as well as target environment, additionally the trait should also have high heritability, genetically variable, simple, non-destructive and fast to measure and not related with any yield reduction under optimum environment. As grain yield under stress condition has poor heritability, genetic improvement or genetic gain may be less through direct selection. In this case, indirect selection via secondary traits is an attractive strategy because during stress conditions, at least some of the traits show high heritability with increased genetic correlation (Lafitte et al., 2003).

The early/late generation testing for stress adaptive traits can be followed depending upon the nature of gene action. Early generation evaluation and selection is preferable for the traits having additive gene action. In maize water-logging tolerance reported to be controlled by additive-dominance genetic model, and with more additive gene effects. Under such conditions reciprocal recurrent selection will be a good breeding approach for enhancing water-logging tolerance (Zaidi et al., 2010). The 12 cycles of recurrent selection method were found effective for the development of waterlogging stress tolerant genotypes with 20% yield advantage (Cairns et al., 2012).

13.5.1 Morphological Traits for Drought and Water Logging Stress Tolerance

Conventional breeding focuses mostly on characterization of the germplasm for stress tolerance related traits in managed stress. Field breeding strategies can be effective when there is knowledge available about nature of gene action involved in the inheritance of target traits. Understanding of inheritance of morphological and physiological traits associated with abiotic stress tolerance is very important. Understanding of yield and morpho-physiological basis of drought tolerance is a pre-requisite for making effective selection strategy for genotypes and breeding for water stress conditions. The study of heterotic expression of the parents would help in the development of desirable genotypes e.g. highly heterotic combination can also be possible when one or both the parents have poor general combining ability (Singh et al., 2010). Major traits associated with maize drought stress tolerance, escape and avoidance are deeper root system, short ASI, erect leaves, earliness, protogyny, stay green, low canopy temperature, high chlorophyll content stability, non-barrenness, osmotic adjustment, pubescence, waxy leaves, more proline, and glycine-betaine accumulation in specific tissue of plant. Similarly, high chlorophyll content stability, low mortality rate of the seedlings, deeper root system, well developed brace roots system, adventitious root formation, short anthesis silking interval (ASI), lodging tolerance, possession of a highly porous aerenchyma, better stalk and middle ear placement are the some of the important traits for water logging tolerance in maize. Screening for all these at various stages can help to identify the materials which can survive better under drought and water logging condition. The seedling, reproductive and grain development stages are the most sensitive to majority of the abiotic stresses in maize. Abiotic stresses conditions in the field results in increased of anthesis silking interval (ASI), respiration, canopy temperature, leaf senescence/rolling, decreases of root development, LAI, photosynthesis, biomass production, and hence ultimately the yield. It also results in heavy tassel firing, nutrient deficiency-N & K, plant mortality, lodging, pest and diseases attack, barrenness due to effect on effective pollinations and seed setting. All these results in significant decrease of plant yield.

13.5.2 Screening Technology for Drought and Water Logging Stress Tolerance in Maize

Choosing of appropriate selection sites and critical growth stage of plant for abiotic stress screening in maize is the crucial step. The sowing should be done in such a way that the sensitive stage must be coinciding with the environmental factors which favour drought and water logging stress. Under climate changing scenario, the 80% of maize rainfed area is mostly facing both drought and water logging stress frequently during the crop duration. Therefore, it is desirable to breed for genotypes having combined drought and water logging stress tolerance. Field Phenotyping should be done preferably in stress prone sites. The soil texture, soil moisture content and proper field leveling are the important point's needs to be taken care while selecting field for screening. Screening for drought stress at flowering stage can be done by withholding irrigation at least 10 days before initiation of flowering and the same can be retain for next 7–10 days during flowering. After that the irrigation can be resumed. Similarly, the stress at early grain filling stage can be induced by restricting the irrigation in trials immediately when the pollination over. The same may be sustained at least for 17–20 days.

It is very much important to have proper categorization of genotypes for their flowering and maturity. It is strongly recommended to have separate trials set on the basis of maturity. The genotypes with similar flowering/maturity groups can be merge to constitute a single trial for evaluation. These points are valid for any types of abiotic or biotic stresses screening. Evaluation under high plant density is a one of the good approaches to select genotypes for abiotic stresses tolerance. The rainout shelter facilities can be availed to evaluate the genotypes for drought. Similarly, screening for water logging stress at seedling stage can be done on flat beds by stagnating water up to minimum level of around 5 cm above the ground surface completely for ten days. Proper bunding should be done around the field to avoid flow of water. Seedling, pre flowering, flowering and grain filling stages are the more vulnerable stages for drought as well as water logging stress. Among all, flowering for drought and seedlings for water logging stress are the most crucial stages, where the occurrence of these stresses can cause maximum yield losses.

13.6 Introgression and Utilization of Distant and Wide Germplasm for Abiotic Stress Tolerance

The genetic variability which reduced over a period of time during artificial selection can be enhanced through the utilization of exotic wild germplasm into the existing germplasm available with any breeding programme for abiotic stress tolerance (Xu et al., 2009). Use of maize wild relatives in pre-breeding programme can also contribute significantly for developing stress resilient germplasm. Various gene banks have conserved approximately 0.05 million maize accessions (Hoisington et al., 1999). CIMMYT also conserved approximately 28,000 accessions http://www.cimmyt.org/germplasm-bank/). Such a huge maize germplasm consisting of pools populations and wild relatives may be the source of large unexploited genetic diversity as well as source of new traits and alleles. This germplasm can also be utilized to develop base populations or broaden the genetic base to exploit beneficial genetic variation. It is necessary to develop trait-specific core set for fully utilization of the genetic variability present in the wild and cultivated maize germplasm. Genetic diversity plays a greater role in the development of stress tolerance pools and populations. The trait contributing towards divergence also equally important. In a study Singh et al. (2020) reported that 100 kernel weight (39.45%) followed by days to anthesis (22.64%) contributed maximum towards total genetic divergence as per cent. Phenotypic characterization of diverse maize germplasm for specific abiotic stress will help in understanding the genetic variability and to develop a stress specific core set. This core set could be exploited through genomic approaches to identify novel genes. A stress specific abiotic stress tolerant population can be developed using genotypes tolerant to that specific stress. For example, a drought-specific pools were developed using anthesis–silking interval, leaf rolling and leaf senescence characters (Monneveu et al., 2006). Teosinte a wild relative of maize has many subspecies of *Zea mays* and some species identified as good source of useful source for improvement of abiotic stress tolerant traits e.g., teosinte *Z. luxurians* and *Z. mays* ssp *huehuetenangensis* can be used as source of genes for adventitious root formation under extreme waterlogging. Introgression lines of maize were developed genomic regions from teosinte (*Z. mays* ssp *nicaraguanensis*) (Mano & Omori, 2007).

13.7 Use of Molecular Markers and Genomic Tools in Abiotic Stresses Tolerance Breeding

Various genetic studies indicated that in general abiotic stress tolerance traits are governed by quantitative traits loci (QTLs). Marker assisted selection methods are effective and save time in breeding programme under high heritability of the trait and also when the cost of phenotypic evaluation of the genotypes is high. MAS methods are also beneficial under the conditions when environmental effects are significant and heritability is low.

13.7.1 Marker Assisted Selection (MAS) and QTL Mapping

Marker assisted selection is a method wherein target trait is selected based on the marker(s) linked to that trait. These markers can be morphological, biochemical or DNA/RNA. Marker-assisted selection (MAS) has become a common method of molecular breeding, where traits are controlled by a small number of major genes. Several markers have been developed and being used in crop improvement. In maize simple sequence repeat (SSR) and single nucleotide polymorphisms (SNPs) markers are being used extensively. Due the availability in large numbers, simple in use and more effectiveness of SSR markers these were used by many maize researchers.

The achievements through MAS technique for abiotic stress breeding depends on the identification of precise Quantitative Trait Loci (QTLs) for the target traits and their introgression. Extensive QTL mapping for drought and waterlogging tolerance was studies by many maize researchers. Five QTL from a drought-tolerant donor line Ac7643 were transferred to susceptible inbred CML247. Zaidi et al. (2015) identified five QTL for grain yield and 13 QTLs for secondary traits associated with waterlogging tolerance in maize.

13.7.2 Marker Assisted Recurrent Selection (MARS)

Marker assisted recurrent selection (MARS) is a breeding method which by the use of molecular markers assembles the desirable alleles in a population. This scheme involve the genotypic selection and intercrossing among best select individuals. This method increases the efficiency of recurrent selection. As this technique used to accumulate the favourable alleles of stress tolerant genes in a population hence, the improved populations can be used as source population to develop next stage of inbred lines with superior performance in terms of abiotic stress traits. In maize MARS was successfully utilized to improve the quantitative traits of drought tolerance. In a study MARS was used to increase the genetic gain for drought tolerance with respect to mean number of combinations of favorable alleles from 114 to 124 in three cycle of population improvement (Abdulmalik et al., 2017) The MARS was also used to increase the frequency of favourable alleles for drought tolerance from 0.510 to 0.515 in C0 to C_2 cycle of population. (Bankole et al., 2017). At CIMMYT researchers achieved remarkable success in developing early generation yellow drought tolerant maize inbreeds using MARS technique.

13.7.3 Rapid Cycle Genomic Selection

Rapid cycle genomic selection (RCGS) improves the genetic gains by decreasing period of selection cycle in stress tolerance breeding programme in maize. RCGS has added advantage over conventional breeding methods in terms of cost and time as three generations per year can be taken under this method. The hybrids derived from the RCGS improved population were yielded 7.3% yield gain than conventional breeding method. Rapid cycle genomic selection (RCGS) was reported an effective breeding technique in multi-parental maize populations for conserving genetic diversity in short period and obtaining high genetic gains (Zhang et al., 2017a, 2017b). In a recent study the realised genetic gain using RCGS in two maize populations MYS-1 and MYS-2 was 0.110 t/ha per year and 0.135 t/ha per year, respectively, under drought stress (Das et al., 2020). Without compromising genetic diversity, the RCGS can also be used as an effective breeding strategy for increasing genetic gains.

13.8 Accelerated Line Development Using Double Haploid (DH) Technology

Development of high yielding maize hybrid or synthetic varieties essentially required homozygous inbred lines as parental lines. The inbred line development is an important component of maize breeding programs. The line development in maize through the conventional pedigree and bulk involve continuous selfing upto 7–8 generations to get the completely almost homozygous inbred lines. This is which is time consuming process. Now doubled haploid (DH) technology has emerged as a powerful technology and become an effective option to the conventional line development method in maize. DH lines can be developed by both in vitro and in vivo methods. However, in maize in *vivo* methods have been found more reliable and are being used for production of in large number of lines of DH lines (Chaikam et al., 2019). By using the facility of off-season nurseries DH technology can produce the homozygous inbred lines in one year in comparison to conventional inbreeding which require three to four years for continuous selfing. Therefore, DH technology is the fastest and efficient method of fully homozygous line development in maize. DH technology not only useful to derive homozygous lines but it can also be used to explore the genetic diversity in maize landraces. (Brauner et al., 2019). The maize breeding programs of some developed countries of Europe, North America, and China already started in vivo DH line development (Molenaar & Melchinger, 2019). CIMMYT, also made the DH technology accessible to public and private organizations. The development of haploid inducer lines with high haploid inducer rate (HIR) opened the way for in *vivo* DH lines development. The In vivo DH technology being followed by may multinational maize companies. After the development of first haploid inducer line "Stock 6," which has the haploid inducer rate (HIR) of 1–3%. Several other inducers were developed using Stock 6 with good field performance and HIR. Most of these inducers like UH400, RWS and PHI were adapted to temperate environments. Earlier CIMMYT has developed temperate inducer lines with HIR 6–9% but recently CIMMYT has developed second-generation Tropically Adapted Inducer Lines (TAILs) with higher HIR (9–14%) and better field performance (Chaikam et al., 2018).

13.9 Rapid Generation Advancement (RGA)

Conventional breeding programme takes much time to complete any breeding cycle of crop improvement. Rapid generation advancement (RGA) and speed breeding are the most promising innovations for increasing the rate of genetic gain and reducing breeding cycle time (Cobb et al., 2019). RGA was first developed by Goulden (1939) and later modified by Grafius (1965). Its latest destination came in the form of speed breeding (Watson et al., 2018). Using different strategies like manipulation of growing environment, nutrients, use of hormones etc., many generations per year can be obtained in RGA method. The RGA not only reduce the generation cycle but also reduce the cost of getting new recombinant as compared to conventional pedigree breeding. RGA technique in maize needs to be standardized to harness its role in abiotic stress tolerant hybrid maize breeding. Specialized facilities like greenhouse, growth champers needs to be developed for rapid generation advancement, to achieve the desired cycle time and generation per year.

13.10 Precision and High Thorough-Put Phenotyping

Phenotyping is an important part of the stress breeding which provide much understanding and information to the breeders of stress tolerance programme. The main purpose of any plant breeding programme is to develop new genotypes which perform better and yield more than the genotype being cultivated in the target environment. Most of the abiotic stress tolerance traits are controlled by quantitative genes, where heritability is poor therefore to enhance the heritability by improving accuracy and achieve more genetic gain, precision phenotyping techniques are required. Advance techniques are required which may accelerate genetic gains by improving selection efficiency. Techniques are also required for automatic monitoring of nutrient and plant health status. In recent years high throughput phenotyping techniques has drown major attention of researchers to develop many protocols for improving desirable plant traits. In phenotyping work of plant breeding, several hundreds or thousands small plots need to be screened for abiotic stress, where use of high precision tools and fast result machines are always required to get instant results. Reliable and automation, high-throughput phenotypic technologies are being considered more and more important tools to get higher genetic gain in stress breeding programmes. Newly developed high-throughput and nondestructive techniques can be used under controlled stress environment for phenotyping of the primary and secondary traits associated with of plant architecture. Ear photometric and imaging techniques are now used for phenotyping of the ear traits in maize. The techniques are available that either used to phenotype the whole or specific part/s of the plant. Proximal phenotyping which done through the ground based small vehicle and sensors mounted on it. They are used to measure the plant canopy temperature, leaf area, chlorophyll fluorescence, nitrogen content, and plant height. Smartphone-based apps and platform were developed for plant phenotyping using pictures and images (Confalonieri et al., 2017). Next generation advanced techniques of automation, robotics, high accuracy sensors, and imaging may provide other opportunities for high-throughput plant phenotyping (HTPP) (Gennaro et al., 2017). Many HTPPs in growth chamber or greenhouse have been developed using digital and advanced technologies like high-throughput, high precision and automation (Zhang et al., 2018).

13.11 Scope of Advanced Genetic Manipulation Technologies in Developing Stress Resilient Maize

As described earlier, identification of molecular initiators and regulators for control of metabolic process leading to stress tolerance is a pre-requisite for designing advanced stress management strategies. Studies have indicated various molecules that play an important role in single- or multiple-stresses. Kimotho et al. (2019) have reviewed the various transcriptional factors that work in ABA-dependent and ABA-independent pathways, and that can be used to imparting tolerance to drought, salinity, heat and cold stresses. Gene manipulation of the key transcription factors can be done to increase or decrease their expression, alter the nature of expression by changing promoter DNA, making the expression respond to a particular regulator etc. This can be done in cis- or trans- approach, depending on whether the gene from a maize itself or from some other species has been taken, respectively. Gene editing involves deliberately introducing knock-in/knock-out mutations, where either the amino acid sequence if changed or the protein expression is shut down. Although many systems exist for introducing specific changes in DNA, one system CRISPR/Cas9 has gained popularity. CRISPR stands for Clustered Regularly Interspaced Short Palindromic Repeats. These are genetic elements, which originate from viral genomes and have got incorporated into the bacterial genome. Endonuclease proteins are associated with these genetic elements, referred to as CRISPR-associated proteins or Cas. The RNA of viral DNA incorporated into bacterial genome, directs Cas proteins to act like specific endonuclease. Whenever, a viral genome is encountered in the cell, bacteria respond by engaging the Cas protein specific for that virus, resulting in cleavage of viral nucleic acid. This phenomenon has been used to design plasmid containing CRISPR Cas systems for precise editing of plant genomes (Fig. 13.2a).

Zhang et al. (2018) identified HKT1 family sodium transporter, as a key factor for imparting salinity tolerance in maize. HKT1 transporter, expressed primarily in root cells, is responsible for exclusion of sodium from the xylem sap. This reduces the flow of sodium to shoot, thereby helping in tolerating saline conditions. The authors used CRISPR/Cas 9 system to create a loss-of-function mutation for HKT1, demonstrating retarded performance of edited plants in saline soil. Shi et al. (2019) reported generation of variants in ARGOS8 gene through CRISP/Cas, which resulted in drought tolerance. ARGOS8 is a negative regulator of ethylene biosynthesis, its

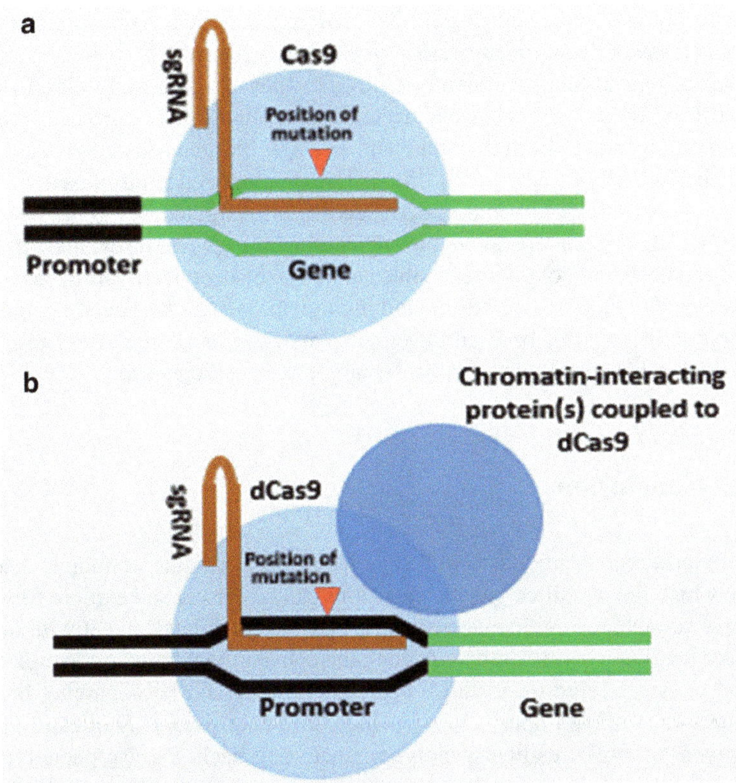

Fig. 13.2 Genome editing technologies for precise gene engineering. **a** CRISPR/Cas9 system for introducing mutations in a specific DNA sequence **b** dCas9 system for introducing epigenetic marks on a specific DNA sequence. The nature of marks created depend on the coupled proteins, which interacts with chromatin to being about desired changes

expression level in maize is low and non-uniform across tissues. Shi et al. (2015) showed that transgenic maize expressing high levels of ARGOS8 is drought tolerant. In this case they used CRISPR/Cas 9 system to generate allelic variation in a key gene ARGOS8. Shi et al. (2015) inserted a GOS2 promoter upstream of the ARGOS8 gene. GOS2 is a maize promoter that imparts a moderate level of constitutive expression. The presence of a GOS2 promoter upstream of ARGOS8 gene resulted in higher and uniform gene expression. This study demonstrates the use of single, native genes for modulating complex traits like drought tolerance.

Now-a-days, technology is available for reversibly altering the expression as well. This field, referred to as epigenome engineering, has opened new avenues in genetic manipulation of plants. This utilizes a particular type of a Cas9 endonuclease, which is capable of recognizing a nucleic acid sequence, but is unable of cleaving it (Pulecio et al., 2017). Due to lack of endonuclease activity, it is also referred to as Cas9 Endonuclease Dead (dCas9). dCas9 can be utilized in various designs to bring about epigenetic changes in the cell (Fig. 13.2b). Through histone modifications, it can be used to increase or decrease expression of particular genes.

In maize gene editing successfully utilized to alter five genes liguleless1 (LIG1), male fertility (Ms26 and Ms45), and acetolactate synthase (ALS) genes (ALS1 and ALS2) through targeted mutagenesis, and site-specific gene insertion using Cas9 and guide RNA (Svitashev et al., 2015). Using integrated multiplexed CRISPR/Cas9-based approaches Liu et al. (2020) successfully targeted 743 genes.

Thus, with the advent of epigenome engineering, the epigenetic events can now be targeted and the benefit of their reversible nature can be beneficially utilized to obtain the dual benefits of stress resistance and high yield. Hence, advanced technologies like genome editing can be used developing maize with abiotic stress resistance. Recent studies have demonstrated this by targeting specific genes.

13.12 Conclusion

There are various conventional as well as molecular and genomic techniques available through which the identified genetic variations can be dissect and explore for identification of suitable recombinants, gene(s)/QTLs and candidates' gene(s) for drought and water logging stress tolerance. Among the conventional breeding, simple selection and classical breeding methods such as pedigree and backcrossing breeding procedures etc. are the important ones which have contributed significantly for the development of stress resilient genotypes since long back. Precise phenotyping at critical crop growth stages under target environments are the keys of success for stress resilient breeding. Generally, seedling, flowering and grain filling stages are the more sensitive stages for drought, heat, cold and water logging stress in maize. Amongst all these stages, the seedling stage for water logging and flowering for drought heat and cold stress are the most sensitive one. Among the molecular and genomic approaches; MAS, MARS, genomic selection, functional genomics, genetic engineering and genome editing are the important ones which can complement the conventional breeding methodologies and therefore can improve the efficiency as well as effectiveness of abiotic stress breeding programme. Developing suitable stress resilient maize genotypes can help to increases the national as well as world average maize productivity.

References

Abdulmalik, R. O., Menkir, A., Meseka, S. K., Unachukwu, N., Ado, S. G., Olarewaju, J. D., Aba, D. A., Hearne, S., Crossa, J., & Gedil, M. (2017). Genetic gains in grain yield of a maize population improved through marker assisted recurrent selection under stress and non-stress conditions in West Africa. *Frontiers in Plant Science, 8*, 841. https://doi.org/10.3389/fpls.2017.0084

Adee, E., Roozeboom, K., Balboa, G. R., Schlegel, A., & Ciampitti, I. A. (2016). Drought-tolerant corn hybrids yield more in drought-stressed environments with no penalty in non-stressed environments. *Frontiers in Plant Science, 7*, 1534. https://doi.org/10.3389/fpls.2016.01534

Anonymous. (2020a). Area and production quantities of maize by country. www.fao.org/faostat/en/#data/QC/visualize. Accessed on September 5, 2020.

Anonymous. (2020b). Fourth advance estimates of production of foodgrains and commercial crops 2019–20: Agricultural statistics division, directorate of economics & statistics). Govt of India. http://www.agricoop.nic.in/acessed on 15–08–2020

Bankole, F., Menkir, A., Olaoye, G., Crossa, J., Hearne, S., & Unachukwu, N. (2017). Genetic gains in yield and yield related traits under drought stress and favorable environments in a maize population improved using marker assisted recurrent selection. *Frontiers in Plant Science, 8*, 808. https://doi.org/10.3389/fpls.2017.00808

Basra, A. (2000). *Crop responses and adaptations to temperature stress: New insights and approaches*. CRC Press.

Bechoux, N., Bernier, G., & Lejeune, P. (2000). Environmental effects on the early stages of tassel morphogenesis in maize (Zea mays L.). *Plant, Cell and Environment, 23*, 91–98.

Brauner, P. C., Schipprack, W., Utz, H. F., Bauer, E., Mayer, M., Schön, C. C., & Melchinger, A. E. (2019). Testcross performance of doubled haploid lines from European fint maize landraces is promising for broadening the genetic base of elite germplasm. *Theoretical and Applied Genetics, 132*(6), 1897–1908.

Cairns, J. E., Sonder, K., Zaidi, P. H., Verhulst, N., Mahuku, G., & Babu, R. (2012). Maize production in a changing climate. *Advances in Agronomy, 144*, 1–58.

Cairns, J. E., Hellin, J., Sonder, K., Araus, J. L., MacRobert, J. F., Thierfelder, C., & Prasanna, B. M. (2013). Adapting maize production to climate change in sub-Saharan Africa. *Food Security, 5*(3), 45–360.

Chaikam, V., Nair, S. K., Martinez, L., Lopez, L. A., Utz, H. F., Melchinger, A. E., & Boddupalli, P. M. (2018). Marker-assisted breeding of improved maternal haploid inducers in maize for the tropical/subtropical regions. *Frontiers in Plant Science, 9*, 1527.

Chaikam, V., Molenaar, W., Melchinger, A. E., & Boddupalli, P. M. (2019). Doubled haploid technology for line development in maize: Technical advances and prospects. *Theoretical and Applied Genetics, 132*, 3227–3243.

Cheikh, N., & Jones, R. J. (2006). Heat stress effects on sink activity of developing maize kernels grown in vitro. *Physiologia Plantarum, 95*(1), 59–66.

Cobb, J. N., Juma, R. U., Biswas, P. S., Arbelaez, J. D., Rutkoski, J., Atlin, G., Hagen, T., Quinn, M., & Ng, E. H. (2019). Enhancing the rate of genetic gain in public-sector plant breeding programs: Lessons from the breeder's equation. *Theoretical and Applied Genetics, 132*(3), 627–645. https://doi.org/10.1007/s00122-019-03317-0

Confalonieri, R., Paleari, L., Foi, M., Movedi, E., Vesely, F. M., & Thoelke, W. (2017). Pocketplant3d: Analysing canopy structure using a smartphone. *Biosystems Engineering, 164*, 1–12.

Cooper, M., Gho, C., Leafgren, R., Tang, T., & Messina, C. (2014). Breeding drought-tolerant maize hybrids for the US corn-belt: Discovery to product. *Journal of Experimental Botany, 65*, 6191–6204.

Das, R. R., Vinayan, M. T., Patel, M. B., Phagna, R. K., Singh, S. B., Shahi, J. P., Sharma, A., Baru, N. S., Babu, R., Seetharam, K., Burgueno, J. A., & Zaidi, P. (2020). Genetic gains with rapid-cycle genomic selection for combined drought and waterlogging tolerance in tropical maize (*Zea mays* L.). *The Plant Genome*. https://doi.org/10.1002/tpg2.20035

Ding, Y., Fromm, M., & Avramova, Z. (2012). Multiple exposures to drought 'train' transcriptional responses in Arabidopsis. *Nature Communications, 3*, 740.

Erdmann, B., Hoffmann, P., & Wiedenroth, E.M. (1986). Changes in the root system of wheat seedlings following root anaerobiosis I. *Anatomy and Respiration in Triticum aestivum L. Annals of Botany, 58*(5), 597–605.

Farooq, M., Hussain, M., Wakeel, A., & Siddique, K. H. (2015). Salt stress in maize: Effects, resistance mechanisms, and management. A review. *Agronomy for Sustainable Development, 35*(2), 461–481.

Gennaro, S. F. D., Rizza, F., Badeck, F. W., Berton, A., Delbono, S., & Gioli, B. (2017). UAV-based high-throughput phenotyping to discriminate barley vigour with visible and near-infrared vegetation indices. *International Journal of Remote Sensing, 39*, 5330–5344.

Goulden, C. H. (1939). Problems in plant selection. In R. C. Burnett (Ed.), *Proceeding of the seventh genetics congress* (pp. 132–133). Cambridge University Press.

Grafius, J. E. (1965). Short cuts in plant breeding. *Crop Science, 5*, 377.

Herman, J., & Sultan, S. (2011). Adaptive transgenerational plasticity in plants: Case studies, mechanisms, and implications for natural populations. *Frontiers in Plant Science, 2*, 102.

Hossain, M. A., & Uddin, S. N. (2011). Mechanisms of waterlogging tolerance in wheat: Morphological and metabolic adaptations under hypoxia or anoxia. *Australian Journal of Crop Science, 5*(9), 1094–1110.

Hoisington, D., Khairall, M., Reeves, T., Ribaut, J. M., Skovmand, B., Taba, S., & Warburton, M. (1999). Plant genetic resources: What can they contribute toward increased crop productivity. *Proceedings of the National Academy of Sciences of the United States of America, 96*, 5937–5943.

Hsiao, T. C., & Xu, L. K. (2000). Sensitivity of growth of roots versus leaves to water stress: Biophysical analysis and relation to water transport. *Journal of Experimental Botany, 51*(350), 1595–1616.

Hussain, H. A., Hussain, S., Khaliq, A., Ashraf, U., Anjum, S. A., Men, S., & Wang, L. (2018). Chilling and drought stresses in crop plants: Implications, cross talk, and potential management opportunities. *Frontiers in Plant Science, 9*, 393. https://doi.org/10.3389/fpls.2018.00393

Joshi, R., & Karan, R. (2013). Physiological, biochemical and molecular mechanisms of plants drought tolerance in plants (pp. 209–231).

Kimotho, R. N., Baillo, E. H., & Zhang, Z. (2019). Transcription factors involved in abiotic stress responses in Maize (Zea mays L.) and their roles in enhanced productivity in the post genomics era. *PeerJ7*, e7211. https://doi.org/10.7717/peerj.7211

Kumar, B., Guleria, S. K., Khanorkar, S. M., Dubey, R. B., Patel, J., Kumar, V., Parihar, C. M., Jat, S. L., Singh, V., Yatish, K. R., Das, A., Sekha, J. C., Bhati, P., Kaur, H., Kumar, M., Singh, A. K., Varghese, E., & Yadav, O. P. (2016). Selection indices to identify maize (*Zea mays* L.) hybrids adapted under drought-stress and drought-free conditions in a tropical climate. *Crop and Pasture Science, 67*, 1087–1095.

Lafitte, R., Blum, A., & Atlin, G. (2003). Using secondary traits to help identify drought-tolerant genotypes. In K. S. Fischer, R. H. Lafitte, S. Fukai, G. Atlin, & B. Hardy (Eds.), *Breeding rice for drought-prone environments* (pp. 37–48). International Rice Research Institute.

Li, P., Cao, W., Fang, H., Xu, S., Yin, S., Zhang, Y., Lin, D., Wang, J., Chen, Y., Xu, C., & Yang, Z. (2017). Transcriptomic profiling of the Maize (Zea mays L.) leaf response to abiotic stresses at the seedling stage. *Frontiers Plant Science, 8*, 290. https://doi.org/10.3389/fpls.2017.00290

Liu, H. J., Jian, L., Xu, J., Zhang, Q., Zhang, M., Jin, M., Peng, Y., Yan, J., Han, B., Liu, J., Gao, F., Liu, X., Huang, L., Wei, W., Ding, Y., Yang, X., Li, Z., Zhang, M., Sun, J., … Yan, R. (2020). High-throughput CRISPR/Cas9 mutagenesis streamlines trait gene identification in Maize. *The Plant Cell, 32*, 1397–1413.

Magorokosho, C., & Tongoona, P. (2003). Selection for drought tolerance in two tropical maize populations. *African Crop Science Journal, 11*(3), 151–161.

Mano, Y., Muraki, M., Komatsu, T., Fujimori, M., Akiyama, F., & Takamizo, T. (2002). Varietal difference in pre-germination flooding tolerance and waterlogging tolerance at the seedling stage in maize inbred lines. *Japan Journal Crop Science, 71*(3), 361–367.

Mano, Y., & Omori, F. (2007). Breeding for flooding tolerant maize using "teosinte" as a germplasm resource. *Plant Root, 1*, 17–21.

Meena, K. K., Sorty, A. M., Bitla, U. M., Choudhary, K., Gupta, P., Pareek, A., Singh, D. P., Prabha, R., Sahu, P. K., Gupta, V. K., Singh, H. B., Krishanani, K. K., & Minhas, P. S. (2017a). Abiotic stress responses and microbe-mediated mitigation in plants: The omics strategies. *Frontiers in Plant Science, 8*, 172. https://doi.org/10.3389/fpls.2017.00172

Meena, H. P., Bainsla, N. K., & Yadav, D. K. (2017). Breeding for abiotic stress tolerance in crop plants. In *Recent Advances in Plant Stress Physiology* (pp. 329–378).

Menezes-Benavente, L., Kernodle, S. P., Margis-Pinheiro, M., & Scandalios, J. G. (2004). Salt-induced antioxidant metabolism defenses in maize (Zea mays L.) seedlings. *Redox Report, 9*(1), 29–36.

Molenaar, W. S., & Melchinger, A. E. (2019). Production of doubled haploid lines for hybrid breeding in maize. In *Advances in breeding techniques for cereal crops* (pp. 143–172). Burleigh Dodds Science Publishing.

Monneveu, P., Sánchez, C., Beck, D., & Edmeades, G. O. (2006). Drought tolerance improvement in tropical maize source populations: Evidence of progress. *Crop Science, 46*, 180–191.

Munns, R. (2002). Comparative physiology of salt and water stress. *Plant, Cell & Environment, 25*(2), 239–250.

Nepolean, T., Kaul, J., Mukri, G., & Mittal, S. (2018). Genomics-enabled next-generation breeding approaches for developing system-specific drought tolerant hybrids in Maize. *Frontiers in Plant Science, 9*, 361. https://doi.org/10.3389/fpls.2018.00361

Prasanna, B. M. (2016). Developing and deploying abiotic stress-tolerant maize varieties in the tropics: Challenges and opportunities. In *Molecular Breeding for Sustainable Crop Improvement*. Springer (pp. 61–77).

Prasanna, B. M., Cairns, J., & Xu, Y. (2013). *Genomic tools and strategies for breeding climate resilient cereals*. https://doi.org/10.1007/978-3-642-37045-8_5

Pulecio, J., Verman, N., Mejía-Ramírez, E., Huangfu, D., & Raya, A. (2017). CRISPR/Cas9-based engineering of the epigenome. *Cell Stem Cell, 21*(4), 431–447.

Ribaut, J. M., Betran, J., Monneveux, P., & Setter, T. (2009). Drought tolerance in maize. In *Handbook of maize: Its biology*. Springer (pp. 311–344).

Rymen, B., Fiorani, F., Kartal, F., Vandepoele, K., Inzé, D., & Beemster, G. T. S. (2007). Cold nights impair leaf growth and cell cycle progression in maize through transcriptional changes of cell cycle gene. *Plant Physiology, 143*, 1429–1438.

Schneider, S., Turetschek, R., Wedeking, R., Wimmer, M. A., & Wienkoop, S. (2019). A protein-linger strategy keeps the plant on-hold after rehydration of drought-stressed beta vulgaris. *Frontiers in Plant Science, 10*, 381. https://doi.org/10.3389/fpls.2019.00381

Shi, J., Habben, J. E., Archibald, R. L., Drummond, B. J., Chamberlin, M. A., Williams, R. W., Lafitte, H. R., & Weers, B. P. (2015). Overexpression of ARGOS genes modifies plant sensitivity to ethylene, leading to improved drought tolerance in both Arabidopsis and maize. *Plant Physiology, 169*, 266–282.

Shi, J., Drummond, B. J., Habben, J. E., Brugire, N., Weers, B. P., Hakimi, S. M., Lafitte, H. R., Schussler, J. R., Mo, H., Beatty, M., Zastrow-Hayes, G., & O'Neill, D. (2019). Ectopic expression of ARGOS 8 reveals a role for ethylene in root-lodging resistance in maize. Summary Significance Statement. *The Plant Journal, 97*(2), 378–390. https://doi.org/10.1111/tpj.2019.97.issue-210.1111/tpj.14131

Singh, S. B., Singh, A. K., & Singh, S. P. (2009). Screening of maize hybrids for drought tolerance using drought susceptibility index for various morpho-physiological and yield characters. *Progressive Research, 4*(1), 11–15.

Singh, S. B., Gupta, B. B., & Singh, A. K. (2010). Heterotic expression and combining ability analysis for yield and its components in maize (*Zea mays.* L) inbreds. Progressive. *Agriculture an International Journal, 10*(2), 275–281.

Singh, S. B., Phagna, R. K., Kumar, B., & Kumar, S. (2017). Determination of selection indices for waterlogging stress tolerance in maize (*Zea mays L.*) under field conditions. *Applied Biological Research, 19*(2), 172–179.

Singh, S. B., Kasana, R. K., Kumar, S., & Kumar, R. (2020). Assessing genetic diversity of newly developed winter maize (*Zea mays* L.) inbred lines. *Indian Journal of Plant Genetic Resources, 33*(1), 68–76.

Singletary, G. W., Banisa, R., & Keeling, P. L. (1994). Heat stress during grain filling in maize: Effects on carbohydrate storage and metabolism. *Functional Plant Biology, 21*(6), 829–841.

Svitashev, S., Young, J. K., Schwartz, C., Gao, H., Falco, S. C., & Cigan, A. M. (2015). Targeted mutagenesis, precise gene editing, and site-specific gene insertion in maize using Cas9 and guide RNA. *Plant Physiology, 169*, 931–945. https://doi.org/10.1104/pp.15.00793

Tang, H., Niu, L., Wei, J., Chen, X., & Chen, Y. (2019). Phosphorus limitation improved salt tolerance in maize through tissue mass density increase, osmolytes accumulation, and Na+ uptake inhibition. *Frontiers in Plant Science, 10*, 856. https://doi.org/10.3389/fpls.2019.00856

Waqas, M. A., Kaya, C., Riaz, A., Farooq, M., Nawaz, I., Wilkes, A., & Li, Y. (2019). Potential mechanisms of abiotic stress tolerance in crop plants induced by thiourea. *Frontiers in plant science, 10*, 1336. https://doi.org/10.3389/fpls.2019.01336

Watson, A., Ghosh, S., Williams, M. J., Cuddy, W. S., Simmonds, J., Rey, M.–D., Asyraf Md Hatta, M., Hinchliffe, A., Steed, A., Reynolds, D., Adamski, N. M., Breakspear, A., Korolev, A., Rayner, T., Dixon, L. E., Riaz, A., Martin, W., Ryan, M., Edwards, D., Batley, J., Raman, H., Carter, J., Rogers, C., Domoney, C., Moore, G., Harwood, W., Nicholson, P., Dieters, M. J., DeLacy, I. H., Zhou, J., Uauy, C., Boden, S. A., Park, R. F., Wulff, B. B. H., & Hickey, L. T. (2018). Speed breeding is a powerful tool to accelerate crop research and breeding. *Nature Plants, 4*(1), 23–29. https://doi.org/10.1038/s41477-017-0083-8

White, J. W., & Reynolds, M. P. (2003). A physiological perspective on modelling temperature response in wheat and maize crops. *Modelling temperature response in wheat and maize* (p. 8).

Xu, Y., Skinner, D. J., Wu, H., Palacios-Rojas, N., Araus, J. L., & Yan, J. (2009). Advances in maize genomics and their value for enhancing genetic gains from breeding. *International Journal of Plant Genomics, 2009*, 30. https://doi.org/10.1155/2009/957602

Zaidi, P. H., Maniselvan, P., Srivastava, A., Yadav, P., & Singh, R. P. (2010). Genetic analysis of water-logging tolerance in tropical maize (Zea mays L.). *Maydica, 55*, 17–26.

Zaidi, P. H., Rashid, Z., Almeida, V. M. T., Phagna, G. D., & RK and Babu R,. (2015). QTL mapping of agronomic waterlogging tolerance using recombinant inbred lines derived from tropical maize (*Zea mays L*) germplasm. *PLoS ONE, 10*(4), e0124350.

Zhang, M., Cao, Y., Wang, Z., Wang, Z. Q., Shi, J., Liang, X., Song, W., Chen, Q., Lai, J., & Jiang, C. A. (2018). Retrotransposon in an HKT1 family sodium transporter causes variation of leaf Na+ exclusion and salt tolerance in maize. *New Phytologist, 217*, 1161–1176.

Zhang, X., Huang, C., Wu, D., Qiao, F., Li, W., & Duan, L. (2017a). High-throughput phenotyping and QTL mapping reveals the genetic architecture of maize plant growth. *Plant Physiology, 173*, 1554–1564.

Zhang, X., Pérez-Rodríguez, P., Burgueño, J., Olsen, M., Buckler, E., Atlin, G., Prasanna, B. M., Vargas, M., San Vicente, & Crossa, J. (2017b). Rapid Cycling genomic selection in a multiparental tropical maize population. *G3 (Bethesda, Md.), 7*(7), 2315–2326.

Zhu, J. K. (2002). Salt and drought stress signal transduction in plants. *Annual Review of Plant Biology, 53*, 247–273.

Zorb, C., Geilfus, C. M., & Dietz, K. J. (2019). Salinity and crop yield. *Plant Biology, 21*, 31–38. https://doi.org/10.1111/plb.12884

Chapter 14
Covid-19 and Anthropause in India: Rediscovering Sustainable Development Policies to Combat Climate Change

Raj Bala, Abhik Ghosh, and Rahul Kumar

14.1 Introduction

SARS-COV-2 (Severe Acute Respiratory Syndrome Corona Virus-2) or Novel virus or Covid-19 was first diagnosed in one of the cities of China named Wuhan in December 2019. That was why this virus also came to be known as the Wuhan virus (Kumar et al., 2020). Lafter on, it spread first in South Korea then Japan followed by Italy, Iran and India (Kachroo, 2020). Covid-19 entered India on 30th January 2020. On this day, three Indian students returned to their native place Kerala from the epicenter of Covid-19, i.e., Wuhan (China). This novel virus is so called because it was a mutation of the animal corona virus that had never been seen before. Till now the origin of this breakout remains undiscovered, but it is suspected that this virus might be associated with seafood or the live animal market in Wuhan. This was because the Wuhan wet market was not observing health and safety regulations and also since that it had been closed down for an unlimited time period (Kachroo, 2020).

Figure 14.1 clearly showed that comparative mortality rate in percentage in ten viral breakouts from 1967 to 2020. Here, the mortality due to the Marburg virus ranked first with 80% and the recent Covid-19 virus had a mortality rate of 2.2% as of January 31, 2020 (WHO et al., 2020).

Figure 14.2 showed that the viral contamination rate of Covid-19 was comparatively high among all other virus breakouts worldwide. Covid-19 has a contamination

R. Bala
Department of Botany and Environment Studies, DAV University, Jalandhar, Punjab, India

A. Ghosh
Department of Anthropology, Panjab University, Chandigarh, India

R. Kumar (✉)
Department of Agricultural Sciences, DAV University, Jalandhar, Punjab, India
e-mail: rahulihbt@gmail.com

S. Sheraz Mahdi et al. (eds.), *Adapting to Climate Change in Agriculture-Theories and Practices*, https://doi.org/10.1007/978-3-031-28142-6_14

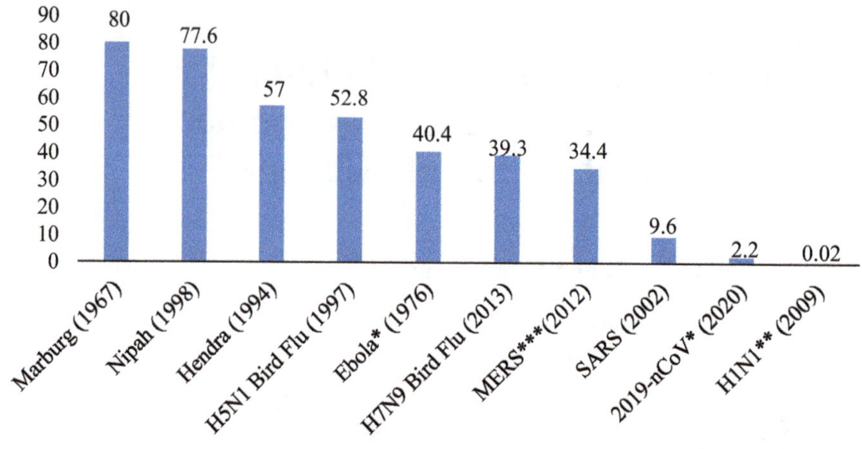

Fig. 14.1 Mortality rate (%) in ten viral breakouts globally. *Source* WHO et al. (2020)

rate of 1.5–3.5 per infected person in March 2020 in comparison to the contamination rate of seasonal flu of 1.3 (ADB et al., 2020). However, this data was subject to change as the corona virus breakout develops.

Covid-19, a new viral disease was considered to proliferate through various modes. It may occur from person-to-person via physical contact or by touching any infected object or surface followed by touching the nose, eyes and mouth. Its symptoms mainly included respiratory infection. This may occur either via direct or indirect exposure with carrier. This infection can be carried by droplets for small distance and aerosols for long distance transference (Moriyama et al., 2020). Following are the residence time periods of the Covid-19 virus which was measured on various objects or surfaces or metal:

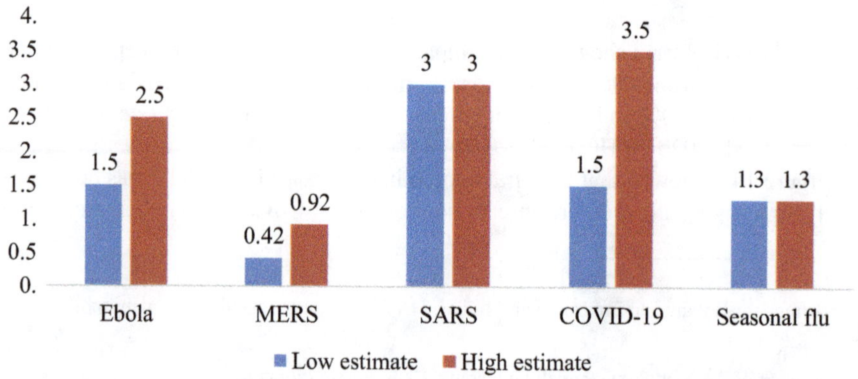

Fig. 14.2 Viral contamination rates (per infected person) breakouts worldwide as of 2020. *Source* ADB et al. (2020)

Table 14.1 Covid-19 current scenario in World and India

	Corona cases	Deaths	Recovered
World	22,069,384	777,751 (5%)	14,807,589 (95%)
India	2,706,450	51,955 (3%)	1,978,747 (97%)

Source Worldometer (2020)

1. Plastic: 2–3 days;
2. Stainless steel: 2–3 days;
3. Cardboard: 1 day;
4. Copper: 4 h (Guo et al., 2020).

The symptoms of this disease are very similar to other viral respiratory infections. Infection in patients vary from person to person. Some patients have mild symptoms while others may suffer from serious medical conditions leading to death. The most common symptoms of Covid-19 viral disease is high fever with or without chills, cough, loss of taste or/and smell, running nose, difficulty in breathing, sore throat, diarrhoea, body ache/fatigue, headache, etc. Aged persons or persons who were already suffering from diabetes, hypertension, lung disease, respiratory problems, etc. were at a higher risk from Covid-19 infection. The incubation period of this virus has not yet been confirmed but it was noticed that it may take from 2 to 14 days to show its appearance. On 18th August, 2020 at 8:36GMT World meter has confirmed the current scenario of corona virus cases in the world as well as in India as follows:

Table 14.1 shows that the percentage of deaths due to Covid-19 is 3% in India as compared to 5% in the world. The recovery rate is also high in India as compared to the world.

Figure 14.3 showed that the Covid-19 disease had now become a pandemic. It was not bounded by any boundary. Earlier, Europe was the most impacted region by Covid-19 but now the Americas have overtaken Europe. Recently, Brazil and United States had the highest Covid-19 confirmed cases as well as deaths worldwide. The highest causality were recorded in Latin America's Mexico, Chile and Colombia due to Covid-19 (WHO, 2020).

As of August 14, 2020, around 21 million cases of Covid-19 had been confirmed worldwide (as shown in Fig. 14.4). Of these, around 757 thousand deaths, 13,924 thousand recoveries and 6390 thousand active infections were reported (Worldometer, 2020).

Figure 14.5 showed that there had been a total of 20.6 million cases of Covid-19 till 13th August, 2020. Figure 14.5 shows a steep rise in Covid-19 cases. Worldwide, many countries were struggling with non-parallel social and health emergencies. Health systems were overburdened and even the developed countries were breaking out in sweat (OWID & ECDC, 2020). To disintegrate the dissemination of novel virus, almost all the countries adopted lockdown procedures that promote anthropause for a time.

In India the Health and Family Welfare Ministry has raised consciousness with respect to the breakout of Covid-19 and suggested control measures to control it to the Central and State Governments. After that the Central and State governments

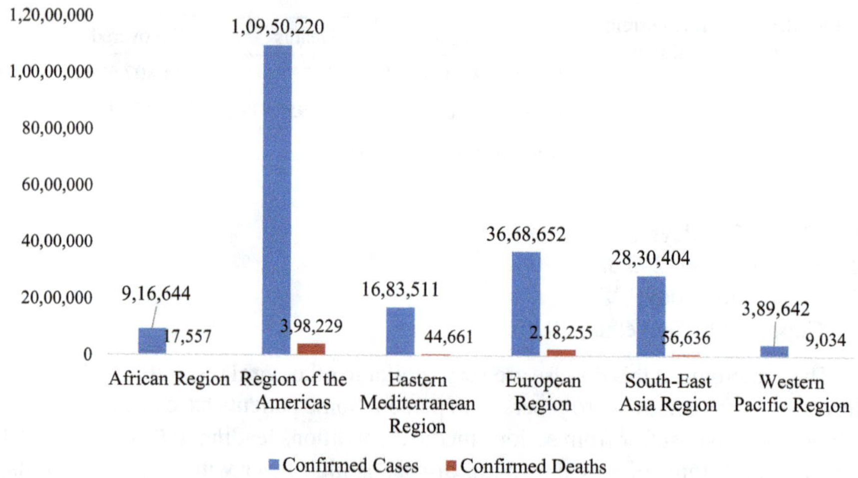

Fig. 14.3 Region wise Covid-19 reported cases and demises (August, 2020). *Source* WHO (2020)

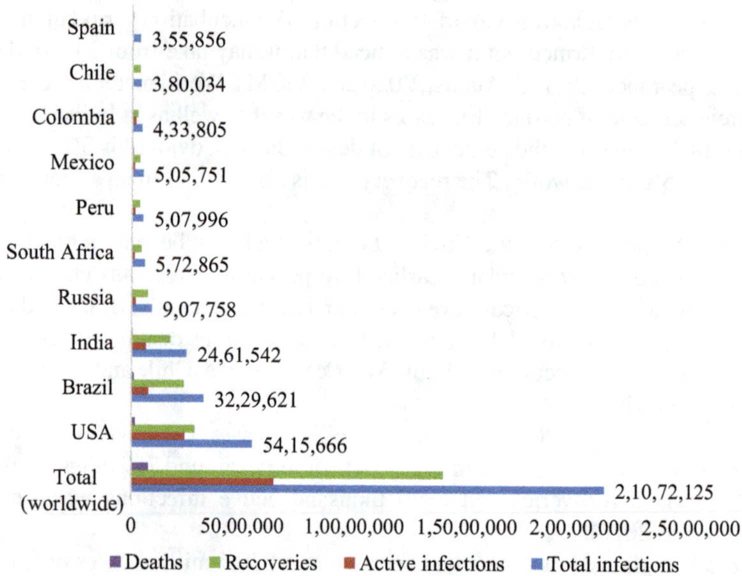

Fig. 14.4 Statistical reported data of Covid-19 globally (till August 14, 2020). *Source* Worldometer (2020)

jointly worked towards formulating several wartime protocols to achieve this goal. Furthermore, when the Covid-19 patients increased to about 500 in India, then on 19th March our Honourable Prime Minister Mr. Narendra Modi asked all the citizens to observe a *Janta* (public) curfew on 22nd March, 2020. On 24th March, 2020 initially

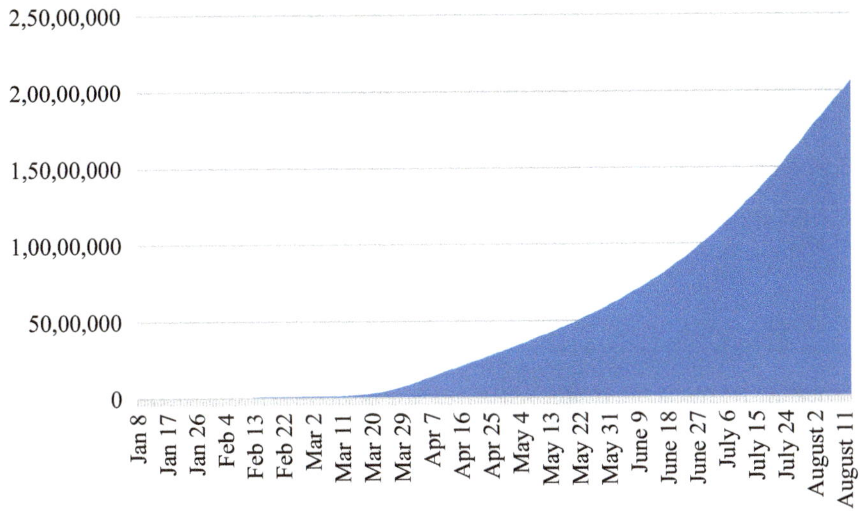

Fig. 14.5 Collective data of Covid-19 on daily basis at world level (from January to August, 2020). *Source* OWID & ECDC (2020)

the nationwide lockdown was announced for a period of 21 days and then again extended till 17th May, 2020 (a total of 55 days). This was done keeping in view the breaking of the transmission cycle of the virus via 'social distancing'.

In India a total of over two million cases of Covid-19 cases were recorded till August 10, 2020. Out of these total cases 1.5 million people recovered and nearly 44.5 thousand died. Figure 14.6 shows that the country had been reporting new Covid 19 cases daily since March 2, 2020 (Johns Hopkins University et al., 2020).

Figure 14.7 shows that on 12th August, 2020 a total of over 2.3 million Covid-19 suffering people were disclosed in which Maharashtra state disclosed the maximum and Andhra Pradesh and Karnataka were disclosed with relatively lower demises due to Covid-19 disease (Covid Tracker India, 2020).

Figure 14.8 shows that people between the age group of 21–40 are mostly affected due to Covid-19 till 27th April 2020. Out of this age group 21–40, people belonging to age group 31–40 were more with approximately 537 cases. If we compare this data to other nations then this trend was significantly lower. Still, in other nations also, like in India, adult age group persons were mostly affected by Covid-19 infection (Covid-19 Tracker India, 2020).

Some conditional relaxations were given to the public after 20th April only for the regions where the spread was minimal. It has been noted that the anthropause caused a decline in the rate of growthof pandemic by 6th April, 2020 to doubling every six days and by 18th April, 2020 to doubling every eight days. Later on, the Government of India after consulting the state governments, categorized all the districts into three zones (green, red and orange) on the basis of the outspread of Covid-19. The lockdown restrictions were eased slowly and slowly in three phase namely Unlock

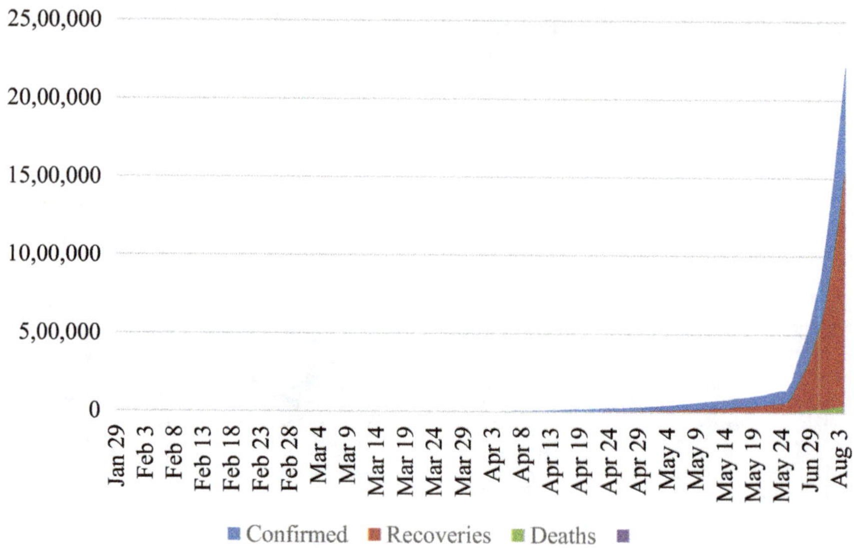

Fig. 14.6 Cumulative number of Covid-19 confirmed, recovered and deceased in India (January–August 2020). *Source* Johns Hopkins University et al. (2020)

1.0 (from 8th June to 30th June, 2020); Unlock 2.0 (from 1st July to 31st July, 2020) and Unlock 3.0 (from 1st August to 31st August, 2020). These lockdown restrictions, no doubt, dramatically impeded industrial sectors and economic growth throughout the world but it has proved to be a rejuvenation of the earth, environment and human health systems. This is a huge positive change in the environment worldwide. So the lock down has been proven to be beneficial for the environment.

Studies show the decrease in air pollution worldwide. All this occurred because of the anthropause. Anthropause refers to the slowing down in the pace of human activities globally, i.e., the coronavirus-induced lockdown period and its impact on other species (Gill, 2020). The atmospheric concentrations of greenhouse gases decreased largely due to man-made sources of carbon dioxide emissions like mobile combustion sources, burning fossil fuels and deforestation and this was the need of the hour to combat climate change. Since 1997, at the onset of the Kyoto protocol, what we needed was a decrease in the concentration of greenhouse gas emissions by limiting our growth. The Covid-19 virus forced us to do it. The results which were eluding us regarding the environment came because of the lockdown during Covid-19. In order to achieve the goals of sustainable development we must learn lessons from the past as well as the present. First, we have to set our priorities in the environment. This Covid-19 scenario showed that our basic needs were very limited. We were polluting our environment in the name of social status. We could easily conserve our environment by taking small steps like reducing unnecessary movements from the home, work from home practice, using local stuff instead of importing from other states or countries, etc.

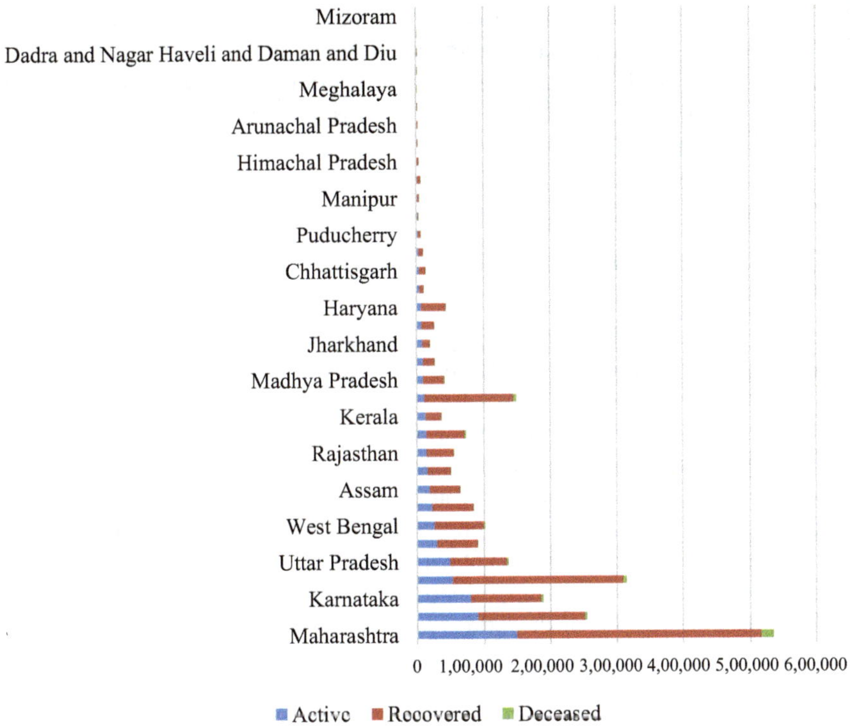

Fig. 14.7 State and Union Territories wise Indian reported data of Covid-19 (August, 2020). *Source* Covid Tracker India (2020)

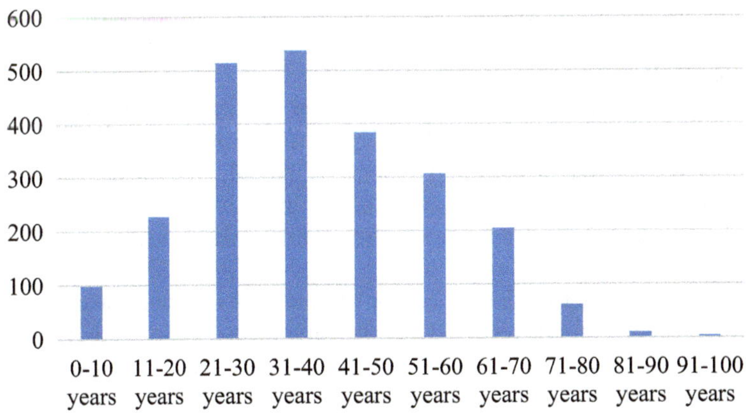

Fig. 14.8 Age-wise Covid-19 cases in India. *Source* Covid-19 Tracker India (2020)

The main goal of this work was to rediscover the sustainable development policies that could combat climate change after experiencing and analyzing the impact of the anthropause which was caused in order to control Covid-19 in India. In order to comprehend the matter in a better way, various journals, the Internet, reference books and other secondary sources were consulted.

14.2 Environmental Impact of Covid-19

Due to the unplanned development, rapid industrialization, increase in automobile industries and rapid economic developments, air pollution has aggravated. The air is considered to be very toxic to breathe because of the contaminants emitted in the air mainly after industrial revolution. All this led to the hike in greenhouse gases in the earth's atmosphere. The earth's temperature is rising which led to global warming. Glaciers were melting which, in turn, caused an increase in the sea level. So, before the initiation of the Covid-19 pandemic, environmental degradation was occurring rapidly due to the over-consumption of natural resources like air, water and soil, but, the lockdown which was enforced to control the breakout of Covid-19 led to a large improvement in the quality of air worldwide. When anthropause was begun in various countries nearly in the same time period, there was less travelling done by people, whether it be by private vehicles or by air, water or land transport. Every industry was closed down. This led to a significant fall in pollution through air and noise. The climate experts of Global Carbon Project (2020) forecasted that the emission of greenhouse gases could drop to proportions which had never reported since post-1945. *Geophysical Research Letters* journal reported the drastic reduction of two pollutants, including nitrogen dioxide, in the air quality. This occurred because of the worldwide lockdown which was imposed to control the outbreaks of Covid-19 pandemic (PTI, 2020a).

Although a high Covid-19 hazard was present worldwide during the months of April to July 2020, but it was noticed that the European and American regions were relatively more affected than tropical regions. In air quality a considerable reduction in NO_2 was observed worldwide. A small cutback in CO and low to moderate cutback in Aerosol Optical Depth (AOD) were also reported worldwide during the month from April to July 2020 (Lal et al., 2020).

If we determine the Covid-19 effects on the environment based on data available from different sources, it is recorded that the anthropause was directly proportional to the positive impact on air quality and water quality. The areas like China and Northern Italy where lengthy lockdown were imposed to control the breakout of Covid-19, have noted remarkable reductions in pollutants and an improvement in their air quality.

In China, there was a record of 40% cutback in NO_2 concentrations in the month of January and February 2020 when compared with 2019 levels. This reduction is equivalent to removing 192,000 cars from the roads (Monks, 2020). Here, isolation leads to reduction in NO_2 by 12.9 $\mu g/m^3$. It was also noted that energy emission and energy consumption was also reduced by 25% in China over two weeks. This, if continued in the future, could also decrease the country's overall annual carbon emissions by one per cent (Saigal, 2020). Wuhan, a city in China from where Covid-19 was identified and transmitted worldwide, also recorded a reduction in the concentration of NO_2 by 22.8 $\mu g/m^3$and Particulate Matter (PM 2.5) by 1.4 $\mu g/m^3$ due to isolation (Monserrate et al., 2020).

Strict isolation measures were adopted by Rome, Madrid, and Paris in Europe. In these areas also Copernicus Sentinel-5P satellite noted remarkable cutback in the concentration of NO_2 (Monserrate et al., 2020). European Space Agency's Sentinel-5P satellite readings show that the concentration of nitrogen dioxide (NO_2) in urban and industrial areas of Asia and Europe dropped by 40% when we compare the time period of late January and early February months of 2019 and 2020 (Monks, 2020).

European Union CAMS (Copernicus Atmosphere Monitoring Service) showed a reduction of PM 2.5 when compared with the data of February months of past three years. Reduction of 20–30% of PM 2.5 in year 2020 was observed in China when comparing the difference between the monthly average for February 2020 and the mean of the monthly averages for February 2017, 2018, and 2019.

After two weeks of nationwide lockdown, nearly 60% drop in the concentration of NO was reported in some cities of UK when compared to the same time period in the year 2019 (Monks, 2020). NASA reported a drop of NO_2 pollution by 30% over New York and major metropolitan areas in north-eastern USA when compared with the monthly average values from the year 2015 to 2019 (Monks, 2020).

In Rio de Janeiro Brazil, 30.3–48.5% drop in CO levels, a remarkable reduction in NO_2 levels, minor drop in PM_{10} levels and hike in ozone concentrations were observed in the first lockdown week. NO_2 and CO median values were lower by 24.1–32.9 and 37.0–43.6% when compared with the same period of 2019 (Dantas et al., 2020).

It was also predicted on the basis of the study that the Northern Hemisphere may be more resistant when compared to the tropical regions from the months of May to July 2020 and these areas may be comparatively more prone to outbreaks in October–November 2020 (Lal et al., 2020). The 55 days complete lockdown announced on 24th March 2020 in India also led to a large improvement in the various segments of environment like air, water, biodiversity, noise, etc. The air quality in urban areas improved from alarming or poor to satisfactory or good. Indian satellite data has also shown a remarkable reduction in the concentration of particulate matter or aerosol levels when lockdown imposed over several parts of the country (PTI, 2020). The main reason behind this was the Anthropause. Mobile combustion sources, industries, crop-residue or waste-burning activities are the main sources of pollution in India. Anthropause halted all these activities and gave positive results to the environment. Drop of 85.1% in $PM_{2.5}$ concentration in Ghaziabad (India's most polluted city) was observed when compared to the $PM_{2.5}$ concentration just three months before. The air quality showed an unprecedented improvement because of the decrease in carbon emissions, PM_{10}, NO_2 levels, and CO levels. All this was the result of restricted human activities (Lokhandwala & Gautam, 2020).

During the 55-day complete lockdown in India, S. K. Satheesh, Professor at Indian Institute of Science (IISc) Bangalore, recorded on an average, 50–60% particulate matter (PM) concentration reduction in southern parts of India and 75% particulate matter (PM) concentration reduction in the Indo-Gangetic basin, including Delhi, UP, Bihar, West Bengal, etc. (PTI, 2020).

Due to the absence of smog and dust haze, residents of district Jalandhar (Punjab) and Saharanpur (UP) were able to see the distant Himalayas for the first time in almost 50 years (Raghunandan, 2020). On April 3rd, residents of district Jalandhar (Punjab) saw the Dhauladhar mountain range. This was very rare because these were 213 km apart from each other (Saigal, 2020).

Not only air but water bodies also recovered and marine life thrived because of the lesser human footfall during the lockdown. The rivers of India (Ganga, Yamuna, and Cauvery, etc.) and Venice have become clean, clear and marine life was visible. In these areas the water became so clear that the fish could be seen and there was better water flow.

After lockdown for one month, CPCB (Central Pollution Control Board), IIT-Roorkee and other agencies tested Ganga water at Haridwar and some other places and it was found that approximately 50% dissolved oxygen (DO) had improved, 500% TDS (Total Dissolved Solids) is reduced, and slightly BOD (Biological Oxygen Demand) is also reduced. These were remarkable improvements in Ganga Water. It was also noted that the colour of the Ganga water looked blue at Rishikesh. Similar improvements in Ganga water was also observed at Varanasi, downstream. All these changes appeared because of the shutting down of industries, hotels, *dharamshalas* and lodges established in Rishikesh and Haridwar. In addition to these the dumping of domestic garbage including plastics into the river drastically decreased during the lockdown time period. Yamuna in Delhi also become blue and clear. It's DO increased and became more than 5 mg/litre, which showed that this water became fit for bathing. The reason behind this was again the shutting down of industries along Yamuna river in Delhi and its neighbouring States (PTI, 2020a).

14.2.1 Effect on Wildlife

Impact of Covid-19 on wildlife is very positive. People saw pumas in downtown Santiago (Chile), dolphins in s in the harbour of Trieste (Italy) and jackals in Tel Aviv, Israel (Rutz et al., 2020). *Mithun*, or the Indian Bison on highways and civet cat sin Kerala was seen moving in India (PTI, 2020a). Critically endangered South Asian River Dolphins had been seen in the Ganga river after 30 years. The South Asian River Dolphins had been spotted at various Ghats of Kolkata. All this was possible because of the reduced pollution. Animals like leopards, deer and some rare species were sighted roaming the streets of some cities. Ten thousand flamingos had gathered in the city of Navi Mumbai. The birds normally migrated to the area every year, but this year their number increased exponentially. Plants were growing better because of less or negligible human interference. With humans as a stand-still, the environment was allowed to self-replenish and grow.

As far as fish were concerned, the lockdown led to a decline in fishing. This resulted in the increase in the biomass of fishes. Overfishing was reduced. Sea turtles were also seen wandering to the areas which they normally avoided in laying down their eggs.

Species like rats, gulls or monkeys which are dependent on humans for their discarded food have reported struggling to stay alive under current conditions.

Risk of poaching or persecution of endangered species like Rhinos or raptors increased because of reduced presence of humans in remote areas (Buckley, 2020). Wildlife poaching incidents in India had increased from 35 (between February 10 and 22) to 88 (between March 23 and May 3) as per a study conducted by TRAFFIC (PTI, 2020b).

It is also predicted that in future economic hardships in economically weak countries may over-exploit natural resources (Rutz et al., 2020). The Covid-19 anthropause taught us that if we make minor changes in our lifestyles we can drastically reduce the disruptive effects on ecosystems (Rutz et al., 2020). Hence, there had been a positive impact on the environment due to the lockdown. Undoubtedly Covid-19 has brought self-cleansing or self-recovery time. We all already knew that the environmental degradation caused by humans was not totally irreversible and this lockdown time period of just 1–2 months proved it practically and this was witnessed by everyone. This was a warning and alerting signal for us to understand and react. Responsible authorities like government, policy makers and general public must take necessary steps to make this healing process regular and permanent in the future.

14.3 Discussion and Suggestions

This virus taught us how we may rediscover our sustainable policies to combat climate change. Covid-19 Anthropause was responsible for healing the Earth to a level which had never been seen in historical times. The noticeable improvements in environment have made us re-think the idea that the Earth could be purified. Anthropause has proved that our movements have a direct effect on the Earth's sustainability. Although, on the one hand, Covid-19 has caused economic limitations and cutback on development activities globally but on the other hand it gave us a glimpse of an environment that we all desire to live in. We should move to a future of eco-friendly lifestyle choices to conserve Mother Earth and for this we have to work only through effective collaboration, communication and action plans among scientists, experts, innovators and policymakers. We have to share our real-time experiences and insights in order to strengthen and contribute towards the solutions we need.

Covid-19 Anthropause brought a significant change in the water quality of some Indian rivers like Ganga, Yamuna, Cauvery, etc. River waters have become clean and clear and marine life was visible. All this happened because of the shutting down of mainly industries. Due to this the drainage of various untreated industrial effluents into rivers has stopped. As industrial and municipal waste water is the major pollutant of river water, so hazardous trading must be alternated with episodic shutting down and resuming to mitigate the effects on water bodies. In future, we must increase water treatment at the industrial source point with the help of municipalities in order to control water pollution in water bodies. Also, the capacity of various treatment plants must be increased in the near future.

The next observation from the Covid-19 Anthropause is the decrease in the carbon dioxide (CO_2) levels in Earth's atmosphere. Approximately 5.5% CO_2 concentration has decreased in 2020 in comparison to 2019. In order to meet the goal of limiting atmospheric temperature increase worldwide to 1.5 °C, we would need to decrease CO_2 emissions by 7.6% each year. To decrease the concentration of air pollutants permanently from the atmosphere some serious blueprints of policies and actions needs to be enacted immediately. In future, in order to control/minimise air pollution we have to adopt some measures to minimise the automobile exhausts. For this we must go for mass public transport instead of adopting private/personal vehicles. We must find ways to incentivise people who ditch their cars and adopt mass public transport. More bicycle lanes must be constructed/marked on roads in order to motivate the usage of bicycles on road. Urban infrastructure centres which include urban commons areas like public parks, water bodies and common green areas must be suitably re-designed. We must also promote pedestrian movement among individuals for shorter distances. Aside from all of this, anthropause may be used as an emergency measure to reduce extreme air pollution episodes in the regions where air quality is low or below quality standards. In the near future, such arrangements may be implemented in the Delhi-NCR area during the winter months.

During the Covid-19 Anthropause we significantly noted large reductions in traffic on roads which results a significant drop in the amount of particulate matter and greenhouse gas emissions in the upper troposphere which is responsible for the recovery of the ozone belt.

When anthropause ends and economic and offices activities resumed, the road traffic steadily increased and the Air Quality index in metropolitan areas observed nearly 250 Lakh of deaths and many respiratory diseases are reported due to air pollution every year. But this situation will become life threatening when Covid-19 is also present. Actually the most common symptoms of Covid-19 is pulmonary ailments, so Covid-19 exaggerates various pulmonary diseases and it leads to more deaths. Some agencies roughly calculated that deaths due to Covid-19 may increase by six per cent with an increase of one micro-gram/cubic metre (1 $\mu g/m^3$) of PM 2.5 particles in the air (Raghunandan, 2020).

So, it is significant that when anthropause ends and people return back to their normal lives, they are not enforced to revert back to their old behavior. A clean environment demands a limit to the growth policy. For this, the poor have to bear the highest burden and this was not considered to be a sustainable development. So, in future, if we wish to achieve sustainable development we should not go back to business with only growth policies. We need to think about the economic activities that may be performed with an eco-friendly approach. We should be either powered by lesser energy or by clean sources of energy rather than fossil fuels. Equality among the different sections of society should be given whether in terms of access to clean electricity or clean cooking fuels, hygienic living conditions, satisfactory jobs at satisfactory wages, social safety and security. This means a total change of systems that create iniquity and benefits for the privileged and the rich, whether in regard to their businesses or in the environment-guzzling facilities that they enjoy.

We have to give due importance to humanity if we want sustainable development instead of social status. In today's world energy consumption is considered as a key indicator of social status but in future the key indicator for social status must be humanity. It is the sense of humanity which teaches us compromises. The individual's commitment and the force of collective action, both must be mobilized to relieve suffering. This ensures respect for human dignity and ultimately creates a more humane society.

Economic growth, social equality and environmental sustainability are the three critical components which promotes sustainable development. For this, all countries especially developed countries, must take leadership to modify or redesign their growth policies and lifestyles. They must learn from this Anthropause and modify their behaviour in a timely manner. India's concern about development and poverty alleviation are admissible but not at the cost of environment. India is promoting clean energy and energy conservation. The necessary technology must be provided at low cost. There should be advancement and coordination in development, application and diffusion of technologies. Practices must be adopted for the mitigation of GHG in all sectors instead of only to the relevant sectors. The Bali Action Plan requires developing countries to formulate and implement Nationally Appropriate Mitigation Actions (NAMAs) which should be supported and enabled by developed countries through finance, technology and capacity building which are 'Monitorable, Verifiable and Reportable (MRV)'. These NAMAs denote that the plans must be prepared by the countries themselves without any external dictation and the financial support for NAMAs of developing countries must be an aid instead of discharging of responsibility by developed countries. The technology must be considered as the key to mitigating and adapting to climate change. Network of regional technology innovation centres is today's need. These centres must be set up in developing countries. Such centres will catalyze collaborative R&D. They also provide reliable information about the cost and performance on available technologies. This also enables capacity building on deployment of clean technologies and their further innovation (Narain et al., 2009). There needs to be a shift from non-renewable to renewable energies, some of which are already cost-effective and can enhance a sustainable energy supply, reduce local pollution and greenhouse gas emissions. Efficient, fast and reliable public transport systems such as metro-railways could reduce urban congestion, local pollution and greenhouse gas emissions.

14.4 Conclusion and Recommendations

It is too early to evaluate and predict the overall consequences of Covid-19 and Anthropause on India's environment. However, some conclusions may be drawn out of this. These are as follows:

1. The improvements in the air and water quality in the environment have shown that our mother earth can be saved if we restrict human actions. Earth's sustainability depends on human actions. It must be admitted that our current economy and political systems are about greed and wealth, rather than about other quality of life issues. They lead not to equality but to a culture of disparity, inequity and privilege for a few.
2. In order to conserve our earth we need to change our lifestyle. Minor changes in our habit can do wonders to the environment. As the environment is common for all so we have to work through effective collaboration, communication and action plans among scientists, experts, innovators and policymakers irrespective of any bias. We have to share our real-time experiences and insights in order to strengthen and contribute towards the solutions that we need without any discrimination. This means the dismantling of a system of wealth-based dissemination of health promoted by governments.
3. We must increase water treatment at the industrial source point with the help of municipalities in order to control water pollution in water bodies. Also, the capacity of various treatment plants must be increased in the near future. Hazardous trading may also be alternated with episodic shut down and resuming to mitigate the effects on water bodies.
4. In future, in order to control/minimise air pollution we have to adopt some measures to minimise the automobile exhausts. For this we must go for mass public transport instead of adopting private/personal vehicles. We must find ways to incentivise people who ditch their cars and adopt mass public transport. More bicycle lanes must be constructed/marked on roads in order to motivate the usage of bicycles on road. Urban infrastructure centres which includes urban commons areas like public parks, water bodies and common green areas must be suitably re-designed. We must also promote pedestrian movement among individuals for shorter distances. In addition to all these, anthropause could be used as an emergency step to mitigate severe air pollution episodes in those areas where air quality is very poor or severe. Such provisions could be adopted in the Delhi-NCR region during the winter months in the near future.
5. We need to think about the economic activities that may be performed with an eco-friendly approach. We should be either powered by lesser energy or by clean sources of energy rather than fossil fuels. Equality among the different sections of society should be given whether in terms of access to clean electricity or clean cooking fuels, hygienic living conditions, satisfactory jobs at satisfactory wages, social safety and security. This means a total change of systems that create iniquity and benefits for the privileged and the rich, whether in regard to their businesses or in the environment-guzzling facilities that they enjoy.

6. To combat climate change effectively a good understanding with coordinated and integrated efforts are required with compromising attitude globally. Thus, efforts need to be at global levels, with sincerity, rather than focusing on individual areas, countries or being national-centric about policies.
7. Network of regional technology innovation centres is today's need. These centres must be established in progressing countries. Such centres will speed up collective R&D. They also provide well-grounded information about the cost and performance on available technologies. This also enables capacity building on deployment of clean technologies and their further innovation (Narain et al., 2009). There needs to be a shift from non-renewable to renewable energies, some of which are already cost-effective and can enhance a sustainable energy supply, reduce local pollution and greenhouse gas emissions. Efficient, fast and reliable public transport systems such as metro-railways could reduce urban congestion, local pollution and greenhouse gas emissions.
8. It would be too early to predict that everywhere protected areas are safe and biodiversity is benefiting due to Anthropause. This may be true area to area and so a regional specificity of programmes in line with broad principles of global effect may be undertaken.
9. Without wasting any time we should help people of our society to realize that human health must be our priorities and remind them the linkage between healthy and resilient ecosystems and human well-being. For this we have to adopt good conservation practices. We are caretakers for the future in terms of the environment rather than of our damaging infrastructures. Damaging onc for the other does not cause development or growth. Rather the economists of de-growth may be more correct in their assumptions.
10. It is important to note that the poor in India are the ones who are most vulnerable to climate change. Regrettably, their properties and livelihoods are dependent on climate-sensitive production factors. To diversify their livelihoods and reduce their dependence, more political and bureaucratic attention is needed. This will necessitate investments in industries other than agriculture. Nutrition services, hospitals, disease prevention, irrigation, rural electrification, rural roads, and other basic investments, especially in central and eastern India, require a substantial increase in targeted investments (Narain et al., 2009).
11. In general, a participatory approach to forest protection, water management, energy management, and rural development will foster long-term development practices while also reducing greenhouse gas emissions or increasing carbon sinks. India's policies should be geared toward adaptation (i.e., increased capacity to defend against and react to serious natural events) and mitigation (aimed at capping the increases in the frequency and severity of the events themselves). To understand the viability and opportunities of low and continued high carbon growth pathways, as well as the results of mitigation mechanisms, a technically sound analysis is required (Shahab, 2015).
12. Lastly there is a requirement of good administrative skills which draw outlines with the aim of various issues like wetlands conservation, water management, equality among the individuals in the society, soil management etc. Also

strengthening agro-ecosystems, fund raising, promoting assurance etc. are other important programmes to be adopted to protect against the calamities caused by climate change (Narain et al., 2009). This world needs a neat and clean environment, sustainable development and social equality, safety and security. All these problems are interconnected. One cannot be solved without the solution of other. Our learning from this lockdown period highlights an urgent need to shift for eco-centric policies. As the whole world is interconnected, so we have to think globally and act locally. One great learning was that we need to really beef up our space, disaster and health systems drastically for any such future disasters.

With the help of a sense of brotherhood in the world we may easily solve big environmental problems in a simple way. The solution to environmental problems is only a matter of compromise and coordination. One of the biggest problem today is climate change. It poses backbreaker problems for various developing countries including India. Current position is sufficient if India is satisfied by its development prospects but if India desire to appear as a global power then India should take a leadership role on these key environmental issues. Today's most important global environmental challenge is climate change associated with food, water, health, energy availability and security. To combat climate change effectively a mutual coordination and cooperation and integrated efforts are required with a globally uncompromising attitude.

References

Asian Development Bank, Lancet and WHO. (2020). Infection rates of viruses involved in outbreaks worldwide as of 2020 (per infected person). *Journal of the American Medical Association*. Retrieved from https://www.adb.org/sites/default/files/publication/571536/adb-brief-128-economic-impact-covid19-developing-asia.pdf

Buckley, R. (2020). Conservation implications of COVID19: Effects via tourism and extractive industries. *Biological Conservation, 247*, 1.

Covid-19 Tracker India. (2020). Number of the coronavirus (Covid-19) cases in India 2020, by age group. Retrieved from www.covid19india.org

Dantas, G., Siciliano, B., França, B. B., & Arbilla, G. (2020). The impact of Covid-19 partial lockdown on the air quality of the city of Rio de Janeiro, Brazil. *Science of the Total Environment, 729*, 1–10.

Gill, V. (2020). Coronavirus: Wildlife scientists examine the great 'human pause'. Retrieved from https://www.bbc.com/news/science-environment-53113896

Global Carbon Project. (2020). Global Carbon Project. Retrieved from https://www.globalcarbonproject.org/carbonbudget/index.htm

Guo, Z., Wang, Z., Zhang, S., et al. (2020). Aerosol and surface distribution of severe acute respiratory syndrome coronavirus 2 in hospital wards, Wuhan, China, 2020. *Emerging Infectious Diseases, 26*(7), 1583–1591.

Johns Hopkins University, GitHub, Various sources and Covid-19 Tracker India. (2020). Cumulative of the coronavirus (Covid-19) confirmed, recovered and deceased numbers across India from January to August 2020. Retrieved from www.covid19india.org

Kachroo, V. (2020.) Novel coronavirus (Covid-19) in India: Current scenario. *International Journal of Research and Review, 7*(3), 435–447.

Kumar, S. U., Kumar, D. T., Christopher, B. P., & Doss, C. G. (2020). The rise and impact of Covid-19 in India. *Frontiers in Medicine, 7*, 1–7. Retrieved from www.frontiersin.org

Lal, P., Kumar, A., Kumar, S., Kumari, S., Saikia, P., Dayanandan, A., Adhikari, D., & Khan, M. L. (2020). The dark cloud with a silver lining: Assessing the impact of the SARS Covid-19 pandemic on the global environment. *Science of the Total Environment, 732*, 1–14.

Lokhandwala, S., & Gautam, P. (2020). Indirect impact of Covid-19 on environment: A brief study in Indian context. *Elsevier Public Health Emergency Collection, 188*, 1–10.

Monks, P. (2020). Here's how lockdown have improved air quality around the world. *The world economic forum Covid action platform*. Retrieved from https://www.weforum.org/agenda/2020/04/coronavirus-lockdowns-air-pollution

Monserrate, M. A., Zambrano, R. M., & Alejandra and Alcalde Luis Sanchez,. (2020). Indirect effects of Covid-19 on the environment. *Science of the Total Environment, 728*, 1–4.

Moriyama, M., Hugentobler, J., Walter, & Iwasaki, A. (2020). Seasonality of respiratory viral infections. *Annual Review of Virology, 7*, 2.1–2.19. Retrieved from https://www.annualreviews.org/

Narain, S., Ghosh, P., Saxena, N. C., Parikh, J., & Soni, P. (2009). Climatechange perspectives from India. UNDP. Retrieved from www.undp.org

OWID & ECDC. (2020). Number of cumulative cases of coronavirus (Covid-19) worldwide from January 8 to August 13, 2020, by day. Retrieved from www.ourworldindata.org

PTI. (2020a). *Covid-19 lockdown-like interventions may help combat air pollution in India, say scientists*. Retrieved from https://www.thehindu.com/sci-tech/energy-and-environment/covid-19-lockdown-like-interventions-may-help-combat-air-pollution-in-india-say-scientists/article31746151.ece

PTI. (2020b). *Coronavirus pandemic: Wildlife poaching in India more than doubled during Covid 19 lockdown*. Retrieved from https://www.hindustantimes.com/more-lifestyle/coronavirus-pandemic-wildlife-poaching-in-india-more-than-doubles-during-covid-19-lockdown/story-Nu0MvySAgVWtZMW3FUiOEL.html

Raghunandan, D. (2020). *Environment and Covid-19 in India: Balancing a clean environment with social justice*. Retrieved from https://www.newsclick.in/World-environment-day-clean-one-with-social-justice

Rutz, C., Loretto, M., Bates, A. E., et al. (2020). COVID-19 lockdown allows researchers to quantify the effects of human activity on wildlife. *Nature Ecology and Evolution*. Retrieved from https://doi.org/10.1038/s41559-020-1237-z

Saigal, K. (2020). *How is covid-19 impacting the environment around us?* Retrieved from https://www.investindia.gov.in/team-india-blogs/how-covid-19-impacting-environment-around-us

Shahab, S. (2015). Challenges of climate change and India's policy options. *Global Journal for Legal Studies, 4*(1), 118–126.

WHO. (2020). Number of novel coronavirus (Covid-19) confirmed cases and deaths as of August 13, 2020, by region, Coronavirus disease 2019 (Covid-19) Situation Report, 206: 1. Retrieved from https://www.who.int/docs/default-source/coronaviruse/situation-reports/20200813-covid-19-sitrep-206.pdf?sfvrsn=bf38f66b_6

WHO, CDC, United Nations, China Global Television Network, Johns Hopkins University and Lancet. (2020). Fatality rate of major virus outbreaks worldwide in the last 50 years as of 2020. Science alert. Retrieved from www.sciencealert.com

Worldometer. (2020). Number of coronavirus (Covid-19) cases, recoveries, and deaths among the most impacted countries worldwide as of August 14, 2020. Retrieved from www.worldometers.info

Chapter 15
Green Consumption Behaviour for Sustainable Development

Swati Garbyal and Ritu Mittal Gupta

15.1 Concept of Green Consumption

Green consumption interchangeably known as sustainable consumption, ethical consumption, and responsible consumption is a form of consumption that is attuned with the protection of the environment for the present and the future generations. It ascribes consumers' concern to address environmental issues through embracing environmentally friendly behavior, such as the use of renewable energy, carrying out organic activities, reduce, recycle, reuse, which leads to zero impact on climate.

Working definition of sustainable consumption by (UNCSD) United Nations Commission on Sustainable Development (1994) "The use of goods and services that respond to basic needs and bring a better quality of life, while minimizing the use of natural resources, toxic materials, and emissions of waste and pollutants over the life cycle, so as not to jeopardize the needs of future generations." There are three principal areas of developed Green consumption: Green energy, Green food, and Green clothing.

Green energy or renewable energy originates from natural sources, such as sunlight, wind, hydro, tidal, plants, algae, geothermal, and biomass can be naturally reloaded. Green consumption focuses on the natural source of energy rather than non-renewable. Green energy does not dispense greenhouse gas and has a slight or no impact on the environment.

Green food is organic food that does not require chemical fertilizers and chemicals to grow. Food like milk, eggs are considered green when it is acquired from animals reared in a natural setting without injecting drugs and antibiotics.

S. Garbyal (✉) · R. M. Gupta
Department of Extension Education and Communication Management, College of Community Science, Punjab Agricultural University, Ludhiana, Punjab 141004, India
e-mail: swatigarbyal341@gmail.com

R. M. Gupta
e-mail: rituhsee@pau.edu

S. Sheraz Mahdi et al. (eds.), *Adapting to Climate Change in Agriculture-Theories and Practices*, https://doi.org/10.1007/978-3-031-28142-6_15

Green clothing is considered organic if its manufacturing process does not consist of any non-biodegradable products, a fabric made up of natural fibers without using chemicals that affect the environment. Fabrics made of recycled products; reused products can be considered green. As the textile industry has contributed a lot to environmental degradation from the waste coming out from looms and factories, must aim towards green clothing using sustainable methods in order to make greener environment (Anonymous, 2020a, 2020b).

15.1.1 Origin and Need of Green Consumption

In the 1960s and early 1970s, western countries faced environmental pollution due to development in the industrial sector, which outgrow their economy and population as well, resulting from the development of a novel idea, i.e. Green consumption.

During the 1980s first-ever American 'green' brands started appearing in the market. Meanwhile, in 1980, the first book 'The Green consumer Guide' was published in the U.K. At the same time in Japan, a movement named 'Buy Green' was started by nearly 700 companies. With the coming year in the 1990s American market faced shortage of demand for green products, but after a decade in the early 2000s, again green product's demand increased.

Thereafter, consumers began to go for sustainability, accepted to spend more money on green products that assure sustainable means of production. Expansion of sustainable form of consumption, organic foods' consumption, high consciousness about the benefit of recycled goods was there (Harrison et al., 2005).

India had been growing rapidly for six decades, after independence attention towards the rapid growth of the Indian economy and relaxation in trade resultant the start of industrial growth. Although Pre-independence India already witnessed the Industrial Revolution in the 1850s resulting start of degradation of environmental resources.

As a result of the population explosion, India faced shrinkage of natural resources too. The resulting need for emphasis to protect the environment. Stringent laws and regulations came into force by the Indian constitution to tackle environmental issues. The Water (Prevention and control of pollution) Act (1974), The Air (Prevention and control of pollution) Act (1981), CPCB (Central Pollution Control Board), and Ministry of Environment, Forest and Climate Change in 1991 had launched the eco-labeling scheme known as the Eco mark. With the rise in per capita income, the buying power of Indians has enhanced too, hence the business enterprise attempted to persuade and motivate the people to use environmentally friendly products. According to Pandey (2017), Regardless, being the third biggest economy on the planet. India is home to one-6th of the world's kin and has the second biggest populace on the planet after China, and the biggest number of individuals living beneath the worldwide neediness line. On account of this sheer size and development, sustainability is a challenge. However, India emerges third largest contributor to the global greenhouse, India raised their concern for green marketing and green consumerism.

Unsustainable consumption practices, massive pressure on the environment, motivate the enterprise and business to work on developing awareness among people regarding green products, environmental issues through advertisement and marketing. Now people are willing to spend on green products and choosing green consumerism (Shamsi and Siddiqui, 2017).

According to an International report on sustainable living (2009) 'Greendex', which measures how consumers respond to environmental issues across the globe. The report measures the scores of housing, transport, food, and goods. India positioned first on this index among 18 countries. Results show that Indian consumers are most recorded concerned about environmental footprint and choosing sustainable products. But within the next 11 years, changes in the economy and socio-economic trends affected the environment directly and indirectly. India as well as the whole globe has a long challenge to deal with environmental issues. All need to realize that development with sustainability is a need of the hour (Greendex, 2009).

15.2 Sustainability and Sustainable Development

Word 'sustainability' came into existence in the 1980s when people became aware of environmental problems such as overpopulation, scarcity of natural resources, and climate change, etc. resultant from industrial expansion. At that time, media and public figures started to draw the focus of consumers towards sustainability, and people eventually came to know that sustainability and development are not mutually exclusive, rather interdependent and go with synergy. The report of Bruntland from the United Nations World Commission on Environment and Development in 1987, under the title 'Our common future' clutch the attention of consumers towards sustainability, it defines sustainable development as "meeting the needs of the present without compromising the ability of future generations to meet their own needs". It stressed the seriousness of pursuing sustainable development, to diminish the abandoned use of the natural resource. Furthermore, the Brundtland commission recognized three major elements of sustainable development: Environment, Economy, and Society. The concept of sustainability is not just the conservation of natural resources to protect the environment, but it is beyond it, consisting of an unchanging and healthy economy and improved quality of life. Brundtland Commission (1983).

(IUCN) International Union for the Conservation of Nature (1991) defined sustainable development as "improving the quality of human life while living within the carrying capacity of supporting ecosystems". Additionally, it explained a sustainable economy as the result of sustainable development, and it documented nine principles that shape it, for example preserving earth's strength and diversity, shortening the consumption of non-sustainable assets, keeping within the earth's carrying capacity, changing individual attitudes and practices, and so on. Consequently, sustainable societies are those that follow the former ideologies, choosing for the sustainable usage of natural and renewable resources.

Thus, sustainability can be mended in multiple areas like energy, wellbeing, economy, welfare, security, financial, family, humanity, personal, etc. Being sustainable means the ability to bear and maintain a process at a desirable level, hence it is also important that managerial, functional, operational, and technical requirements are worked out. To achieve sustainability, we must convert the consumers into green consumers.

15.2.1 Green Consumption and Sustainable Development

Consuming green products, whether it is food, clothing, or anything else, not only improves health, ecosystem but also leads to sustainable development. The raw materials required for making green products are biodegradable, organic, and renewable which a minimum effect on the environment. Hence, it steps towards sustainability without compromising the present needs. Similarly, the use of green energy i.e., renewable energy, leads to sustainable development.

15.3 Greendex Research on Consumer Behavior

A survey was done by National Geographic/GlobeScan Greendex, an international research approach to measure and monitor consumer progress towards environmentally sustainable consumption. Findings of the study done in 2014 showed positive results, as countries were more concerned about environmental issues compared to the previous survey in 2012. Greendex comprehensive ranking system, regulate each country's Greendex score, researchers surveyed more than 18,000 consumers. The researchers inquired consumers about their habits in a realm of different areas, including conservation of energy, food buying behavior, and transportation practices, likings in terms of organic and conventional products, and environmental knowledge and attitude. The rank of 18 countries was uncovered based upon data analysis. The countries like Australia, India, China, the United States, the United Kingdom, Brazil, South Korea, Argentina, Hungary, Russia, Mexico, Sweden, Spain, South Africa, Germany, France, Japan, and Canada were included in the survey.

India got the first position, drawing in a Greendex score of 61.4. In precise, the country gets high marks for its housing, transportation, and food selection, resulting from positive reinforcement, as India seems to be the most easily influenced among, with Mexico, Brazil, and Argentina to change when they were informed about their personal influence on the environment, while British, German, and Swedish consumers were the least influenced countries. China was the second most sustainable country. South Korea has moved into third place before Brazil and Argentina. Canadians and Americans remained lowest. Their main demise was their falling score in the area of housing. Various sub-indexes based on which different countries were ranked on Greendex are as under:

- ***Housing***

Consumers in India, Mexico, and China, scores have increased since 2012 regarding sustainable consumption at the domestic level. While countries like the USA, Canada, and Japan have decreased scores in domestic consumption. Germans and Canadians record the major cuts in their housing sub-index scores, compelled by household energy consumption.

- ***Transportation***

Consumers in India still stood the highest on the transportation sub-index, whereas consumers in North America rank at the lowest. Europeans are more inclined towards using public transportation than the USA, Brazil, and China. Since 2012, Chinese consumers have adopted unsustainable habits, like possessing at least one car or truck and driving alone often, they have declined their scores in transportation.

- ***Food***

Results revealed that food consumption habits are improving, as food sub-index scores have increased in 11 of the 17 countries since 2012. Scores are low in countries like Sweden and Spain, determined by factors such as high chicken and seafood consumption in both countries. Indian consumers were still in the first position of sustainable consumption in terms of food, because of fairly little non-vegetarian consumption. Mexican consumers continue to score lowermost on the food sub-index, accompanied by Japanese, American, and Spanish consumers.

- ***Goods***

Since 2012, Greendex goods scores had reduced for consumers in 11 of the countries surveyed, while increased in five, resulting in unsustainable behavior in consuming goods. South Korea is at the top in green consumption behavior regarding goods. Canadians and Americans at the bottom of the index, while Swedish consumers registered the good improvement since 2012, by repairing broken items and using their own durable biodegradable shopping bags instead of plastic.

- ***Intent to change behavior***

Overall results of the study showed that after knowing the environmental footprints, Mexicans, followed by Brazilians, Indians, and Argentineans, are more determined to improve their green consumption behavior, as consumers in these countries are easy to be influenced, while in countries like British, German, and Swedish, consumers are not liable to change their behavior towards green consumptions even after knowing their environmental footprint. Consistently, consumers in the large emerging economies of India and China have scored highest in green consumption behavior. Change in behavior the only way to achieve sustainability in the coming time.

15.4 Green Consumer and Their Behavior

- ***Green consumer***

According to Valentini (2011), "Green consumer is a consumer who, for every kind of consumer choice, choose something less hazardous for the environment and human health."

According to Kutloano (2011), "green consumer shops with awareness, in other words, they understand the implications of their purchases. Basically, they know the origin, source of their product and more or less where it will end up after consumption."

An in-depth awareness and concern about environmental issues are essential, but the readiness to spend higher prices for certified and less detrimental products seems to be more important.

- ***Non-green consumer***

According to Erica, a non-green consumer is a person who does not pay attention to his purchase, who does not care about health, who does not care about the environment, and who is selfish. Basically, non-green consumers are those who don't care about organic or inorganic, who is ignorant about environmental issues, and consume a lot, beyond their needs. High consumption levels are considered clear indicators to classify a non-green consumer.

Non-green consumers neither recycle nor look at the quality of the products they buy, don't pay attention when purchasing consumer goods.

In the words of Belk (1988), "people are what they buy and possess, and non-green consumers' indifference and non-consciousness about environmental issues emerge from their uncontrolled consumption."

Non- green consumers neither show respect for the environment when s/he goes walking in the mountain nor s/he pay attention to preserving and not spoiling.

15.4.1 Behavior and Green Consumption Behavior

Behaviors are deeply entrenched in social and institutional settings. We are directed much by what others around us say and do. We often find ourselves 'locked in' to unsustainable behaviors, despite our own best intention (Jackson, 2005).

Human behavior is the response of an individual to internal and external stimuli. Behavior plays an important role in individual life, some behavior takes us in the right direction in lifelike politeness, tidiness, timeliness, eating healthy, hydrating oneself, etc. But some behaviors are needed to change as creates problems in one or other way like smoking, poor diet, lack of exercise, aggressive, stealing, telling lie, etc. Getting out into the new behavior and doing something new is helpful to foster our life. Change is good because you have the opportunity to embrace newly experienced. It is an important prerequisite for making progress.

Green Consumption Behavior: Pro-environmental Behavior

Numerous environmental issues cause a threat to the sustainability of the environment, in which air pollution, water pollution, climate change, drought, famine, global warming, water scarcity are common. Most of the problems have their root cause in human behavior and can be coped with by performing significant behavior. Modification in human behavior regarding sustainable consumption can be useful in coping with environmental issues.

Green indicates the conservation of natural resources, while consumption means destruction. Consumption is usually a physical, social, economic process that is influenced by nature, surroundings, psychology of consumer, geography, culture, norms, and societal arrangements.

Green consumption together generates an idea of a set of practices of environment-friendly activities. It is one form of pro-environmental behavior that make little harm to the environment. According to Kollmuss and Agyeman (2002), pro-environmental behavior is defined as "behavior that consciously seeks to minimize the negative impact of one's actions on the natural and built world".

Individuals knowingly choose this type of behavior an individual to get a minimal effect on the environment. Some research in this field with empirical support claim pro-environmental behavioral multidimensional construct:

- ***Public sphere***: Participating in the environmental campaign, rallies, and members of NGOs working for environmental issues.
- ***Private sphere***: Purchasing green products, like energy-efficient electric items, doing organic composting, using public transport instead of private vehicles, recycling, etc.

Many influencing effects impact the pro-environmental behavior or green consumption behavior.

15.5 Factors Affecting Green Consumption Behavior

Certain factors affect the behavior of individuals towards green consumption, some factors trigger the behavior while other factors restrict the behavior. After analyzing various researches on factors affecting green consumption behavior, the authors identified that all the factors that influence green consumption behavior can be broadly categorized under four types: personal, societal, governmental, and functional factors.

15.5.1 Personal Factors:

- *Perception:*

Perceived consumer effectiveness—When consumers estimate that any step taken by them leads to impact the environment positively, and in being able to bring about outcomes that he or she values and wants to achieve.

Perceived seriousness of environmental problems—It is the perception of consumers towards environmental issues and how much the consumer is serious about the issues.

Perceived benefit—Consumers' perception towards the product will provide benefit to them and the environment.

Perceived risk—Consumer's perception towards the risk with the product after purchasing it is a negative correlation with green behavior.

Perceived usefulness—Consumer's perception towards the product that usage of that product will be useful in terms of minimal environmental and health effect.

Perceived environmental responsibility—Consumer perception towards their responsibility for environmental problems and contribution in combating the problems.

Conditional value—It is a perceived usefulness of an alternative green product in a specific situation. It is influenced by the physical and social value that enhances the product's social and functional value.

Epistemic value—It is perceived usefulness attained by a certain product that will stimulate curiosity and able to satisfy the desire for knowledge among consumers. Epistemic value also contains innovation in an existing product that keeps consumers stick to the existing brand.

Self-image—Self-image is basically how an individual sees himself/herself, be it abilities, appearance, and personality.

Self-identity—Self-identity is a whole picture of who we believe we are, while self-image is one piece of that picture.

Social identity—Social identity refers perception of an individual about himself based on his membership, religious group, etc.

- ***Attitude***—It refers to the feeling of someone towards something, it can be positive, negative, or maybe neutral. Our positive attitude is a prerequisite to our green behavior.
- ***Belief***—This consists of Control belief, Outcome belief, Citizenship belief, and Response efficacy. Control belief means when consumers believe that they can control the environmental problems by being green. Outcome belief means when an individual belief in a predicted future outcome to be beneficial. Citizenship belief when the consumer thinks he should be a responsible citizen for whatever happening to the environment. Response efficacy when a person belief the recommended action leads to reduce the risk and avoid threats.

- ***Knowledge***—Is how many consumers know about environmental issues, different consumption practices? Whether they know about global environmental, policy knowledge, and waste knowledge or not is detrimental to green consumption.
- ***Behavior***—Behavior have certain elements like intrinsic motivation, emotion, and moral. Intrinsic means internally, if the consumer is self-motivated to conserve the environment, he'll go for the green. Consumers who are more emotionally attached to environmental issues tend to become green consumers. If the consumer has moral values to save the environment, it will be visible in his/her behavior.

15.5.2 Situational Factor

Service provision or consumers' job profile also acts as a major influencer while purchasing a product. Other situational factors are socio-demographics, which include age, education level, and income. Studies show that the more the age more concerned about the environment a person is, the more the consumer is educated, the more he'll be concerned about environmental issues, while in the case of income, the consumer who can spend more is more likely to act green.

15.5.3 Societal Factors

- ***Normative beliefs***—Consumers incline to behave according to the norms he/she believes. Consumers sometimes change their behavior, thoughts, or values to be liked and accepted by others.
- ***Subjective norms***—It is a perceived social pressure from others within a society. It is of two types, Descriptive norms (commonly practiced) and Injunctive norm (morally right) both have a great influence on green behavior.
- ***Social value***—It is a means and ends that people have in their life to live it like status, health, peace, enlightenment, and power and influence.
- ***The socio-cultural***—This factor is the combination of society and culture, which affect thought, behavior and feeling.

15.5.4 Functional Factors

- ***Quality***—Quality of products plays a major role, some consumers opt for quality irrespective of the price, while others consider price ignoring product quality.
- ***Price***—Consumers are sometimes willing to go green, but price acts as a major barrier, price-conscious consumers always consider the price before purchasing any product.

15.5.5 Governmental Factors

- ***Availability of green products*** - If greener products will be available in sufficient amounts, then there are more chances that consumers will buy them.
- ***Accessibility of green products***—Along with availability, accessibility must be considered. In a market, green products may be available, but because of the high price, they may not be accessed by consumers.
- ***Policy intervention***—The government, along with NGOs and other private agency must develop policies regarding green products. There must be a subsidy on purchasing green, and a relaxation in percentage of tax to the shopkeeper who sold greener products. Implementation of good policy can enhance green consumption practices.

15.6 Barriers to Green Consumption

- ***Price***: First thing that acts as a barrier to be greener is price. Most of the average consumers purchase their products only based on price, which reflects in a consumers' mind while purchasing products. People choose to spend less without considering how that product can cause problems to their health and environment. It seems too expensive to buy organic products for them.
- ***Low availability***: As more companies develop non-sustainable products than eco-friendly products, people usually go for products that are abundant around them.
- ***Less awareness***: There are more advertisements for non-green products than green products, as media plays an important role in opening the horizon among consumers. Consumers purchase those products which are mostly seen by them on television, videos, etc. What they see goes into their subconscious mind, and while purchasing a product subconsciously, they buy what they have seen or heard about.
- ***Lack of concern***: Some consumers are aware, have accessibility, and can afford green products, yet they have less concern about the environment. The only way to change their level of concern is by making them more and more aware, adding them to the environmental campaign, telling them about the dreadful effects of unsustainable consumption, and the use of various intervention strategies to alter their behavior.

These barriers must be thoroughly looked upon, and strategies can be worked out to overcome these. There are certain models which can be used to change the behavior of consumers, accompanied by overcoming barriers.

15.7 Behavior Change and Behavior Change Model

- ***Behavior change***: It refers to any change and alteration in present behavior of human. Interventions has huge ability to change the current behavior pattern. Making change in behavior is not an easy task, however researchers have discovered numerous methods to make desirable changes in human behavior, these are being utilized by doctors, educators, extensionist, etc.

15.7.1 The Elements of Change *Cherry k (2019)*

There are three basic and most important component in changing human behavior these are as follows:

- ***Readiness to change***: Do individual have knowledge so that they can be ready to make long lasting change in behavior?
- ***Barriers to change***: is there any blockages that are preventing individual to change the behavior?
- ***Expect relapse***: what condition might occur that make individual to perform early behavior?

15.7.2 Behavior Change Models to Trigger Behavior Change

Models of behavior change elucidate why the particular behavior change. In behavioral determinations, the major factors such as personal, environmental, and behavioral attributes are cited by these models. In the past few years, these models are being used in various fields, like health, education, energy, criminology, and international development. Some of the behavior change model are given below:

- ***Health belief model (1950s)—***

To predict health-related behaviors, especially regarding health services, a socio-psychological health behavior model, i.e., the health belief model (HBM), has been developed.

The HBM was developed by social psychologist Irwin M. Rosenstock, Golfrey M., Hochbaum, S. Stephen Kegeles, and Howard Leventhal at the U.S. Public Health Services in the year 1950s. This model is still known as the most outstanding and utilized theory in human behavior study. Figure 15.1 shows that the HBM advocating the individual's engagement (or lack of engagement) in health-promoting behavior depends on the individual's belief regarding their health problems, perceived benefits of action, and barriers to action and self-efficacy. A stimulus or cue to action should be available to trigger the health-promoting behavior (Morris et al., 2012a, 2012b).

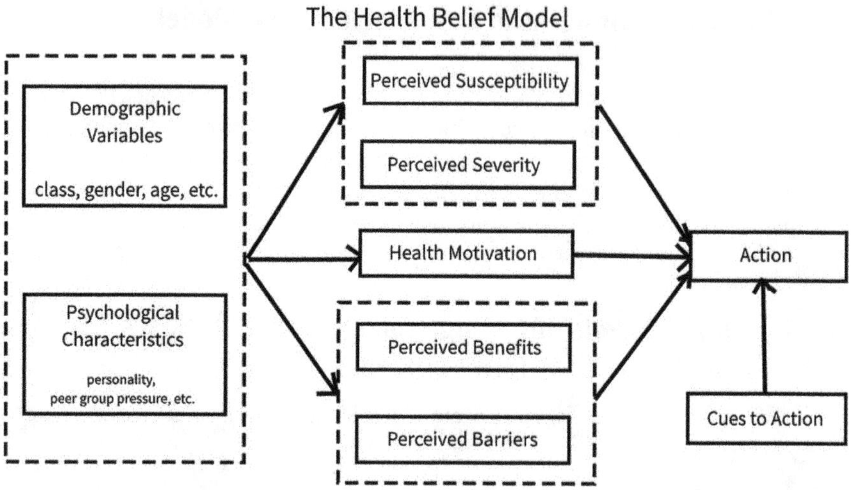

Fig. 15.1 Health belief model

- ***The 4 E's model of behavior change (2005)—***

The '4e's model arisen from the study of Jackson (2005). The study focused on strategies of behavior change and promoted behavior change under four categories. The Center of this model has the behavior and attitude of individuals, and interventions such as information, education, and incentives are intended to affect the behavior. The core area of this model is to understand the behavior in an individual's lifestyle context, starting from their current position, and then understanding the way they live. As there is no single solution to influence behavior, but in several ways. So in Fig. 15.2 the 4 Es framework has come up with the mixed interventions, such as Enable, encourage, exemplify, and engage to promote the behavior change (Allen, 2020; Jackson, 2005)

- ***Theory of reasoned action (TRA, 1967)—***

TRA was developed by Martin Fishbein and Icek Ajzen in the year 1967, which explained in Fig. 15.3 that the association between human behavior and attitude. Primarily, theory predicts the behavior of an individual based on their pre-existing attitudes and intentions. The theory says the decision of an individual to engage in a particular behavior is depending on the expected outcome that is assumed by that individual (Anonymous, 2019).

- ***Theory of planned behavior (1980)—***

The Theory of Planned Behavior (TPB) is an extension of the theory of reason action (TRA). In Fig. 15.4 TPB understands and foresees the behaviors which postulate that behavior is determined by behavioral intentions and under certain circumstances by perceived behavioral control. Behavioral intentions are blended with three

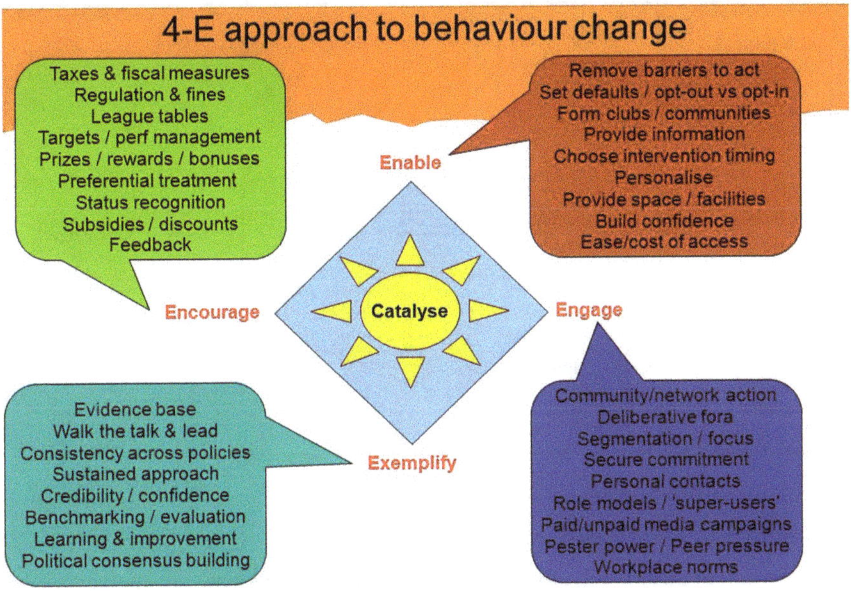

Fig. 15.2 4E's model

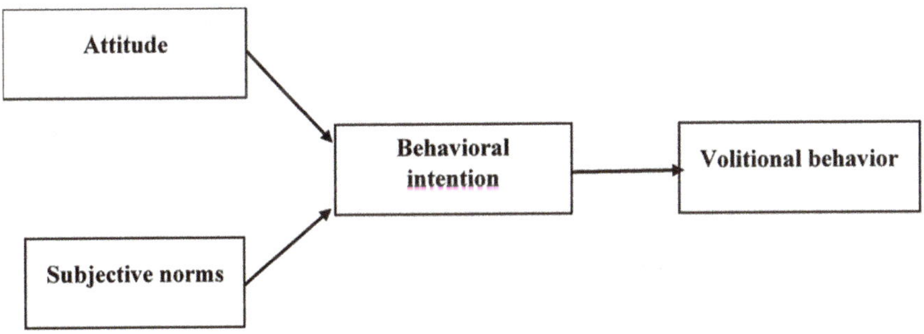

Fig. 15.3 Theory of reasoned action

factors: subjective norms, attitudes toward the behavior, and perceived behavior control (Kan & Fabrigar, 2017).

- ***Trans-theoretical or stages of change model (1970)—***

The Trans-theoretical Model, also known as The Stages of Change (SoC) model was developed in the late 1970s by investigators James Prochaska and Carlo Di Clemente. This model is used extensively as a cognitive model, which categorizes individuals in five categories that signify different milestones or levels of motivational readiness in a continuum of behavior change.

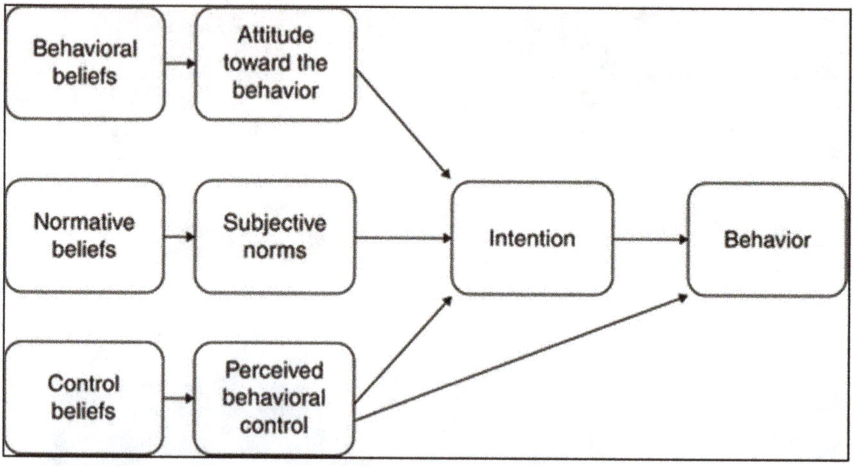

Fig. 15.4 Theory of planned behavior

According to this model, changes happen slowly, and relapses are an unavoidable part of the process to make an enduring change. In the beginning, the individual often reluctant to change, but sooner or later they develop a proactive and dedicated approach for behavior change.

This model elucidates that a steady movement of little steps requires regularly to achieve a bigger objective.

Figure 15.5 shows the six stages (i) pre-contemplation, (ii) contemplation, (iii) preparation, (iv) action, and (v) maintenance (vi) Relapse stage (Morris et al., 2012a, 2012b).

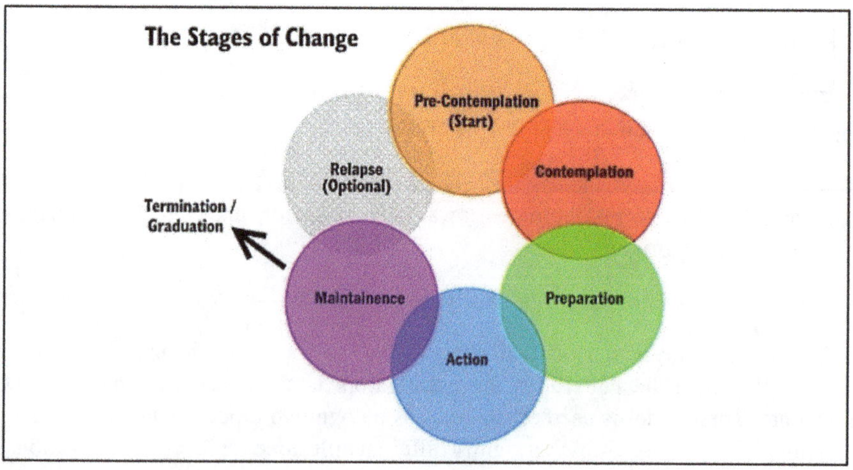

Fig. 15.5 Trans-theoretical change model

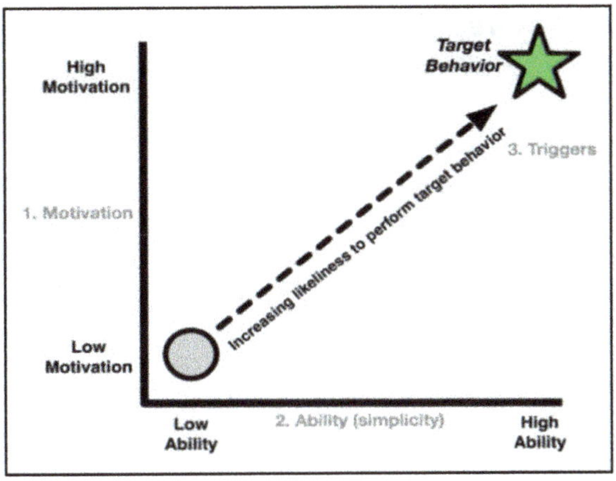

Fig. 15.6 Fogg behavior model

- ***BJ Fogg's behavior model (FBM) (2009)—***

B. J. Fogg proposed the Fogg Behavior Model (FBM) in the year 2009, to analyze the behavior change. 'Behavior' is a word, Fogg used to depict an activity that somebody performs. He raised the question that how we can skill an individual to do the right actions. Fogg said three elements are needed to change the behavior. These are 'B = MAT' i.e., Motivation, Ability, and a Trigger (prompt). Collaboration of these three elements leads to the desired change in the behavior (Anonymous, 2017).

Figure 15.6 shows the first element that is 'Motivation' further includes three drives, namely 'sensation', 'anticipation', and 'belonging'. 'Sensation' has two sides: pleasure and pain. Individuals are motivated to seek pleasure and avoid pain. 'Anticipation', a second core motivator, has two sides: hope and fear. 'Third core element, i.e., 'Belonging' is a social dimension that has two sides: social acceptance and social rejection. People usually perform socially acceptable things.

The second element "Ability" includes promoting the target behavior for which the users have the high ability. BJ Fogg lists dimensions that characterized the high ability or simplicity of performing a behavior, i.e. money, time, physiological effort, brain cycle/thought process, and social deviance.

Thirdly, "Trigger", later termed as Prompt, are reminders that may be direct or indirect to the performance of a behavior. It is the starting gun to fire, and they'll get going. There are three types of Trigger 'spark', 'facilitator', and 'signal'.

Thus, to inculcate green consumerism in behavior, any of these models can be used by educationists, extensionists, and environmentalists.

15.8 Green Marketing: A Precursor to Green Consumerism

Green consumption not merely signifies purchasing green products, but also act, behave and being green. Green consumerism comes from a desire to protect scarce natural resources for the next generation, without compromising the need of the present generation. Green marketing plays an important role in spreading green consumerism. Nayyar (2015) "Green marketing refers to holistic marketing concept wherein the production, marketing, consumption and disposal of products and services happen in a manner that is less detrimental to the environment."

According to American Marketing Association (2020), "green marketing is a market of products that are presumed to be environmentally safe. Thus, green marketing incorporated a broad range of categories like product modification, changes in production process, green packaging, as well as changes on advertisement."

15.8.1 Strategies for Successful Green Marketing

There are certain green marketer need to take into consideration while doing green marketing.

- ***Identify your consumer***: First and foremost, it is to gather information about your consumers to whom you are going to give your goods and services, know their awareness towards green products, their knowledge, their needs, demand, interest, etc.
- ***Educate your consumers***: Second thing, after knowing their awareness level, provide them information about different green products, how these products are beneficial for their health and the environment as a whole.
- ***Credibility and transparency***: It means the company should carry all the things in their product which they have mentioned and claim to be. They should always be transparent to their consumer about the ingredients, packaging of the products.
- ***Assure the buyer***: Consumers should be assured that the product is environment friendly and true to the claim, and will not lead to an unsustainable cord in the future too.
- ***Consider their price***: Price is one of the barriers while purchasing the product. Consider the pocket of buyers, if the product is of high price, then tell consumers about its importance, how it functions differently from other products.

15.9 Green Products and Their Characteristics

Green products refer to the product which has an environmental friendly nature, various characteristics of green products are:

- Products with natural ingredients.
- Originally grown products.
- Recyclable, reusable, and biodegradable products.
- Products with eco-friendly packaging.
- Energy-efficient products.
- Products using recycled contents, non-toxic chemicals.
- Not affect the health of humans and animals badly.
- Not developed from unnecessary cruelty to animals.
- Not developed by killing endangered species.

15.10 Suggestions to Enhance Green Consumption Behaviour

Consuming green products is one of the major aspects in the direction to attain sustainable development. Changing the behaviour of consumers towards green consumption is one of the major challenges for environmentalists, governments, NGOs, agencies. If they get success in changing the behaviour then soon we can achieve sustainable development goals. Some of the suggestions are:

- ***Awareness stage***: First and most important step to be considered, without awareness no consumer will go for the green. Forceful backing is required to make the consumer aware of environmental issues. Rallies, billboards, street plays, advertisements, short stories can work for raising awareness. Mass media can be used successfully for this purpose.
- ***Information and knowledge buildup stage***: Providing information to build up their knowledge must be the next crucial step. Awaken consumers must be exposed to overall information regarding prevailing environmental problems, like climate change, global warming, other disturbances, and its global, local individual effect. Strategies like lectures, demonstrations, short videos, and short films must be shown to impart knowledge regarding problems. Research results must be exposed to the consumers that how badly the things have affected the environment so far.
- ***Cause of the problem and alternative solution***: It is important to talk about different causes, like high energy consumption, the surge in areas like transportation, industrial sector, fossil fuel industries, etc., and its leading effect on the earth. Alternate solutions must be given, like use LEDs instead of incandescent light bulbs, use of solar lights, avoiding wastage of water, water waste management, indulging in the best of waste, minimize the use of plastic at the domestic and industrial level. Provides them the examples of countries and states who are in direction to sustainable consumption, and those that have zero or minimum carbon

emission, like Bhutan. More emphasis must be given to a conventional method of livelihood, organic farming instead of chemical fertilizers, riding a bicycle for short distance instead of motor vehicles, using handmade rugs in the home instead of plastic one, bucket bath instead of a shower, washing utensils through bucket water rather opening running water tap, etc. Such activities can be demonstrated through videos to the consumers, with minimize the use of natural resources and sustaining it.

The government, NGOs, other private agencies, schools, universities can come up together to mitigate the environmental issues. All must work hand-in-hand to combat the situation, all the above steps can be followed to make consumers conscious and alert. Various models can be developed by different institutions that aim for sustainability without compromising today's needs. Subsidies and loans with minimal interest can be provided by government banks to the ventures and ideas related to green energy. Positive reinforcement must be there by providing price and honorarium to the NGOs, or other private agencies which follow green instructions, which engages the community towards being green by various ways and means. The yearly survey must be done to find out the cleanest and organic area, village, city, and state across the nation. The survey can be done based on their green consumption, conservation of natural resources, and waste-water management. Village, city, or state which ranks first in green consumption behaviour must be awarded, their working model must be replicated in other areas of the country. Media must be involved in this process as media act as a major influencer among consumers. Government must direct media to advertise green products, and also direct manufacturers to keep the green products affordable to the general population.

Every individual of the society must realize the value of sustainable consumption to secure their natural habitat for now and for their next generations. Promoting green consumerism with proper planning, implementation and follow-up can work as a ladder towards sustainable development.

References

Air (Prevention and control of pollution) Act. (1981). Retrieved from: https://www.jspcb.nic.in/upload/uploadfiles/files/AirAct.pdf. Accessed on August 14, 2020.

Allen, W. (2020). *Six ways that agencies can influence behavior through public policy*. Retrieved from: https://learningforsustainability.net/post/6es-policy-framework/. Accessed on July 19, 2020.

American marketing association. (2020). *Green marketing*. Retrieved from https://www.ama.org/the-definition-of-marketing-what-is-marketing/. Accessed on August 2, 2020.

Anonymous. (2017). *B J Fogg behavior model*. Retrieved from https://www.growthengineering.co.uk/bj-foggs-behavior-model/. Accessed on July 30, 2020.

Anonymous. (2019). *Theory of reasoned action*. Retrieved from: https://en.wikipedia.org/wiki/Theory_of_reasoned_action. Accessed on July 29, 2020.

Anonymous. (2020a). *Green consumption*. Retrieved from: https://en.wikipedia.org/wiki/Green_consumption. Accessed on July 30, 2020.

Anonymous. (2020b). *Green consumption.* Retrieved from https://en.wikipedia.org/wiki/Green_consumption#:~:text=The%20three%20main%20reasons%20that,Ethical%20concerns. Accessed on July 24, 2020.

Belk, R. W. (1988). Possessions and the extended self. *Journal of Consumer Research, 15*(1), 139–168.

Brundtland Commission. (1983). *Sustainable development.* Retrieved from https://simple.wikipedia.org/wiki/Sustainable_development. Accessed on August 30, 2020.

Fog, B. J. (2009). *A behavior model for persuasive design.* Retrieved from https://www.mebook.se/images/page_file/38/Fogg%20Behavior%20Model.pdf. Accessed on July 30, 2020.

Greendex. (2009). *Consumer choice and the environment—A worldwide tracking survey, consumer choice and environment*, Greendex Report.

Greendex. (2014). *Consumer choice and the environment—A worldwide tracking survey, consumer choice and environment*, Greendex Report.

Harrison, R., Newholm, T., & Shaw, D. (2005). Studying the ethical consumer: A review of research. *Journal of Consumer Behaviour., 6*(5), 253–270.

IUCN (International Union for the Conservation of Nature). (1991). "Caring for the Earth: A Strategy for Sustainable Living", with the UN Environment program and the World Wildlife Fund, Gland, Switzerland.

Jackson, T. (2005). Motivating sustainable consumption: A review of evidence on consumer behavior and behavioral change. *Sustainable Development Research Network* (Vol. 15) University of Surrey, England, UK.

Jain, J., & Gupta, S. (2015). *Green marketing. Green marketing: A changing Paradigm in contemporary marketing.* Kalyani Publishers (pp.146–151).

Katie young. (2018). *The rise of green consumerism: What do brands need to know?* Retrieved from https://blog.globalwebindex.com/chart-of-the-week/green-consumerism/. Accessed on August 1, 2020.

Kollmuss, A., & Agyeman, J. (2002). Mind the gap: Why do people act environmentally and what are the barriers to pro-environmental behavior? *Environmental Education Research, 8*(3), 239–260.

Kutloano. (2011). *Green consumer.* Retrieved from: research.cbs.dk› files› stefania_valentini (p. 26). Accessed on July 25, 2020.

Kan, M. P. H., & Fabrigar, L. R. (2017). Retrieved from https://link.springer.com/referenceworkentry/10.1007%2F978-3-319-28099-8_1191-1. Accessed on August 2, 2020.

Morris, J., Marzano, M., Dandy, N., O'Brien, L. (2012a). *Forestry, sustainable behaviours and behaviour change: Theories.* Retrieved from https://www.forestresearch.gov.uk/documents/1409/behaviour_review_theory.pdf. Accessed on July 31, 2020.

Morris, J., Marzano, M., Dandy, N., & O'Brien, L. (2012b). Health belief model. Forest research: Sustainable behaviours and behaviour change: Theories. *Journal of Forest Research, 1*, 6–7.

Nayyar, A. (2015). A comprehensive scenario of Indian it organizations regarding green marketing: Initiatives and adoptions. *Green marketing: A changing Paradigm in contemporary marketing.* Kalyani Publishers (pp. 1–8).

Pandey, A. (2017). *What India can teach the world about sustainability? World economic forum*, Bhopal, India.

Shamsi, M. S., & Siddiqui, Z. S. (2017). Green product and consumer behavior: An analytical study. *Pertanika Journal of Social Sciences and Humanities, 25*(4), 1545–1554.

United Nations Commission on Sustainable Development (UNCSD). (1994). *Sustainable consumption.* Symposium on Sustainable Consumption.

Valentini, S. (2011). *The role of green consumption practices in defining self-identities: A qualitative research in Denmark and Italy.* Copenhagen Business School (pp. 25–26).

Water (Prevention and Control of Pollution). (1974). Retrieved from http://Extwprlegs1.Fao.Org/Docs/Pdf/Ind2085.Pdf. Accessed on August 6, 2020.

Chapter 16
Strategies for Sustainable Climate Smart Livestock Farming

Bilawal Singh, Amandeep Singh, Y. S. Jadoun, Pragya Bhadauria, and Gurpreet Kour

16.1 Introduction

Agriculture and livestock in India is inextricably linked in the rural domain of India (Ahmad et al., 2019). Any possible impact on either of the two affects the other substantially. Rising temperature, floods, droughts, crop failures, pest attacks, etc. are the common constraints associated with agriculture due to climate change. On the other hand, climate change is also affecting livestock in a number of ways. The direct implication on livestock being reduction in the availability and yield of fodder and grasslands for livestock production. Livestock being a living system is prone to risks which inevitably become evident in due course of time. Any natural calamity occurring due to climate change like disease outbreaks, accidental fires, cyclones, floods, etc. may lead to loss of animal life, ultimately distressing the poor farmer whose sole source of income perishes (Singh et al., 2020a, 2020b, 2020c) which is a serious implication of climate change. Climate change also affects the livestock production as it leads to competition for land and water, and food security at a time when it is most needed (Thornton, 2010). Climate change has significant impacts on both the natural resources and the ecosystems upon which the livestock sector

B. Singh · Y. S. Jadoun (✉)
Department of Veterinary and Animal Husbandry Extension Education, Guru Angad Dev Veterinary and Animal Sciences University (GADVASU), Ludhiana, Punjab, India
e-mail: ysvet1203@gmail.com

A. Singh
Guru Angad Dev Veterinary and Animal, Sciences University (GADVASU), Ludhiana, Punjab, India

P. Bhadauria
ICAR-Agricultural Technology Application Research Institute Ludhiana, Ludhiana, Punjab, India

G. Kour
Department of Animal Genetics and Breeding, Guru Angad Dev Veterinary and Animal Sciences University (GADVASU), Ludhiana, Punjab, India

S. Sheraz Mahdi et al. (eds.), *Adapting to Climate Change in Agriculture-Theories and Practices*, https://doi.org/10.1007/978-3-031-28142-6_16

depends. Food and Agriculture Organization of United Nations have also stated that livestock food chains are major contributors to greenhouse gas emissions (FAO, 2006).

The vital contribution of climate smart animal agriculture in Indian context is par visualization. Being resource poor, Indian farmers in totality rely on animal agriculture for meeting their economic and nutritional needs. Therefore, there is an imminent need to develop strategies for climate smart livestock farming. Livestock is not only integral to mixed farming systems, but also provides food and income to large number of people, often those living in rural setups. In the global context, livestock production represents 40% of agriculture's gross value and makes a key contribution to global food security. Its contribution is especially important to landless and marginal farmers where livestock represents a unique source of energy, protein and micronutrients.

Greenhouse gas emissions primarily leads to global climate change that result in warming of the atmosphere (Pearce et al., 2014). The livestock sector contributes 14.5% of GHG emissions worldwide, and thus may increase land degradation, air and water pollution, and decrease in biodiversity (Gerber et al., 2015). At the same time, climate change will affect livestock production through affecting the quantity and quality of feeds, disease outbreaks, heat stress and biodiversity loss while the demand for livestock products is expected to increase by 100% by mid of the twenty-first century (Garnett, 2009). Therefore, the challenge is to maintain a balance between productivity, household food security, and environmental preservation. Before understanding the strategies for CSLF, let us understand basics of being climate smart.

16.2 Climate Smart Farming (CSF)

CSF is a novel approach for agricultural development under the new realities of climate change (Kumar et al., 2020). As per the Food and Agricultural Organization of the United Nations, CSF is defined as "farming that sustainably increases productivity, enhances resilience (adaptation), reduces/removes green-house gases (GHGs) (mitigation) where possible, and enhances achievement of national food security and development goals". In the definition provided by FAO, the principal goal of CSF is identified as food security and development (FAO, 2013a, 2013b); while productivity, adaptation, and mitigation are identified as the three interlinked pillars necessary for achieving this goal. CSF is neither new agricultural system, nor a set of practices rather than an innovative methodology which devises the development pathways for betterment of agriculture sectors more productive, efficient and sustainable and makes the agriculture sectors more productive and sustainable. Therefore, further it can contribute to climate change adaptation and mitigation.

16.3 Climate Smart Agriculture (CSA)

CSA is an approach for developing agriculture by taking appropriate transformation and reorientation measures (Lipper et al., 2014). According to Food and Agriculture Organization of United Nations, CSA is defined as "agriculture that sustainably increases productivity, enhances resilience (adaptation), reduces/removes GHGs (mitigation) wherever possible, and enhances helps in achieving national food security and development goals". Provision of food security and development is the primary goal of CSA while productivity, adaptation, and mitigation are the three interlinked pillars for achieving this goal (FAO, 2013a, 2013b; Lipper et al., 2014). The CSA has three main objectives: sustainably enhancing agricultural productivity to support reasonable increase in income, food security and development; augmenting adaptive capacity and resilience to shocks at multiple levels, from farm to national; and decreasing greenhouse gas emissions and increasing carbon sequestration wherever possible.

16.4 Climate Smart Livestock Farming (CSLF)

According to Kadzere (2019), there is an urgent need to alleviate the negative effects of climate change on livestock production to meet growing demand for livestock products worldwide. Livestock depends on environment and natural resources which are affected by climate change. Increased ambient temperature, altered rainfall and shifts in precipitation patterns, heat waves, heat stress, droughts, floods and erratic changes in seasonal patterns are emerging challenges livestock production which have become evitable due to climate change. Reduction in yields and quality of feeds and fodder, possible increase in incidences of diseases and competition for the available resources are some indirect effects of climate change on livestock production systems (Behera et al., 2019). In Indian context, where almost 70% of livestock are owned by small and marginal farmers and landless labourers is at high risk (Singh et al., 2020a, 2020b, 2020c) due to climate change. Several approaches like genetic improvement of animals, use of molecular genetic markers for heat tolerance in selection programme, provision of scientifically designed housings, improving diets, better herd management to improve output, better management of grassland, utilization of digital technologies like establishment of weather forecasting system for earlier warning of farmers regarding alarming weathers etc. will help in combating with the changing climate (Behera et al., 2019) leading to CSLF. Livestock waste management plays a significant role in CSLF as the ill effects of livestock waste can be converted into profitable products (Singh et al., 2020a, 2020b, 2021). CSLF also involves moderations in extension services, researchers and trainers to collaborate and upgrade their skills in order to keep pace with ever-changing sector norms. Climate smart livestock services must be integrated and involve all stakeholders, including public, private and non-government sectors, industry value chains, and educational institutions at a local, national and international level. This holistic approach is required because of the fact that climate change has local and global implications, as GHGs emitted in one country can influence production in another (Kadzere, 2019).

16.5 Contribution of Livestock to Climate Change

Livestock sector is responsible for 14.5% of greenhouse gas emissions which includes 9% CO_2 from land-use change caused by demand for feed grains, grazing land and agricultural energy, 37% methane from enteric fermentation and manure management and 65% N_2O from animal manure. Ruminants are less efficient in converting forage into useful products than monogastrics (pigs and poultry). Therefore, a large share of GHG emissions is contributed by livestock. The ruminants, both small and large, are the main contributors (98%) to the enteric methane emission in India. In India more than 90% of the total methane emission from enteric fermentation is being contributed by the large ruminants (cattle and buffalo) and rest from small ruminants and others (Patra, 2017). The major contributors to methane emissions in Indiaare indigenous, crossbred cattle, buffalo and sheep and goat accounting 40, 8, 40 and 10% respectively. On an average total emission of methane from Indian livestock was 10.08 MT considering categories of ruminants and different type of feed resources available (Singhal et al., 2005). The contribution of milch buffaloes was 59.6%, crossbred cows 11.4% and indigenous cows 28.9% to the total emissions from dairy animals (Upadhaya et al., 2009). Contribution of goats to methane emission was only 4.5%. The total global methane emissions from livestock manure management have been estimated as 9.3 Tg/year as compare to India are estimated to be 1.27 Tg in the year 1994 (Scheehle, 2002). As per the estimates of Jha et al. (2011), total methane emission from Indian livestock (cattle, buffalo, sheep, goat, horses and ponies, camel and pig), has been estimated as 9.92 ± 2.37 Tg. Key livestock species like cattle, buffalo, sheep and goat were contributing about 9.14 ± 2.21 Tg of methane. Contribution of enteric fermentation was about 8.97 ± 2.22 Tg and manure management practices was 0.95 ± 0.15 Tg. In India, the states of Andhra Pradesh, Bihar, Madhya Pradesh, Maharashtra, Rajasthan and Uttar Pradesh were having major share of about 5.8 Tg of methane and were considered to be major states of methane emission due to large number of animals. Mizoram, Goa, and Sikkim were lowest methane producers.

Let us understand the emissions through an example. Quantity of feed consumed and its digestibility are two important factors, which decide the total methane production. On an average, 70 and 120 kg of Methane is released per year per cow. Methane is a greenhouse gas as carbon dioxide (CO_2). But the negative effect of methane on climate is 23 times higher than the effect of CO_2. Therefore, the release of approximately 100 kg methane every year for each cow is equivalent to about 2300 kg CO_2 per year. The livestock traits (age, weight and species), health and living conditions influence the energy requirement. Production of higher methane gas results from higher feed intake and energy requirement. On average Indian cattle produces about 35 kg per annum methane as compared to 95 kg per annum for dairy cows in Germany (Sirohi & Michaelowa, 2007) due lower energy requirement. The lowest annual methane production for dairy (180 kg/herd) and non-dairy cattle was reported in Indian subcontinent while comparing with other regions of the world (Sharma et al., 2006).

16.6 Impact of Climate Change on Livestock

Domestic animals maintain their body temperature within a relatively narrow zone to remain healthy and productive and are classified as homeotherms. The fluctuations in ambient temperature above or below thermo-neutral zone produce stress to the animal and untimely hampers production, reproduction and growth. Air temperature, relative humidity, air movement and solar radiations are four environmental factors which influence effective temperature. The temperature-humidity-index (THI) is used usually to indicate the degree of stress on dairy cattle. With increase in THI more than 74, high producing dairy cows are affected adversely. Livestock Weather Safety Index (LWSI) was given by Eigenberg et al. (2007) to classify the combined intensity of temperature and humidity into four categories of THI values:

(a) THI less than or equal to 74 is Normal,
(b) THI 75–78 is Alert,
(c) THI 79–83 is Danger, and
(d) THI value 84 and above is Emergency condition.

Reduction in feed intake, increased water intake, changed metabolic rate and maintenance requirements, increased evaporative water loss, increased respiration rate, changed blood hormone profile and increased body temperature are some of the effects noticed in animals due to climate change. The comprehensive details of impact of climate change on livestock can be understood below.

16.6.1 Impact on Milk Production

Climate change on livestock has a direct impact on the milk yield. Increase in number of stressful days (THI more than 80) and their frequency impact yield and production of cattle and buffaloes (Upadhaya et al. 2009). High heat environment is a major factor that can negatively affect milk production in dairy animals, especially in animals of high genetic merit. At all India level an estimated annual loss due to direct thermal stress on livestock is about 1.8 million tonnes of milk (Rs. 2661.62 crores), that is, nearly 2% of the total milk production in the country.

16.6.2 Impact on Animal Reproduction

Intensity of estrus period decreases, therefore less conception rate occurs. So, heat stress may reduce the fertility of dairy cows in summer by poor expression of behavioural signs of estrus due to a reduced estradiol secretion from the dominant follicle. In these situations, the calving interval becomes longer. So, lifetime production of dairy animal decreases. Heat stress during pregnancy slows down growth of the foetus due to decreased blood supply to the uterus which causes placental insufficiency to provide maternal nutrient which leads to decreased fetal growth and calf size or sometimes leads to early embryonic death in animals exposed to heat stress. Reproductive processes in male animal are very sensitive to disruption by hyperthermia with the most pronounced consequences being reduced quantity and quality of sperm production and decreased fertility.

Heat stress also leads to reduced sperm concentration and seminal volume. It is reported that ejaculate volume, concentration of spermatozoa and sperm motility in bulls are lower in summer than in winter season (Krishnan et al., 2017).

16.6.3 Impact on Feed and Fodder Availability

The quality and quantity of feed and fodder also changes due to climate change. Droughts and extreme rainfall variability can trigger periods of severe feed scarcity, especially in dry land areas, with devastating effects on livestock population. Reductions in the quantity and quality of feed (leading to less feed intake and higher mortality) could make the impacts of climate change on livestock systems severe in certain places.

16.6.4 Impact on Livestock Health

Animal Diseases

Climate change affects by increasing spectrum of pathogens which further increases the disease susceptibility of the animal and thus, supports the pathogenicity of the causative agent. The pattern and distribution of livestock diseases along with changes in disease severity have major impact on livestock systems.

Effect on Vectors

Vectors play a significant role in disease transmission among livestock. Ticks, lice, mites, mosquitoes and flies, the developmental stages are major vectors and often heavily dependent on temperature and humidity. Alterations in the pattern of rainfall and temperature regimes affect both the distribution and the abundance of disease-causing vectors, as can changes in the frequency of extreme events (Thornton et al., 2009). Cattle ticks like, *Boophilusmicroplus, Haemaphysalisbispinosa* and *Hyalommaanatolicum* have found to aggravate the vector borne diseases in livestock (Kumar et al., 2004).

Effect on Pathogens

Changes in precipitation, soil moisture, humidity, temperature, etc. play a vital role in the multiplication and spread of pathogens and parasites. Air flow or wind pattern also affect the distribution of pathogens and spread from one place to another. Excessive rainfall leads to spread of vectors and also help them multiply.

16.7 Need of Climate Smart Livestock Farming (CSLF)

Climate change significantly effects the ecosystems on which livestock sector depends. Increased temperature, changes in rainfall and precipitation patterns effect the livestock systems directly. Indirect impacts are experienced through alterations in ecosystems, changes in the yields, quality and type of feed crops, possible increases in animal diseases and increased competition for resources. At the same time, livestock food chains are major contributors to greenhouse gas (GHG) emissions (FAO, 2006). Usually, livestock respiration is not added as a net source of carbon dioxide (CO_2) discharges because they are part of the global biological cycle. Kyoto Protocol suggests that the consumed amounts of carbon dioxide in vegetative form are equivalent to those emitted by the livestock. Therefore, livestock acts as carbon sink as the excess carbon is used by livestock in building tissues for meat and secreting milk, thereby making a vital contribution in food security and mitigation of climate change. Heat stress and availability of water in arid and semi-arid region will affect livestock production. Commercial chicken and pork production has created massive environmental and social collateral damage. They mostly depend on imported soybeans grown as high input monoculture from South America, using land formerly under rainforests. The concentration of industrial production system in the US (Tyson) and Germany (PHW Group) dominate the world market and can have a devastating effect on local markets of developing countries. Monoculture crop production reduces biological diversity, grassland, and conversion of forests and release of CO_2, nitrous oxide emission from synthetic fertilizers, nitrification of soils and ammonia load in the atmosphere. Shifting to the diversification of livestock animals, crops and crop varieties, mixed crops, agroforestry and leaves of *Zizyphus* (ber), acacia and chickpea and integration of cattle production with agriculture had been the most promising adaptation measures under climate change. Keeping in mind the collateral damage done during the process of livestock production, it becomes important at this point of time to adopt various strategies for CSLF.

16.8 Strategies and Working Models for Climate Smart Sustainable Livestock Production (CSLF)

The CSLF aims to strengthen livelihoods and food security, especially small and marginal farm holders through improving the management and judicious use of natural resources through adoption of appropriate technologies and methods for production, processing and marketing of livestock and livestock products. In livestock sector, CSLF can be achieved through the followings strategies and working models which are currently being used in different parts of the world for livestock production.

16.8.1 Strategies for Climate Smart Livestock Farming (CSLF)

There are many strategies for adapting livestock production systems to a changed climate. These include: technological options (e.g., using species with better drought tolerance); behavioral modifications (e.g., changes in dietary pattern, such as consuming less meat); managerial choices (e.g., different farm management practices); and policy alternatives (e.g. planning regulations and infrastructural development). Generally, livestock are more resistant to climate change than are crops because they are mobile and can travel to find feed (Hoffman & Vogel, 2008). Possible barriers to adaptation within the livestock sector may include: poor local infrastructure; lack of access to credit; lack of market access; poor access to water and water management technologies; poor or absent animal health services; and lacking access or knowledge of climate information (Nayak et al., 2018). The following are the strategies which can be acted upon to reach at CSLF.

Nutritional Interventions

The production of the animal depends on the type of feed it consumes. Over and under feeding should be avoided. Over feeding leads to heat loss and under feeding results in decreased production. Economic feed processing techniques such as wetting of grasses, cropping and chopping of green, grinding, pelleting, use of urea-molasses will reduce the energy loss in the digestion and decrease the heat loss for maintenance of body temperature. Use of available green fodder during summer or efficient use of non-conventional feed resources or newer feed resources will help to negotiate the fodder scarcity produced due to adverse climatic condition (Behera et al., 2019).

Nutritional technologies for improvement in rumen efficiency like diet manipulation, direct inhibitors, feed additives, propionate enhancers, methane oxidisers, probiotics, defaunation and hormones can help in reduction of methane production (Moss, 1994). Dietary manipulation through increased green fodder decreases methane production by 5.7%. Increasing the concentrates in the diet of animals reduces methane by 15–32% depending on the ratio of concentrate in diet (Singhal & Mohini, 2002). The methane mitigation from molasses urea supplementation was found to be 8.7% (Srivastava & Garg, 2002) and 21% from use of feed additive monens in (De & Singh, 2001). Improvements in feed efficiency and milk production can significantly reduce GHG emissions and land use of the dairy herd (Bell et al., 2011). It was noticed that highly digestible high-energy rations are an effective form of summer diet to help animals to control body temperature by reduction of excess heat. Providing low fiber diet and a cool drinking water low fiber diet renders comfort to the animals. Bypass fat was proved in a study to reduce heat stress in cattle and buffalo around 18–20%. Increasing the concentration of minerals and vitamins in the diet to compensate for the reduction in feed intake, particularly sodium, potassium, magnesium and niacin levels in the diet have proved to be useful (Bell et al., 2011). Supplementing cows with 1.5–1.6% DM of potassium and 0.5–0.6% DM of sodium will potentially improve milk yield in heat-stressed cows. Include magnesium at 0.35–0.4% DM to help to avoid metabolic problems (grass tetany) when feeding higher amounts of potassium. Including niacin (6 g/cow/day) may also be beneficial. It has been reported to decrease skin temperature and increase milk yield. Enhancement in milk yield has also been reported by feeding 150–200 g/cow/day of sodium bicarbonate during hot weather to help buffer the rumen. The use of antioxidants such as Vit. E, Vit. A and selenium help in reducing the impact of heat stress by oxidant balance, resulting in improved reproductive efficiency and animal health (Behera et al., 2019).

Reproductive Interventions

Progesterone supplementation during early pregnancy has proven beneficial in some studies. Supplementation of exogenous progesterone during summer heat stress has the potential to improve fertility. Synchronization of heat in dairy animals with GnRH and PGF2α also improves fertility. Utilizing embryo transfer technology (ETT) is considered a potential strategy for minimizing the negative effects of heat stress on bovine reproduction.

Manure Management

In India, most of the animal manure is extensively used as fuel in the form of dry dung cakes or spread in field (Singh et al., 2020a, 2020b, 2020c). Animal wastes including manure account for more than 25 million tonnes methane emission globally per year. Better management of animal excreta through various interventions can reduce the methane emission (Singh et al., 2020a, 2020b). Manure solids separation and anaerobic degradation pre-treatment can mitigate CH_4 emission from subsurface-applied manure, which may otherwise be greater than that from surface applied manure. The GHG emission from manure is dependent on the temperature, timing of application and duration of the storage. Moreover, manure have the residues of certain chemicals which are noxious for humans as well as for environment. Further, the excretions of diseased animals may have certain zoonotic pathogens which are very harmful for humans and can remain in soil for several days to weeks. The active compounds in the animal excretions and the effluents erupting from the livestock products and processing industries pose a greater threat to all the components of environment. The fusion of traditional managemental procedures with climate smart manure management can better serve the purpose of manure management (Singh & Rashid, 2017). Manure management via recycling is an important step in sustainable livestock waste management as well as to reduce the negative environmental impact associated with its mismanagement. Production of Biogas from animals by anaerobic digestion has been traditionally a common practice in Asia, particularly in tropical areas such as Indonesia, India and Vietnam (Henuk, 2001). Biogas is used as fuel for cooking and lighting purposes and in diesel engines to substitute diesel-oil. The left-over decomposed slurry is a good source of manure for agricultural lands as it contains 80% carbon, 1.8% nitrogen, 1% phosphorous and 0.9% potash which makes the slurry an excellent source of humus and micronutrients for crops. Since ages, the livestock manures have been used effectively as organic fertilizers. Animal dropping contain all essential plant nutrients and have been well documented to be an excellent fertilizer (Bell, 2002). Due to high nitrogen content in poultry manure, it has been recognized as the most desirable of these natural fertilizers because of its high nitrogen content (Sloan et al., 2008). Further, animal manure can be potentially converted into vermin-cast and vermin-meal (protein meal) via low cost vermiculture system (Singh & Rashid, 2017).

Housing and Managemental Interventions

Good house ensures proper design, height and orientation with good open and covered space. Adequate ventilation and comfortable floor space per animal will provide cooler microenvironment inside the house. Proper housing ensures stress free environment which leads to better productivity among livestock, thus building resilience among them. While constructing animal house, heat ameliorative measures like foggers, sprayers, drinkers and shady areas should be properly built.

Current risks of climate change impacts can be reduced by adoption suitable adaptation strategies. Soin hot and humid climates housing and management of dairy animals may be improved in two ways:

(i) Adaptation by the animal to the climate through selection, breeding and acclimatization,
(ii) Adaptation of climate to the animal by provision of protective structures and cooling devices.

The long axis with north and south orientation will expose the area under the shade to the morning and afternoon sun and assist in keeping it dry. Therefore, this type of orientation of shed is suitable in sub-temperate to temperate climate of hilly region. The height for cattle sheds would be 10–12 ft. It is advisable to paint white the top of a shelter, so as to reflect much of the incident solar radiation. The underside of the shelter should be painted black, so that the radiant heat reflected from the ground onto the animals should be reduced. Crops provide cooler atmosphere for cattle housing than bare ground. Plantation of tree around the animal shed produces long term cooling effect on the animal. To reduce effects of temperature on livestock in summers, it is advised to take feed to livestock rather than livestock to feed. Walking to feed increases heat load which also leads to heat stress.

Reduction in Livestock Population

Livestock are one of the chief contributors to today's most serious environmental problems. Urgent action is required to remedy the situation. Global analyses have clearly shown that non-CO_2 greenhouse gas (GHG) emissions i. enteric methane (CH_4) and nitrous oxide (N_2O) etc. are inversely related to animal productivity (Gerber et al., 2013). Animal productivity can be increased through improvements in animal genetics, feeding, reproduction, health, and overall management of the animal operation. In many parts of the world, reduction in animal numbers was the single most influential mitigation strategy that significantly reduced the carbon footprint. But in countries like India, where majority of the livestock ownership is with poor landless and marginal farmers, the viability of this strategy comes under question.

Precision Livestock Farming

The optimum utilization of resources in a farm can be obtained through precision livestock farming (Tripathi & Bisen, 2019) along with cutting direct and indirect greenhouse gas emissions. Smart tags, drones, cameras, sensors and computers are available for specific interventions in livestock management. The software based digital tools cuts the supply of inputs after sensing the threshold.

Integrated Farming System

Any type of farming is threatened due to degradation of natural resources and climate change. Climate change impacts crops, livestock, forestry, fisheries and aquaculture, and can cause grave social and economic consequences in the form of reduced incomes, eroded livelihoods, trade disruption and adverse health impacts. If we wish to reduce the vulnerabilities, better we should change our farming practices according to climate or we integrate various types of farming systems so that the loss in one can be compensated by other. Integrated Farming Systems (IFS) employ a unique resource management strategy to help achieving economic benefit and sustain agricultural production without undermining the resource base and environmental quality. IFS ensures inclusive growth in agriculture, pro-poor, and environmentally sustainable (Altieri, 2002). IFS can bring along economic growth and poverty reduction, with enhanced resilience of small farmers to disasters (Altieri et al., 2012). Livestock can be integrated with agriculture, horticulture, sericulture, fisheries, etc. for better returns.

Improving Resource Use Efficiency

In livestock production, the optimum utilization of resources not only help in cost-cutting but also ensures sustainability. For example, feed conversion ratio or feed use efficiency should be enhanced in livestock sector for optimum utilization of feed which will lead to better utilization of feed resources and less depletion.

Building Resilience in Livestock

Research should be oriented to develop such varieties of livestock which will be more adaptive to changing climate. The resilience of the present livestock should also be enhanced by selective exposures to vulnerabilities. It also includes the development of new technologies in livestock production which are not available at present.

Breeding Management

Selective breeding of the livestock with stress tolerant breeds should be done. Identification and breeding of animals through marker assisted techniques can bring significant progress in breeding management. Setting of mother bull farms for selected areas and breeding them with the local livestock to enhance resilience should be practiced. Farmers should be provided with incentives for breeding their animals to the better breeds mandated for climate resilience.

Using Digital Technologies

Digital technologies have multifarious role in building resilience and mitigation of climate change in livestock production. The network of farmers can be created using social media whereby timely information can be passed onto them which will also strengthen extension and advisory services. Early weather forecasting systems can help farmers to be ready with preventive measures after the passage of calamity. Disease surveillance and monitoring systems can aware the farmers well before disease outbreak so that the preventive and control measures can be strengthened.

Further, livestock monitoring can be done using digital technologies like tags and collars which helps in better management.

Better Extension Advisory Services

Extension advisory services are the backbone of any livestock enterprise. Farmers requiring any sort of information turn to an extension agent. Therefore, it becomes mandatory for extension services throughout the world to focus on climate smart aspect of farming. The new strategies should be discussed with the farmers and the literature regarding same shall be distributed. Focus group discussions, mass media talks, popular articles in mass media, blogs on social media can be effective in spreading the knowledge about climate smart activities.

16.8.2 Working Models for Climate Smart Sustainable Livestock Production (CSLF)

In this sub-section, the working models which are instrumental globally in reaching at climate smart livestock farming have been discussed.

Climate Smart Approaches for Smallholder Dairy Development in East Africa

The climate smart project running in East Africa named as the East Africa Dairy Development Project (EADD) focuses on Uganda, Rwanda and Kenya. In its initial phase, EADD provided 179,000 farming families with extensive training on dairy husbandry, business practices and operation, and marketing of dairy products. Presently in its second phase, the program aims to work with over 200,000 farmers on improving dairy production and access to markets. EADD has currently incorporated climate-smart agriculture as an overarching objective. This includes supporting farmers to intensify milk production by transitioning to fewer but more productive cattle per household, which will reduce emissions per unit of milk. Key climate-smart activities include improved feeding using crop by-products, fodder banks, improved manure management, agroforestry, improved pasture species and planted legumes. EADD is financed by the Bill and Melinda Gates Foundation and has received a grant of USD 25.5 million for the second phase of the program.

Kuroiler Chicken: A Better Backyard Broiler for Rural India

Commercial meat production often means using lots of resources—feed, land, water, energy. But animals like chickens and pigs have traditionally been raised on discarded food scraps and not much more. In rural Indian villages, many women with little or no land keep a few hardy backyard chickens that don't need expensive care or feed. But compared with modern commercial breeds, these birds are very slow to produce eggs or meat. Kegg farms, an Indian poultry breeder, set out to develop a better chicken for rural India. Then the company built a grassroots distribution network, enabling villagers to turn their small flocks into profitable businesses. Introduced in the early 1990s, the hybrid Kuroiler breed eats the same diet of insects, vegetation, seeds and scraps as a native chicken but puts on weight 5 times as quickly and produces 2–3 times as many eggs per year. The surplus allows households to sell more chickens and eggs as well as consume them. Kuroiler chicks reach remote villages through a series of hatcheries and distributors—8000 small enterprises supplying at least 800,000 farmers, each representing a new income source for rural communities. The popular breed has also been successfully exported to countries such as Uganda. The commercial operations related to this breed of chicken has now been taken care by the rural women, also leading to their empowerment. Better yet, the system is relatively climate-proof: the birds are bred for drought and disease tolerance and they do not depend on plentiful grain harvests. The impact of Kuroilers can be stated in a manner that these birds are extraordinary solution to a very context-specific problem: breeding hardy, high-yield chickens adapted to the conditions and available resources in poor rural Indian households. Concentrating investment in this narrow niche proved highly successful. The distribution system leverages thousands of small rural enterprises to make Kuroilers as accessible as possible—literally delivering chicks to the doorsteps of some of the most isolated people in India. With 20 million chicks per year being distributed across 13 Indian states, the annual profits for the chicken farmers are estimated at over USD 12 million.

Farmer Field Schools (FFS) Supporting Climate Smart Cattle Ranching in Nicaragua

Learning by doing is an effective way to spread climate-smart practices, as shown by a Central American programme of farmer field schools for sylvo-pastoralism, the practice of integrating forestry with grazing land. In FFS, the farmers share lessons and design their own management practices in the process. As part of the learning process, 7 sylvo-pastoral dual-purpose cattle farms (producing both milk and calves) in central Nicaragua were compared with an equal number of traditional ranches. The farmers had added trees to their pastures—which can improve soil and productivity—and the animals ate a combination of grass and woody fodder. On conventional farms, milk production and income were more than 50% higher, and the wooded pastures sequestered twice as much carbon. Higher adoption of these innovative approaches depends on participatory processes like the farmer field schools, as well as incentives such as certification and demand for greener products in value chains.

The Incentive System for Low-Carbon Agriculture in Brazil

Agriculture sector of Brazil is the second largest source of greenhouse gases after the energy sector. But there are clear opportunities in the country to mitigate agriculture's contribution to climate change. For instance, Brazil has about 40 million hectares of degraded pasture. Reinstating these pastures could increase beef yields sixfold, from around 30 kg/ha per year to 180 kg/ha, and reduce the pressure to expand agriculture into the Amazon region. More carbon is sequestered by well managed pasture than degraded pasture. Low-Carbon Agriculture (ABC) plan of Brazil includes a credit initiative called the ABC Programme. This programme provides low-interest loans for sustainable agricultural practices such as no-till agriculture; restoration of degraded pasture; integration of crops, livestock and forest; planting of commercial forests; biological nitrogen fixation; and treatment of animal wastes. The program's ambitious goals include rehabilitating 15 million ha of degraded pastures, 4 million ha of integrated crop-livestock-forest, planting 3 million ha commercial trees, and treating 4.4 million cubic meters of animal waste. The target is to reduce greenhouse gas emissions by 160 million tonnes of carbon dioxide equivalent annually by 2020.

Index-Based Livestock Insurance (IBLI) to Increase Climate Resilience of Pastoralists in Kenya and Ethiopia

Pastoral livelihoods are undermined by many natural calamities such as droughts accompanied by widespread livestock mortality. Index-based livestock insurance (IBLI) provides pastoralists the option to insure themselves against these events. In comparison with the traditional insurance products, Index-based livestock insurance traces local forage conditions using real-time, publicly available satellite data ('greenness maps') to decide the severity of drought, predict area-average livestock losses and calculate policyholders' indemnity payments. If the contractual threshold or 'strike' point of forage loss or predicted livestock mortality is reached, the IBLI contract is triggered and policyholders receive a pay-out proportionate to the number and type of animals insured and the severity of vegetative loss and expected herd loss in the policyholder's geographic area. Thus, IBLI aims to deliver a productive safety net for households affected by livestock loss and help them effectively manage the resulting shock. IBLI may also incentivize investments that enhance productivity, as farmers feel safe investing in the purchase and keeping of livestock as knowledge that insurance payouts would cover losses due to extreme weather events. The Partners like ILRI, Cornell University and the CGIAR research programs for Dryland Systems and Climate Change, Agriculture and Food Security, etc. are associated with IBLI. Donors that have supported the project include the Australian Agency for

International Development (AusAID), the UK Department for International Development (DFID), the European Union, the Global Index Insurance Facility (GIIF), the US Agency for International Development (USAID) and the World Bank.

Milking the Demand for Dairy in Kenya

Fast-growing cities worldwide are thirsty for milk. Demand for milk and milk products in developing countries are growing at 2.5% year on year, and small-scale farmers are already trying to increase production but processing and packaging of milk is major challenge. Milk bound for far-flung urban markets must be pasteurized, but an FAO study in Kenya found that heating and chilling equipment costs up to USD 50,000—a dubious investment for dairy cooperatives dealing in small volumes. A better substitute is the low-cost technology known as Milk-Pro, originally developed in South Africa. Whereas conventional pasteurizers require specialized electrical supplies, Milk-Pro runs on standard outlets or generators. Milk is vacuum-packed in plastic sachets for pasteurization and shipping, which prevents contamination and extends the refrigerated shelf life to 15 days. Cutting the number of steps for processing fresh milk also reduces the water used for cleaning equipment and containers. The system's price, just under USD 10,000, is within range for a Kenyan dairy cooperative. The payback period is one year when 750 L of milk is processed per day.

Carbon Finance to Bring Back Grasslands in China

When overgrazing strips pastures bare, productivity suffers. Yet despite the threat to their livelihoods, many poor herders cannot afford the short-term costs of switching to a more sustainable system. Carbon credits could be a way to fund that shift, but only if local people are able to demonstrate that their new practices lower greenhouse gas emissions and store more carbon in the soil. The Food and Agriculture Organization (FAO), working together with the Chinese Academy of Agriculture Sciences (CAAS), the World Agroforestry Centre (ICRAF) and China's Northwest Institute of Plateau Biology (NWIPB), has developed a carbon accounting methodology to make this easier for small-scale sustainable grazing projects. Yak and sheep herders in the Three Rivers region of northern China will be the first to test out the approach.

16.9 Conclusion

Agriculture and allied sectors require significant transformations to meet challenges of food security and climate change. To meet the challenges of thermal stress on livestock production systems, the focus is primarily on cattle in warm or hot climates, for which animal housing and management is the high priority adaptation strategy to minimize the effects of climate and climate change on livestock production. Factors affecting variability in enteric CH_4 production requires urgent attention and efforts to decrease the uncertainty in GHG emission inventories for which viable GHG reduction strategies needs to be identified. Increasing resource efficiency is necessary both to increase and safeguard food security on the long term and to contribute to mitigate climate change. It is essential to build resilience to every type of risk and be prepared for uncertainty and change. Efficiency and resilience have to be considered together, at every scale and from environmental, social and economic perspectives. In order to operationalize green economy, there is need to implement climate-smart agriculture. All the stakeholders should come together with coordinated action to achieve long term climate change perspectives. To alleviate some of the complex challenges posed by climate change, farming has to become "climate smart", that is, sustainably increase agricultural productivity and incomes. Climate-Smart Farming contributes to the achievement of sustainable development goals. It integrates the three dimensions of sustainable development (social, environmental and economic) by jointly addressing food security and climate challenges. The extension and advisory service (EAS) can play a major role in technology development and information dissemination, strengthening farmers' capacity, facilitation and brokering, and advocacy and policy support.

References

Ahmad, S., Kour, G., Singh, A., & Gulzar, M. (2019). Animal genetic resources of India—An overview. *International Journal of Livestock Research, 9*(3), 1–12. https://doi.org/10.5455/ijlr.20181025013931

Altieri, M. A. (2002). Agro-ecology: The science of natural resource management for poor farmers in marginal environments. *Agriculture, Ecosystems & Environment, 93*(1), 1–24.

Altieri, M. A., Funes-Monzote, F. R., & Petersen, P. (2012). Agro-ecologically efficient agricultural systems for smallholder farmers: Contributions to food sovereignty. *Agronomy for Sustainable Development, 32*(1), 1–13.

Behera, R., Rai, S., Sathpathy, D., Sahu, A., Karunakaran, M., Talokdar, A., Singh, A., & M & al, A. (2019). Climate smart livestock production. *Innovative Farming, 4*(1), 015–018.

Bell, D. D. (2002). Waste management. In D. D. Bell & W. D. Weaver, Jr., (Eds.), *Chicken meat & egg production* (5th edn). Kluwer Academic publisher.

Bell, M. J., Wall, E., Russell, G., Simm, G., & Stott, A. W. (2011). The effect of improving cow productivity, fertility, & longevity on the global warming potential of dairy systems. *Journal of Dairy Sciences, 94*, 3662–3678.

De, D., & Singh, G. P. (2001).Monensin enriches UMMP supplementation on in vitro methane production in crossbred calves. In *Proceedings of the Tenth Animal Nutritional Conference (Abstract papers)*, (p. 161). Animal Nutrition Society of India, ICAR-NDRI, Karnal, India.

Eigenberg, R. A., Brown-Brandl, T. M., & Nienaber, J. A. (2007). Development of a livestock weather safety monitor for feedlot cattle. *Applied Engineering in Agriculture, 23*(5), 657–660.

FAO. (2006). Cereal Supply & Demand Brief. http://www.fao.org/worldfoodsituation/csdb/en/

FAO. (2013a). *'Module 8: Climate-smart livestock' in climate-smart agriculture sourcebook*, FAO, Rome, Italy.

FAO. (2013b). *Climate-smart agriculture; source book*. Food & Agriculture Organization of the United Nations, Rome, Italy.

Garnett, T. (2009). Livestock-related greenhouse gas emissions: Impacts & options for policy-makers. *Environmental Science Policy, 12*, 491–503.

Gerber, P., et al. (2013). *Tackling climate change through livestock: A global assessment of emissions & mitigation opportunities*. FAO.

Gerber, P. J., Mottet, A., Opio, C. I., Falcucci, A., & Teillard, F. (2015). Environmental impacts of beef production: Review of challenges & perspectives for durability. *Meat Science, 109*, 2–12.

Henuk, Y. L. (2001). Nutrient adjustments of the diets fed to cage & barn laying hens to decrease waste. Ph.D. Thesis, University of Queensland.

Hoffman, M., & Vogel, C. (2008). Climate change impacts on African Rangelands. *Rangelands, 30*, 12–17.

Jha, A. K., Singh, K., Sharma, C., Singh, S. K., & Gupta, P. K. (2011). Assessment of methane & nitrous oxide emissions from livestock in India. *Journal of Earth Science & Climatic Change, 2*(1).

Kadzere, C. T. (2019). Towards climate-smart livestock farming. *Farmer's Weekly, 19016*, 6–7.

Krishnan, G., Bagath, M., Pragna, P., Vidya, M. K., Aleena, J., Archana, P. R., & Bhatta, R. (2017). Mitigation of the heat stress impact in livestock reproduction. *Theriogenology, 8*, 8–9.

Kumar, P., Singh, A., & Kumar, D. (2020). An overview of working models & approaches to climate smart livestock farming. *IJLSAS, 2*(1), 28–36.

Kumar, S., Prasad, K. D., & Deb, A. R. (2004). Seasonal prevalence of different ectoparasites infecting cattle & buffaloes. *BAU Journal of Research, 16*(1), 159–163.

Lipper, L., et al. (2014). Climate smart agriculture for food security. *Nature Climate Change, 4*, 1068–1072.

Moss, A. R. (1994). Methane production by ruminants—Literature review of Dietary manipulation to reduce methane production & Laboratory procedures for estimating methane potential of diets. *Nutrition Abstracts and Reviews Series B, 64*, 786–806.

Nayak, P. K. et al. (2018). Integrated rice-based crop livestock agroforestry system: An coefficient & climate resilient agricultural practice for small & marginal farmers. In: *Extended Summaries in 3rd ARRW International Symposium*, Feb 6–9 at NRRI, Cuttack (pp. 235–236).

Patra, A. K. (2017). Accounting methane & nitrous oxide emissions, & carbon footprints of livestock food products in different states of India. *Journal of Cleaner Production, 162*, 678–686.

Pearce, W., Holmberg, K., Hellsten, I., & Nerlich, B. (2014). Climate change on Twitter: Topics, communities & conversations about the 2013 IPCC Working Group 1 report. *PLoS ONE, 9*(4), e94785.

Scheehle, E. (2002). Emissions & projections of non-CO_2 Green House gases from developing countries 1990–2020 (p. 73).

Sharma, S., Bhattacharya, A., & Garg, A. (2006). Greenhouse gas emission from India: A perspective. *Current Science, 90*(3), 326–333.

Singh, A., Kumar, P., Kumar, H., Neeraj, A., Kumar, P., & Kour, G. (2020). Status of livestock insurance in India & a complete guide: An evidence-based review. *International Journal of Livestock Research, 10*(5), 8–19. https://doi.org/10.5455/ijlr.20200224090417

Singh, A., & Rashid, M. (2017). Impact of animal waste on environment, its managemental strategies & treatment protocols to reduce environmental contamination. *Veterinary Science Research Journal, 8*(1&2), 1–12. https://doi.org/10.15740/HAS/VSRJ/8.1&2/1-12

Singh, A., Tiwari, R., & Dutt, T. (2020). Augmentation of farmers' income in India through sustainable waste management techniques. *Waste Management & Research, Online First*. https://doi.org/10.1177/0734242X20953892

Singh, A., Tiwari, R., & Dutt, T. (2021). An ICT driven intervention for transforming waste to wealth: Methodic development & assessment of IVRI-Waste Management Guide App. *Journal of Material Cycles & Waste Management, Online First*.https://doi.org/10.1007/s10163-021-01236-1

Singh, A., Tiwari, R., Joshi, P., & Dutt, T. (2020c). Insights into organic waste management practices followed by dairy farmers of Ludhiana District, Punjab: Policy challenges & solutions. *Waste Management & Research, 38*(3), 291–299. https://doi.org/10.1177/0734242X19886632

Singhal, K. K., & Mohini, M. (2002). *Uncertainty reduction in methane & nitrous oxide gases emission from livestock in India* (p. 62). Project report, Dairy Cattle Nutrition Division, National Dairy Research Institute, Karnal.

Singhal, K. K., Mohini, M., Jha, A. K., & Gupta, P. K. (2005). Methane emission estimates from enteric fermentation in Indian livestock: Dry matter intake approach. *Current Science, 88*(1), 119–127.

Sirohi, S., & Michaelowa, A. (2007). Sufferer & cause: Indian livestock & climate change. *Climatic Change, 85*, 285–298.

Sloan, D. R., Kidder, G., & Jacobs, R. D. (2008). *Poultry manure as a fertilizer*. University of Florida.

Srivastava, A. K., & Garg, M. R. (2002). Use of sulfur hexafluoride tracer technique for measurement of methane emission from ruminants. *Indian Journal of Dairy Science, 55*, 36–39.

Thornton, P. (2010). Livestock production: Recent trends, future prospects. *Philosophical Transactions of the Royal Society B, 365*, 2853–2867.

Thornton, P., van de Steeg, J., Notenbaert, M. H., & Herrero, M. (2009). The impacts of climate change on livestock & livestock systems in developing countries: A review of what we know & what we need to know. *Agriculture Systems, 101*, 113–127.

Tripathi, R., & Bisen, J. P. (2019). *Climate resilient agricultural technologies for future. Training manual, model training course on climate resilient agricultural technologies for future*, ICAR-National Rice Research Institute, Cuttack (pp 1–102).

Upadhaya, R. C., Ashutosh, A. K., Gupta, S. K., Gupta, S. V., Singh, S. V., & Rani, N. (2009). Inventory of methane emission from livestock in India. In P. K. Aggarwal (Ed.), *Global climate change & Indian agriculture. Case studies from the ICAR Network project*. ICAR (pp. 117–122).

Chapter 17
Improving Agricultural Carbon Sequestration Strategies by Eco Friendly Procedures for Managing Crop Residues and Weeds

A. Muni Kumari and Jitendra Bhaurao Durge

17.1 Sustainable Agriculture

Sustainable agriculture aims in preservation of the ecological system by adopting ecofriendly procedures in agriculture. Long-term production of crops and livestock for attaining required food production and saving the environment is possible with sustainable agricultural practices. Sustainable agriculture concerns about the environment and focuses on practices that support improving the required food production and do not have adverse effects on the ecological system, soil, water, and air (Palm et al., 2014). To solve natural resource problems in agriculture, productive use of scientific knowledge and approach must be inculcated in agricultural practices enhancing the skills of farmers towards sustainable agriculture. Agricultural sustainability can be achieved by redesigning the agricultural approaches for long term benefits (Wang, 2009). Agricultural approaches for achieving agricultural sustainability are called as sustainable agricultural practices. Sustainable agriculture practices aim at improving agriculture productivity, the addition of nutritional values to the crops, maintaining the soil health, reducing soil degradation/erosion, water-saving, maintaining biodiversity, providing a healthy environment for beneficial organisms, and reducing environmental pollution.

A. Muni Kumari (✉)
Department of Genetics and Genomics, Yogi Vemana University, Kadapa, Andhra Pradesh 516003, India
e-mail: munikumari29@gmail.com

J. B. Durge
Department of Agronomy, Shri Shivaji Agriculture College, Amravati, Maharashtra 444602, India

S. Sheraz Mahdi et al. (eds.), *Adapting to Climate Change in Agriculture-Theories and Practices*, https://doi.org/10.1007/978-3-031-28142-6_17

17.2 Climate Change and Greenhouse Gases

Atmosphere on earth surface blocks most of the light, radiation from the universe and makes life possible on earth. It allows visible light, radio waves and small wavelength infrared light to reach the earth. Atmospheric Carbon dioxide and water vapor absorbs most of the infrared light coming to earth from the Universe. Natural mechanisms such as changes in solar activity and large volcanic eruptions can disrupt global energy balance. In addition to natural process man driven mechanisms such as greenhouse gases emission is another major reason for Global warming. Human activities from last 150 years like fossil fuel burning as a source for electricity, heat and transportation is major cause for an increase in greenhouse gases in the atmosphere. Greenhouse gases from agriculture come from livestock, agricultural soil and rice production. Greenhouse gases absorb and emit some of the radiation energy with in the thermal infrared range on earth. Greenhouse gases cause the greenhouse effect on plants. Heat-trapping greenhouse gases causes raise in temperature of Earth's climate called as Global warming. Global warming results in Climate change. Climate change includes changes in arrival of seasons, unseasonal rains or no rains when expected, extreme weather conditions etc. Industrial activity, agricultural activity and deforestation are also leading to Earth's climate change. Greenhouse gases causing Global warming mainly include Carbon Dioxide (CO_2), Nitrous Oxide (N_2O), Methane (CH_4), Water Vapor and Ozone (O_3) (Langmuir, 1999). Some industrial process emits synthetic greenhouse gases in smaller quantities such as Hydro fluorocarbons, per fluorocarbons, sulfur hexafluoride and nitrogen trifluoride having high global warming potential. Some of these are gases such as chlorofluorocarbons, hydro chlorofluorocarbons and halons are ozone-depleting substances. Fossil fuel burning, solid waste, biogas production, cement production in industries, organisms respiration, decomposition, carbonate rocks weathering, forest fire, eruption of volcanoes etc., can release CO_2 in to atmosphere. Methane is released from livestock, biogas production, decay of organic waste in solid waste landfills etc. Factors releasing nitrous oxide in to atmosphere include use of nitrogen fertilizer in agriculture, industrial activities, fossil fuels combustion, solid waste and wastewater treatment, factories and automobiles. Due to greenhouse gases, land and ocean surface temperatures are warming referred as global warming. The increase in temperature or global warming is continuing in an alarming stage. There is decrease in cold seasons and increase in warm season. In many parts of the world including Europe, Asia and Australia number of heat waves and its occurrence has increased. The Greenland and Antarctic ice sheets are melting due to increased temperatures and continuously losing mass resulting in an increase in ocean levels. Significant increase in mean maximum temperature is observed in many countries in the last 60 years and rainfall variability is increasing. Climate resilient agriculture makes a way to transform current systems and has a wider perspective than increase in production only (Lopez-Ridaura et al., 2018). It supports food production systems in local, regional and global level that are socioeconomically and environmentally sustainable.

17.3 Carbon Sequestration

Carbon sequestration is a process of collecting and isolating CO_2 into stable terrestrial (land, vegetation and forests), Geo (underground), Ocean and other carbon pools in a way that it will not be re-emitted in to atmosphere in near future. Managing the CO_2 emission is by two ways—reducing emission of CO_2 and increasing carbon sequestration. Carbon sequestration is nothing but Carbon dioxide capture and storage or removal of CO_2 from atmosphere. It is a technique of CO_2 capturing and storing in the soil. It aims at mitigating global warming and climate change by slow down or reducing the atmospheric accumulation of greenhouse gases. This process is aimed at long-term storage of carbon dioxide or other forms of carbon, separated from industrial sources or/and other energy sources and storing inside earth (geological storage) and inside ocean (ocean storage) for long term isolation of CO_2 from atmosphere. Oceans, Terrestrial systems (forests and agriculture) are C sequestering potential systems. In biological carbon cycle plants sequester or absorb atmospheric CO_2 for photosynthesis. Forestlands and agricultural lands can be used for Carbon sequestration. Lands absorb CO_2 from atmosphere act as sink and retain the CO_2 inside the land. This retained CO_2 is allowed to stay inside the land and will not be allowed to come out of the land and release into atmosphere due to agricultural activities. Deforestation can result in increase in CO_2 as land areas can absorb CO_2 from atmosphere act as sink. Deforestation and use of forest land for residential purpose and other commercial use leads to reduced forest land and reduction in CO_2 absorption and retention. The Intergovernmental Panel on Climate Change (IPCC) recommends for 80% reduction in CO_2 emission globally by 2050. In order to meet this industries needs to reduce CO_2 emissions. To meet the target industries must adopt Carbon dioxide (CO_2) capture and storage (CCS). The removal of CO_2 released from industries and energy-related sources and transporting for storing in deep-sea or deep under the soil isolating away from the atmosphere (Leeson et al., 2017).

17.4 Carbon Trading

Due to its important role in climate change, greenhouse gas CO_2 got strategic importance in developed as well as developing nations. For reducing the emissions of CO_2 by the polluting industries, carbon trading came in to existence. The Kyoto protocol found a way to give market value to carbon dioxide for trading globally as carbon units. For trading carbon units, emission of carbon must be reduced, then carbon budgeting is calculated as the sum of all exchanges between earth's reservoirs and atmosphere in the carbon cycle. By carbon trading a trading certificate is achieved representing the reduction in emission of carbon in various ways. Countries which reduce carbon emissions get carbon emission reduction (CER) certificates which are traded as carbon trading. Countries earn carbon emission reduction (CER) certificates through completion of clean development mechanism (CDM) projects or in the form of emission reduction units (ERU) which are earned on successful completion of joint implementation projects. If countries continue with carbon emissions above the targets set by Kyoto protocol, they are required to buy 'carbon credits' from other participant countries. A tradable certificate called carbon credit represents the right to emit one ton of carbon dioxide. An industry has to adopt a new technology or improve an existing technology to reduce the emission of carbon dioxide. Countries whose carbon emissions are less when compared to their assigned (tradable allowances), then the country can sell the excess amount in the form of carbon units to the countries whose emissions have exceeded the assigned amount. Carbon trading is done by selling the carbon credits of the developing country to developed country treating carbon dioxide as commodity this free market is known as "CARBON MARKET". Carbon trading results in trading of units of carbon dioxide reduced in the environment. Different Carbon regulatory markets existing are Kyoto market, EU ETS, RGGI (regional greenhouse initiative) in the US, voluntary markets, New South Wales, Chicago climate Exchange (CCX). Paris Agreement of United Nations Framework Convention on Climate Change (UNFCCC) deals with reduction of greenhouse gases (GHG), the goal of Paris Agreement is to reduce global average temperature to well below 2 °C and attain the temperature to pre-industrial levels. The Agreement was negotiated by 12 Dec, 2015. As of Feb 2020 all UNFCCC members have signed the agreement and 189 countries have become part of it. Different Policies, emissions levels and energy prices are the main influencing factors on carbon price (Ji et al., 2019). Agricultural systems have a vast potential to sequester carbon. Management practices that can enhance carbon sequestration and agricultural productivity need to be identified and harmonized. It is also important to include quality characters of grains in sustainable cropping systems (Grahmann et al., 2016). Worlds food needs can be reached by reducing the carbon footprint of agriculture, using the resources efficiently in farming, changing the diets, reducing the crop loss by insects pests, changing agricultural practices according to climate change, reducing post-harvest loss and food wastage.

17.5 CO_2 and Zero Tillage

Agriculture and climate are closely related. The mechanization in agriculture made 78 billion metric tons of carbon stored in soil to get lost in to atmosphere. Soil organic carbon (SOC) is a reserve source of CO_2 isolated from atmosphere. As SOC is derived from CO_2 in the atmosphere, soil can become the sink for CO_2 to remove CO_2 from atmosphere. Scientists are focusing on the study of soil organic carbon its distribution and behavior. The SOC is essential in maintaining good physical condition of soil and absorbing and retaining water and also in supply of nutrients to crops. Crop rotation, Crop diversification, integrated farming system, Tillage and residue management, conservation agriculture, Crop residue mulching, Nutrient management integrated nutrient management, balanced fertilization. Agroforestry and grasslands are some of the crop management methods for carbon sequestration. Conventional tillage results in soil erosion. Research has shown frequent tillage not a beneficial rather a detrimental practice. Zero tillage contributes in mitigating climate change by minimizing emissions of GHGs from soils (Mangalassery et al., 2014). Cultivating the crops without disturbing the soil by reducing plowing and crop residue retention got importance to rescue the global degrading soils and strengthen the agricultural practices to improve the food supply to reach the continuous demand (Derpsch et al., 2010). Minimum tillage involves less or reduced soil disturbance and considerable soil disturbance when compared to conventional tillage. It is aimed at minimizing the tillage to required extent for making a soil suitable for seed germination and crop cultivation. Zero tillage is an environmentally friendly farming technique. It can reduce the carbon emission from the agricultural land (Ogle et al., 2012). In this method of farming no-tillage is applied in agriculture land between harvest and sowing. Weeds are controlled by using herbicides. Mulching reduces the soil erosion stubble on the soil surface improves the soil structure and improve water retention. Crop residues retained on the soil surface as stubble mulch. In Stubble mulch tillage or stubble mulch farming is done by keeping the crop residues on the surface of the soil to protect the soil at all times, with zero disturbance to soil (Derpsch et al., 2014). Uncut residue helps in loosen the soil and kill weeds. Mulching protects against evaporation and erosion losses. In addition zero tillage is part of carbon sequestration procedure. In zero tillage seeding and fertilizer applications are combined and instruments like happy seeder a machine mounted to a tractor is used. It is used to cut and liftthe straw and sowing crop plants into the soil and depositing the straw over the field as mulch. In zero tillage system surface seeding method is used for placement of seeds onto the soil surface without any land preparation. This technique is followed for cultivation of legumes, oilseeds and wheat. To hasten the decomposition in mulching technique, soil surface is cut to 12–15 cm depth using sweeps or blades to mix some residues with soil leaving enough residues on surface. In stubble mulch farming strip is cleared using wide trash-bars and narrow furrow of 5–10 cm is made with planter for seeding leaving residues on soil surface. After two to three years of zero-tillage practice advantages like improvement in soil condi tions and structure favoring root growth, formation of channels in soil due to the decomposition of dead roots results in infiltration of nutrients. Zero tillage and crop residue retention together can result in soil structure improvement and minimized soil erosion (Honsdorf et al., 2020). Adoption of zero tillage in long time results in higher macro porosity of the soil results in nutrient distribution, healthy roots, gas and water cycling (Galdos et al., 2019). Farmers adopting the conservation agriculture have to use zero tillage and some farmers are still using some tillage. To address this issue, International Maize and Wheat Improvement Center (CIMMYT), in a study reported that in a well irrigated condition for different wheat lines of different bred, yields were better in zero tillage compared to conventional tillage. Their results state that zero tillage or conventional tillage systems did not produce lines having adaption specific for any tillage system, variation in tillage systems didn't impact on selection traits such as plant height, water lodging tolerance and earliness. Many studies conducted to find the difference in yield when compared between zero-tillage and conventional tillage. Pittelkow et al. (2015), in Meta-analysis of 678 such studies, including 50 crops and 63 countries, reported 5.1% of reduction in yield under zero-till condition. Performance of zero-till systems was best in dry rain-fed production systems compared to tilled systems. In order to overcome the gap between zero-till and conventional till systems, meta-analysis study show the need to improve zero tillage systems in crop and site-specific conditions.

17.5.1 Constraints of Zero Tillage

Zero tillage has some drawbacks that need to be addressed to improve and implement zero tillage methods in agriculture. Seeds experience difficulty in germination. Due to less decomposition of organic matter, nitrogen fertilizer supply to soil is required. Due to reduced aeration in soil root nodule formation is reduced in leguminous crops. Swing operation cannot be implemented with conventional equipment and procurement of latest equipment becomes a necessity and difficult for poor farmers. Higher soil bulk density of soil in zero tillage conditions may hinder root development of seedlings and high disease pressure. Residue retention on soil surface and slow decomposition of residues may increase nitrogen immobilization.

17.6 Zero Tillage and Herbicides

Weeds are undesired plants in agro-eco-systems causinghuge losses directly or indirectlyby reducing the yields and increasing the cost of cultivation, reduce the potentiality of productive lands, decrease input use capacity, heavy erosion, blockage in irrigation canals, paths and water ponds. Heavily invaded areas results in fires, reduction in grazing areas and loss of space for livestock production, harbors insects and pests, infections, damages the ecosystem, reduction in biodiversity and also affect human and cattle health. Weeds reduce the crop production by 45% or severe crop loss. One-third of the production cost may be spent for weed management. Manual method of weed control is difficult and requires more manpower. The herbicide kills unwanted weeds and plants with out damaging the crop. To maintain soil with zero disturbance, zero tillage system uses herbicides. Usage of herbicides resulting in pollution and ecological problems. Inhibitory activity of herbicide is based on plant hormones. Herbicides are classified based on usage, specificity to control types of weeds, time of application in the fielded.

17.6.1 Types of Herbicides

Translocate herbicides: These herbicides move through Xylem and phloem transport mechanisms in the plant. Depending on herbicide concentration, plant growth conditions, and species it may take two weeks to develop the symptoms in target weeds. Example: Glyphosate.

Contact herbicides: Theseherbicides require direct contact with plant move little within the plant. Some of these herbicides become inactive after long contact with soil. They exhibit their activity of killing target weeds with in 24 h. Example: diquat, bromoxynilparaquatand oxyfluorfen.

Selective herbicides: These herbicides are formulated to control specific weed categories. When applied they target to kill specific weeds being non-toxic or less toxic to other-plants. Example: 2, 4-D (Selectively toxic to broadleaf weeds).

Non-selective herbicides or knockdown herbicides: These herbicides–havea broad spectrum of herbicidal activity to control broadleaf and grass weeds. They can exhibit their activity for some months and prevent weed growth. Example: Parquet, Glyphosate.

Non-residual herbicides: These herbicides get de activated on contact with soil exhibiting no activity after few days of application. They are applied in combination with residual herbicides to extend their activity and inhibit weed germination. Example: non-selective parquet and glyphosate.

Post-emergent and pre-emergent: It refers to weed germination and time of application. Post emergenceherbicides are applied any time after the target weed emerges from the soil. Example: 2, 4-DB, bromoxynil. Pre-emergent herbicide is applied any time before the weeds have emerged in the field. Example: Simazine.

Pre plant herbicides: These are applied several days just before planting the crop Example: EPTC, Glyphosate.

Established stands: Herbicides applied after the root systems have developed enough to allow selective use. Example: 2diuron, herbal, hexazino.

Bio herbicides: Some microbial species can be used as bio herbicides. Some microorganisms involved in herbicide decomposition indicate changes in soil composition after herbicide application. Number of bio indicator microorganisms decrease after herbicide application. Example: Population of Nitrogen-fixing bacteria decrease in 7–14 days of herbicide application.

17.6.2 Adverse Effects of Herbicides

Microorganisms support the plant growth there by help the environment by harmful chemical detoxification, suppress the growth of pathogens, and produce plant growth promoting substances, increase crop productivity, encourage plant growth, stimulate stress related responses and increase plant resistance (Kumar & Sai Gopal, 2015). Understanding the impact on soil micro biota is essential in sustainable agricultural strategies. Herbicides targeting to kill herbs or weeds and pesticides targeting to kill pests are being used extensively in agriculture to decrease the loss in production by suppressing the pests and weeds population. The herbicides, insecticides, and pesticide inputs enter streams, rivers, and other water bodies through leaching called agricultural run-offs. Intensive usage of pesticides and herbicides damages the soil biodiversity, disturbs microorganisms living in field communities and soil ecosystems, kills soil microbes and reduces microbial carbon biomass. Soil microbes are very much essential for maintaining soil structure, function, and fertility. Soil microorganisms help plants to grow by providing essential growth nutrients. Sustainable farming practices and organic farming must be focused for attaining food security (Muller et al., 2017) without causing damage to the microbial community. More than 95% of the applied herbicides and 98% of insecticides reach the soil and react with soil micro-organisms. Herbicides effect the microbial population and reduction in number can be observed within 7–30 days after application. Herbicides adversely affect microbial population by damaging the structure, biosynthetic mechanisms and enzyme activities, cellular membrane composition, protein biosynthesis resulting in the death of many susceptible microbes (Meena et al., 2020). This affects biodiversity of soil and disruption of plant growth regulators. The extent of adverse effects depends on adsorption, decomposition, and bioactivity, desorption, concentration, and toxicity of agrochemicals. Herbicides when used in combinations, microbial communities get more effected.

The reduction in soil microbes is less in no-till (NT) system compared to tillage. A decrease in soil microbes can be noted about 12 days after application of herbicides and combination of herbicides. Eg. Fomensafen, Fluazifap-butyl + Fomensafen. Some herbicides when applied in combination with inorganic fertilizers reduces soil microbial activities. Due to physical and biochemical transformations of herbicides in the field several secondary metabolites are released which are lethal to non-target microbial communities. Along with herbicide solvent and surfactant used also increases the toxicity of the herbicide. Example: surfactant polyoxyethylene amine, glyphosate. Herbicidal effect depends on soil type as in coarse-textured soils like sandy clay loam soils the effect can be more severe. Some herbicides are more hazardous and reduce the soil microbial population when applied-for a long time or repeated use. Example: herbicides belonging to the Triazines group. Atrazine and metolachlor act on actinomycetes and bacteria in the soil. Glyphosate inhibits the growth of soil microbes and mycorrhizal fungi. Single herbicide usage for long time on the same land develop resistance in weeds. Herbicides on reaching the pond cause reduction in dissolved oxygen and effect biological oxygen demand. Depletion of non-target bacteria, fungi, and protozoa results in increase in disease-causing microorganisms, Decrease in number of beneficial organisms allows the growth of opportunistic disease-causing organisms. Weed management in agriculture by ecofriendly approach is essential to balance ecological systems.

17.7 Eco-friendly Weed Management

Modern sustainable agricultural practices must include ecological weed management in place of traditional weed management systems. Eco-friendly weed management methods must be followed as part of sustainable agriculture practices (Owombo et al., 2014). Many eco-friendly weed management methods are established and well described by many scientists (Bahadur et al., 2015). Ecological weed management systems work better by integrating alternative weed control- measures, cropping systems unfavorable for weeds growth, fallowing methods of crop rotation using legumes fallowed by non-leguminous plants and planting cover crops can reduce weed population. Cover crops have allopathic effects to minimize the impact of weed (Vukicevich et al., 2016). Using efficient cultivars, adopting new techniques for sowing and planting, making changes in sowing and planting times, usage of organic residues in mulching, manuring and adopting reduced or zero tillage can prevent weed seed germination and growth, It minimizes the mineralization process leading toreduction in energy consumption and carbon oxide emission (Hobbs et al., 2008). Cultural practices includes soil management, crop management strategies like managing the time and methods of fertilizer applications, effective water management, irrigation, crop rotation (Liu et al., 2016). Allelopathicis the production of biochemical known as allelochemicals by microbes that effect growth of other microbes it can be beneficial or detrimental to the target organisms and their community. For implementation of new approaches in weed management systems and weeds management, participation of multidisciplinary personnel is essential (Liebman et al., 2016). Non-chemical based weed management methods or technology/innovation methods are a very essential part of organic farming and must be adopted by farmers. Farmers must gain awareness by dissemination of knowledge by innovators. Agriculture is exponentially increasing and started using information and communication technology (Walter et al., 2017).

17.7.1 Weed Management Methods in Organic Farming

Preventive management methods to prevent weeds from being established in the field is by using weed free certified seeds, maintaining weed free hay, cleaning the equipment before reaching the field, and testing the irrigation water for contaminated with weed seeds, maintaining weed free irrigation canals. Other methods of management includes pulling and cutting to control some weed species relatively small in population. Following thermal management methods such as—using flame for burning, using hot water steam, blowing hot air, infrared weeders, using microwaves, laser radiation and UV light. Spot burning of the weeds should be practiced after gathering the weed materials from patches with in the field. Effective weed management methods includes cultural weed management techniques such as crop rotation, selecting competitive crops, mulching, promoting vigorous crop growth by fallowing agronomic practices. Integrated weed management involves combination of methods for weed control. Pure seeds, crop variants, different methods of planting maintaining space, application of heat, crop rotation along with intercropping, water testing and maintenance, using manure, biological weed control, and usage of herbicides. Maintaining appropriate crop rotation with legumes, nonlegumes and growing cover crop, selecting competitive cultivars, modifications in planting methods, varying sowing and changing planting time, using organic residues for mulching, adding manure and adopting reduced or zero tillage.

Factors such as appropriate plant population, method of plantation, proper spacing, adequate seed, selecting suitable/competent variety, appropriate time of sowing are significant in weed growth. Crop varieties that grow faster having bigger leaves cover soil between the crop rows with their shade and can limit the growth of weeds. Crops having fast growth compete over the weeds. Crops with narrow space between the rows and less spacing, planting patterns improves efficiency in resource use and restrict weed growth by reducing the amount of light and nutrients required for weed growth and improves crop productivity.

Green Manure Crops

These can suppress weeds by secreting and releasing some weed seed inhibiting chemicals in to the soil. These crops are cultivated in organic farming and incorporated into the soil as manure. It enriches soil, plant diversity and crop rotation and helps in reduction of weeds adapted to a particular crop.

Intercrops

The cultivation of two or more variety of crops in distinct row arrangements simultaneously on the same land is known as intercropping. It increases the crop varieties and enhances the proper usage of resources and also helps to reduce weed growth and limit herbicide usage in farming there by reducing the crop loss.

Competitive Cultivars

Weed suppressing—competitive cultivars have early growth and seed production. It is environmentally safe benign and low-cost procedure for weed management and can suppress weed germination and growth. Transplanted crops with healthy growth, faster seedlings production have reduced crop loss due to competitive growth suppressing weed population.

Crop Establishment Methods

Crop establishment methods such as Zero-till and furrow irrigated raised bed systems can suppress weed seed germination and subsequent growth as moisture for germination and growth will not be available. Thus these systems have reduced weed growth with increase in grain yield compared to conventional tillage system.

Seed Rate

Maintaining optimum plant population improves, showing the excellent smothering effect on weeds and improves the crops yield and profit by thick and faster growth.

Irrigation

Irrigation is one of the methods for controlling weeds, maintaining irrigation timing and irrigating for proper number of times can decrease the weed density as weed seeds cannot germinate and grow without water. Making good land leveling and flooding the water continuously in the field till the water covers the soil completely can reduce weed establishment.

Cover/smother Crops

Cover crops, having shadowing effects, faster growth and more density are cultivated in crop rotation and also inter cropping. Cover crops suppress weed growth due to allelopathic effects, reduction in the sunlight, water, nutrients and space availability for weeds (Vukicevich et al., 2016).

Crop Rotation

Crop rotation is the process of growing plants having different lifecycles by maintaining variation in plantation and harvesting days. Weed management with this procedure can affect the population dynamics of weeds and reduce their population.

Nutrient Management

Fertilizers in agro-ecosystems alter the nutrient levels. Competitions for nutrients between crop and weed may benefit weeds more than crops. Weed interference in crops can be reduced by altering the fertilizer dosage, application time etc., It significantly reduces the weed population and results in higher grain yield. A by product having sufficient nutrient capacity produced by decomposing the plant and animal matter by using primary decomposers is called bio-fertilizer.

Organic Manures

Organic manure is a leftover by product of animal or plant matter after making use of the primary product. Sugar industry by product manure press mudis obtained 2 percent of the weight of sugarcane crushed. It contains macro and micronutrients and organiccarbon. It can destroy the weeds by preventing weed seedgermination when applied in the soil by reducing soil pH and releasing some allelochemicals by the action of microbes present in press mud. Applying press mud in combination with neem seeds is more effective in eliminating weeds.

Weed Management Methods

Depending on the weed type and crop human or mechanical methods or in combination is applied to removeand destroyweeds. Methods like hoeing, tillage, harrowing, torsion weedingare used at various weed growth stages. Manual method of weeding is eco-friendly but tedious and labor-intensive.

Soil Solarization

In the process of soil solarization is a proved and efficient method of weed control by covering the wet soil with polyethylene sheet. It helps in increasing the temperature of soil by 8–12 °C, an increase in temperature, in 4–6 weeks of time this method can inhibit the growth of solar heat sensitive weeds of the soil with out any damage to the crop.

Hand Pulling and Digging

Hand pulling is a procedure of manually removing the entire weed plants from inside the soil when infections are less severe. Manual pulling of weed is convenient when the soil is wet after rain plant roots, underground buds come out of the soil easily.

Mulching

It's an Eco-friendly weed management system in which straw and leaves spread over the ground as mulch. Mulches cover the surface of the soil and obstructing the light reaching the weeds and inhibiting their growth. Mulching reduces soil deterioration, water evaporation by retaining soil moisture and helps in regulating temperature fluctuations. It prevents runoff, soil loss and reduces weed growth. Dry leaves, plant residues, soil, straw, stems, bark, composted material etc. can be used as mulch to control weeds. Mulch improves the soil texture, nutrient value, chemical composition and biological properties of soil. It improves crop growth and yield. Infrared mulches that let infrared light to warmth soil and stop photosynthetically active radiation are effective in inhibiting the weed growth and improve crop production.

Bio-herbicides

Using insects or pathogens to kill weeds is termed as Biological weed control. Bio-herbicide is inoculums of plant pathogens formulated and applied as herbicide. It involves insects, nematodes, bacteria, viruses, and fungi that infects weeds and inhibits their growth. Fungi is widely applied when compared to bacteria, viruses, and nematodes.

Allelopathic Plants

Some plants exhibit allelopathic properties release allelochemicals as a crop-environmental or ecological processes. Various parts of these plants produce and release chemicals that effect growth of neighboring plants. The chemical compounds with allelopathic potentiality are called allelochemicals or allelochemicals. Produce allelochemicals or act as allelochemicals. These plants with allelochemicals are capable of competing with neighboring plants. The Allelopathic mechanisms have significant role in controlling weeds. Some allelopathic plants effect on aquatic weeds. Examples include Acalypha indica, Trianthema portulacastrum, Eichhornia crassipes Parthenium hysterophorus, Cyperus rotundus. Argemone maxicana Lantana camara Eupatorium adenophorus Cyperus sp., Imperata cylindrical Mikania micrantha Echhinochloa colonum, Phalaris minor Imperata cylindrical Portulaca oleracea.

Bio-fertilizers

Bio-fertilizers consists of living microorganisms that have a potential to encourage plant growth by improving the nutrient uptake. Biofertilizers are known as plant growth promoters. They promote growth of plants by enhancing natural process of nitrogen fixation, phosphorus solubulization, synthesizing growth promoting substances. Bio fertilizers enrich soil organic matter enhancing health of the soil. Use of bio fertilizers reduces the necessity for synthetic fertilizers and pesticides.

17.8 Conclusion

Human activities have resulted in an imbalance of carbon dioxide flow in the atmosphere. Raise in CO_2 and other greenhouse gases results in global warming, raise in sea-level, changes in ecosystems. Climate change can be mitigated by reducing global energy usage or developing low carbon fuels, bringing down CO_2 levels by Carbon sequestration. Collected CO_2 from atmosphere can be stored in the ocean, terrestrial, and soil carbon pools of Earth. Fallowing carbon sequestration practices in agriculture such as zero tillage agriculture soils can become carbon storage pools. This can bring financial benefit to any country involved by carbon trading. Biotic removal CO_2 from atmosphere involves higher plants and micro-organisms. Numerous microorganisms still need to be evaluated for their beneficial pathways which can serve the future generations. Knowingly Unknowingly we are destroying the soil microorganisms making soil as just a material to hold the plant. In thirst of benefits and improving the productivities, we are losing the natural ecological balance. Carbon sequestration has no meaning without soil microorganisms. Microorganisms are important participants in carbon sequestration. Using pesticides and herbicides cause great damage to soil microorganisms. We should find alternatives to herbicides and insecticides to continue life on the planet. Sustainable agricultural practices and carbon sequestration should run hand in hand without damaging the natural ecosystem. Many farmers are unaware of the negative effects of chemicals used in farming on health and environment and believe that non-chemical technologies are ineffective. Non-chemical weed control relies primarily on practices that are tedious and cost effective. Farmers may not fallow nonchemical methods due to various reasons like lack of required time and money. Adoption of Non-chemical weed control can be achieved by educating farmers on eco-friendly techniques, making aware of benefits of ecofriendly weed management, focusing on the reasons that inhibit farmer'sadoption of ecofriendly techniques, fulfilling the farmer's requirements, their aspiration, interests, and desires got its importance.

References

Bahadur, S., Verma, S. K., Prasad, S. K., Madane, A. J., Maurya, P., Gaurav, V. K., Verma, & Sihag, S. K. (2015). Eco-friendly weed management for sustainable crop production-A review. *Journal Crop and Weed, 11*(1), 181–189.

Derpsch, R., Franzluebbers, A. J., Duiker, S. W., Reicosky, D. C., Koeller, K., Friedrich, T., Sturny, W. G., Sá, J. C. M., & Weiss, K. (2014). Why do we need to standardize no-tillage research? *Soil and Tillage Research., 137*, 16–22.

Derpsch, R., Friedrich, T., Kassam, A., & Hongwen, L. (2010). Current status of adoption of no-till farming in the world and some of its main benefits. *International Journal of Agricultural and Biological Engineering, 3*, 1–25.

Galdos, M. V., Pires, L. F., Copper, H. V., Calonego, J. C., Rosolem, C. A., & Mooney, S. J. (2019). Assessing the long term effects of zero-tillage on the macroporosity of Brazilian soils using X-ray computed tomography. *Geoderma, 337*, 1126–1135.

Grahmann, K., Govaerts, B., Fonteyne, S., et al. (2016). Nitrogen fertilizer placement and timing affects bread wheat (Triticum aestivum) quality and yield in an irrigated bed planting system. *Nutrient Cycling in Agroecosystems, 106*, 185–199.

Hobbs, P. R., Sayre, K., & Gupta, R. (2008). The role of conservation agriculture in sustainable agriculture. *Philosophical Transactions of the Royal Society B: Biological Sciences, 363*, 543–555.

Honsdorf, N., Verhulst, N., Crossa, J., Vargas, M., Govaerts, B., & Ammar, K. (2020). Durum wheat selection under zero tillage increases early vigor and is neutral to yield. *Field Crops Research, 248*, 107675.

Ji, C., Li, X., Hu, Y., et al. (2019). Research on carbon price in emissions trading scheme: A bibliometric analysis. *Natural Hazards, 99*, 1381–1396.

Kumar, B. L., & Sai Gopal, D. V. R. (2015). Effective role of indigenous microorganisms for sustainable environment. *3 Biotech 5*, 867–876.

Langmuir, C. (1999). Climate change and greenhouse gases. *Eos. 80*(39), 453–458.

Leeson, D., Mac Dowell, N., Shah, N., & Petit Fennell, P. S. (2017). ATechno-economic analysis and systematic review of carbon capture and storage (CCS) applied to the iron and steel, cement, oil refining, and pulp and paper industries, as well as other high purity sources. *International Journal of Greenhouse Gas Control, 61*, 71–84.

Liebman, M., Baraibar, B., Buckley, Y., Childs, D., Christensen, S., Cousens, R., Eizenberg, H., Heijting, S., Loddo, D., Merotto, A. Jr., Renton, M., & Riemens, M. (2016). Ecologically sustainable weed management: How do we get from proof-of-concept to adoption? *Ecological Applications, 26*(5), 1352–1369.

Liu, C., Cutforth, H., Chai, Q., & Gan, Y. (2016). Farming tactics to reduce the carbon footprint of crop cultivation in semiarid areas. A review. *Agronomy for Sustainable Development, 36*, 69.

Lopez-Ridaura, S., Frelat, R., van Wijk, M. T., Valbuena, D., Krupnik, T. J., & Jat, M. L. (2018). Climatesmart agriculture, farm household typologies, and food security An ex-ante assessment from Eastern India. *Agricultural Systems, 159*, 57–68.

Mangalassery, S., Sjögersten, S., Sparkes, D. L., Sturrock, C. J., Craigon, J., & Mooney, S. J. (2014). To what extent can zero tillage lead to a reduction in greenhouse gas emissions from temperate soils? *Scientific Reports, 4*, 4586.

Meena, R. S., Kumar, S., Datta, R., Lal, R., Vijayakumar, V., Brtnicky, M., Sharma, M. P., Yadav, G. S., Jhariya, M. K., Jangir, C. K., Pathan, S. I., Dokulilova, T., Pecina, V., & Marfo, T. D. (2020). Impact of agrochemicals on soil microbiota and management: A review. *Land, 9*, 34.

Muller, A., Schader, C., Scialabba, N.-H., Brüggemann, J., Isensee, A., Erb, K.-H., Smith, P., Klocke, P., Leiber, F., Stolze, M., & Niggli, U. (2017). Strategies for feeding the world more sustainably with organic agriculture. *Nature Communications, 8*, 1290.

Ogle, S. M., Swan, A., & Paustian, K. (2012). No-till management impacts on crop productivity, carbon input, and soil carbon sequestration. *Agriculture, Ecosystems & Environment, 149*, 37–49.

Owombo, P. T., Aregbesola, O. Z., & Adeloye, K. A. (2014). Eco-friendliness of weed management methods in organic farming: The need for extension education. *Journal of Agriculture and Sustainability., 6*(2), 179–199.

Palm, C., Blanco-Canqui, H., DeClerck, F., Gatere, L., & Grace, P. (2014). Conservation agriculture and ecosystem services: An overview. *Agriculture, Ecosystems & Environment, 187*, 87–105.

Pittelkow, C. M., Linquist, B. A., Lundy, M. E., Liang, X., Groenigen, K. J., Lee, J., Gestel, N., Six, J., Venterea, R. T., & Kessel, C. (2015). When does no-till yield more? A global meta-analysis. *Field Crops Research, 183*, 156–168.

Vukicevich, E., Lowery, T., Bowen, P., Úrbez-Torres, J. R., & Hart, M. (2016). Cover crops to increase soil microbial diversity and mitigate decline in perennial agriculture. A review. *Agronomy for Sustainable Development, 36*, 48.

Walter, A., Finger, R., Huber, R., & Buchmann, N. (2017). Opinion: Smart farming is key to developing sustainable agriculture. *Proceedings of the National Academy of Sciences U S A, 114*(24), 6148–6150.

Wang, Y.-W. (2009). Sustainable agricultural practices: Energy inputs and outputs, pesticide, fertilizer and greenhouse gas management. *Asia Pacific Journal of Clinical Nutrion, 18*(4), 498–500.

Chapter 18
Innovative Improved Skip Row Sowing Technique (IISRST) for Sole Soybean Crop Under Rainfed and Irrigated Situation, by Using Conventional Sowing Implements, for Sustainable Yield Advantage

Jitendra B. Durge, A. Muni Kumari, Savita A. Kanse, and S. Sheraz Mahdi

18.1 Need to Overcome Problems in Conventional Sowing Method

In Maharashtra soybean cultivators have to face many problems during cultivation in kharif season. There is an urgent need to minimize problems of soybean cultivators. Major problem of sole soybean crop is, it covers all the field once vegetative growth stage is over. Means up to flowering stage, all the field get covered by soybean crop. It happens because of huge plant population. As per recommendation soybean is sown at a spacing of 30 cm × 7.5 cm or 45 cm × 5 cm. Seed rate recommended is 75 kg ha^{-1}. Accordingly, plant population maintained is 444,444 plant ha^{-1}. Once crop covers all the field, there is a problem of crop monitoring. Farmer has no space available for entry in the field. Because of restricted entry (i.e., overcrowding), there is more incidence of insects—pests and diseases with severe crop weed competition. It all happens during reproductive growth phase of soybean crop. It causes heavy losses in the yield every year with improper quality of the produce.

J. B. Durge (✉) · S. A. Kanse
Department of Agronomy, Shri Shivaji Agriculture College, Amravati, Maharashtra 444603, India
e-mail: jitendradurge@gmail.com

A. Muni Kumari
Department of Genesis and Genomics, Yogi Vemana University, Kadapa, Andhra Pradesh 516003, India

S. Sheraz Mahdi
Division of Agronomy, SKUAST-Kashmir, Shalimar, Srinagar, Jammu & Kashmir 190025, India

S. Sheraz Mahdi et al. (eds.), *Adapting to Climate Change in Agriculture-Theories and Practices*, https://doi.org/10.1007/978-3-031-28142-6_18

Within crop competition is also having an adverse effect on soybean crop. Under normal rains excessive growth of crop takes place. During scanty rains or during long dry spell at vegetative growth phase and during post flowering stage, crop growth gets affected adversely. It causes stunted growth and improper grain filling respectively. Under both situations there is crop failure. Maintaining optimum growth of soybean crop, it seems to be difficult task. Due to inconvenience in crop monitoring, there is a severe problem of stem fly, girdle beetle, semilooper, tobacco leaf eating caterpillar, leaf roller etc. pests and infestation of leaf rust, root rot, collar rot, stem blight, pod blight etc. diseases. This also causes heavy losses because of improper and untimely spraying. This is another cause of crop failure. Many times, during long dry spell in rains, if farmer has an irrigation facility, he is not able to provide furrow irrigation in time to crop. Even if he is having sprinkler unit, it becomes difficult to farmer to provide irrigation. Because of huge plant population, there is within crop competition in addition to severe crop—weed competition. It leads to competition for soil moisture, soil nutrients, vertical space, solar radiation, etc. Following are the problems of soybean crop, when grown by conventional bullock drawn and tractor drawn sowing implement.

18.2 Problems

- Heavy plant population.
- No space for entry in the field.
- Overcrowding of crop.
- Difficulties in crop monitoring and inspection.
- Untimely and improper spraying.
- Inconvenience during spraying.
- Heavy infestation of insect, pest and diseases.
- Heavy crop—weed competition.
- Stunted growth under scantly rains.
- Water logging during excessive rains.
- Excessive growth under normal rains.
- Within crop competition for growth factors.
- Poor aeration.
- Improper humidity.
- Uneven distribution of sunlight.
- Moisture stress during long dry spell.
- No space for providing protective irrigation, during reproductive growth phase.

These problems highly affect productivity and quality of produce, of soybean crop every year. To overcome all these problems, there is an urgent need to change sowing method of soybean crop. To overcome this situation farmer can very easily switch from conventional sowing method to Innovative Improved Skip Row Sowing

Technique (IISRST), that too by using conventional bullock drawn and tractor drawn sowing implements.

By using same conventional sowing implements farmer can perform sowing by skip row method. In Maharashtra State conventional farmer use bullock drawn three or four tined implement and tractor drawn five or six or seven or nine tined implement for sowing of sole soybean crop. In case of Innovative Improved Skip Row Sowing Technique (IISRST), while sowing soybean crop, farmer has to skip one row after every specific number of rows. Means after every three rows—4th row, after every four rows—5th row, after every six rows—7th row, after every seven rows—8th row have to be skipped. This decision of skipping row, totally depends upon the type of conventional sowing implement available with the farmer. It may be bullock drawn or tractor drawn implement. In this case sowing of soybean is done as per recommended Row to Row (R × R) and Plant to Plant spacing (P × P). Seed rate required will ultimately governed by the row skipped. If every 4th row is skipped, 25% less seed rate will be required. If every 5th row is skipped, 20% less seed rate will be required. If every 6th row is skipped, 16.67% less seed rate will be required. If every 7th row is skipped, 14. 29% less seed rate will be required. If every 8th row is skipped, 12.5% less seed rate will be required. So, depending upon the row skipped, up to that extent seed rate required will be less. Recommended seed rate for sole soybean is 75 kg ha^{-1}. If every 4th row is skipped, seed rate required will be 56.25 kg ha^{-1}. If every 5th row is skipped, seed rate required will be 60 kg ha^{-1}. If every 6th row is skipped, seed rate required will be 62.5 kg ha^{-1}. If every 7th row is skipped, seed rate required will be 64.30 kg ha^{-1}. If every 8th row is skipped, seed rate required will be 65.625 kg ha^{-1}. In addition to this there is a saving of fertilizer also in proportionate with the row skipped. This is an additional advantage, while skipping the row (i.e., every 4th or 5th or 6th or 7th or 8th row), farmer has to keep empty space, double as that of R × R spacing, to form strips of three rows or four rows or five rows or six rows or seven rows. If R × R spacing is 45 cm, farmer has to skip 90 cm empty space to form strips of specific number of rows in the field. Accordingly farmer has to skip that particular row, so that strip of 3 rows, 4 rows, 5 rows, 6 rows or 7 rows, will be formed in the field. This space of empty (i.e., unsown) row ultimately helpful for entry in field, for crop monitoring, spraying, better aeration, proper humidity, even distribution of sunlight, for avoidance of excessive growth etc. At the time of hoeing operation (locally called Dawarani or Kolapani), farmer has to open a furrow at empty row, by tying a rope to the tines of the hoe, to form a dead furrow. Ultimately Broad Base and Furrow (BBF) like structure will get created in the field. These furrows will play a key role of In-situ rain water conservation (Pendke et al., 2019) and also it will be useful for providing furrow irrigation to the crop, if irrigation facility is available with the farmer. This empty space can also be utilized for placing pipes of sprinkler unit at the time of irrigation. Table 18.1 indicates seed rate required with skipped row.

Table 18.1 Seed rate required in kg ha^{-1} in accordance with row skipped

Sr. No.	Row skipped	Less seed rate required (%)	Seed rate required (kg ha^{-1})
1	Every 4th row	25.0	56.25
2	Every 5th row	20.0	60.00
3	Every 6th row	16.6	62.50
4	Every 7th row	14.2	64.30
5	Every 8th row	12.5	65.625

18.3 Innovative Improved Skip Row Sowing Techniques

In Maharashtra State conventionally sole soybean crop is sown by 3 or 4 tined bullock drawn implement. In Maharashtra locally it is called as Tiffan, Chaufan, Kakari (i.e., marker) with Sarata's and conventional tractor drawn sowing implements used are with five or six or seven or nine times. By using these conventional tractor and bullock drawn sowing implements, sowing of soybean crop can be successfully done by Innovative Improved Skip Row Sowing Technique (IISRST) as under.

18.3.1 Bullock Drawn Implement with Three Tines—(i.e., Tiffan or Kakari with Sarata's.)

While performing sowing with Tiffan or Kakari farmer can skip every 4th row or 7th row. If farmer wants to skip every 4th row, then every time while returning and going back, farmer has to skip one row (i.e., every 4th row) to from strips of 3 row in the field, with every fourth row unsown (Fig. 18.1) (Sharma et al., 2019).

If the farmer wants to maintain strips of six rows of soybean crop in the field, then while returning from the other end, again he will sow adjoining three rows but while going back (end of the field from where sowing started), he will skip one row (i.e. every seventh row he will keep vacant) to from strips of six rows each in the field, with every 7th row unsown (Fig. 18.2).

18.3.2 Bullock Drawn Implement with Four Tines—(i.e., Chaufan or Kakari with Saratas)

While performing sowing with Chaufan or Kakari (i.e., marker) with four tines, every time while returning and going back, farmer has to skip one row (i.e., every 5th row) to form strips of four rows each in the field, with every 5th row unsown (Fig. 18.3) (Gupta et al., 2017).

Fig. 18.1 Soybean—three rows skip row planting

Fig. 18.2 Soybean—six rows skip row planting

18.3.3 Tractor Drawn Sowing Implement with Five Tines

While performing sowing with tractor drawn implement having five times, farmer has to skip every 6th row, while returning and going back, to from strips of five row each in the field with every 6th row unsown (Fig. 18.4).

Fig. 18.3 Soybean—four rows skip row planting

Fig. 18.4 Soybean—five rows skip row planting

Fig. 18.5 Soybean—six rows skip row planting

18.3.4 Tractor Drawn Sowing Implement with Six Tines

While performing sowing with tractor having six tines, farmer can maintain strips of six rows or four rows in the field. To maintain six row structure (i.e., strips) in the field, every time while returning and going back, he has to keep every 7th row vacant to form strips of six rows each in the field (Fig. 18.5) (Verma et al., 2018).

To maintain strips of 4 row in the field, farmer has to close the tube from the seed unit and fertilizer unit by closing it with a lid or by putting cloth in it. For this he has to close one tine each from both corners of seed and fertilizer unit, so that sowing can be performed in strips of 4 rows each. While doing this every time corner row will remain vacant. Now every time while returning and going back farmer has to keep last corner tine in the empty furrow itself, which is left vacant earlier, to form strips of four rows each, with every 5th row unsown (Fig. 18.6) (Asawar et al., 2019).

18.3.5 Tractor Drawn Implement with Seven Tines

While performing sowing with seven tined tractor drawn implement, strips of 3 rows or 5 rows or 6 rows or 7 rows can be maintained in the field. To form strips of 3 row, the central i.e., 4th hole of seed and fertilizer unit should be closed with a lid or by putting cloth in the opening of the tube. Every time while returning and going back, if farmer keeps one row vacant, there will be formation of strips of 3 rows in the field with every 4th row unsown (Fig. 18.7).

Fig. 18.6 Soybean—four rows skip row planting

Fig. 18.7 Soybean—three rows skip row planting

Fig. 18.8 Soybean—six rows skip row planting

While returning and going back every time, if adjoining rows are sown with normal R × R spacing, there will be sowing in strips of six rows each with every 7th row unsown (Fig. 18.8).

If farmer closes mouth of the tube from the seed and fertilizer unit, one each of both the ends, ultimately sowing will be performed in the strips of 5 rows each with corner row vacant. Every time while returning and going back last tine of the sowing implement have to keep in empty furrow itself, keeping every 6th row unsown (Fig. 18.9).

If farmer keeps every 8th row vacant while returning and going back, strips of seven rows each, will be formed in the field with every 8th row unsown (Fig. 18.10).

18.3.6 Tractor Drawn Sowing Implement with Nine Tines

While performing sowing with tractor drawn sowing implement with nine tines, middle i.e., mouth of 5th tube has to be closed with a lid or cloth. Now while performing sowing, every time while returning and going back, one row has to be kept vacant (i.e., every 10th row) which means, there will be sowing in strips of 4 rows each in the field, keeping every fifth row unsown (Fig. 18.11) (Dhakad et al., 2020).

In this way while performing sowing with conventional sowing implement, with slight change, farmer can easily switch to Innovative Improved Skip Row Sowing

Fig. 18.9 Soybean—five rows skip row planting

Fig. 18.10 Soybean—seven rows skip row planting

Technique (IISRST) without any additional investment, by using his mental skill. Thus, by giving appearance like BBF in the field, ultimately farmer can minimize problems of sole soybean crop. Farmer can achieve following advantages through adoption of Innovative Improved Skip Row Sowing Technique (IISRST).

Fig. 18.11 Soybean—four rows skip row planting

18.4 Advantages

- Saving of seed.
- Saving of seed cost.
- Saving of fertilizer.
- Saving of fertilizer cost.
- Space for entry in the field.
- Space for crop monitoring.
- Timely diagnosis of insect, pest and diseases.
- Reduced crop weed competition.
- Proper aeration.
- Help to maintained proper humidity in field.
- Proper and even distribution of sunlight.
- Even growth of crop.
- Proper monitoring and management of individual strip.
- Overcrowding of crop can be avoided.
- Spraying can be done timely, properly and conveniently to individual strip.
- In-situ rain water conservation.
- Provision for irrigation through already opened furrows.
- Space to place pipes of sprinkler unit.
- Minimization of incidence of inset—pest and diseases.
- Border effect can be achieved through space of vacant row.
- Improvement in quality of produce with sustainable yield advantage.

- Saving of inputs and inputs cost like Insecticides, Fungicides etc.

18.5 Disadvantage

- Reduced plant population.
- Weed infestation may be more at space of empty rows (i.e. unsown row).

18.6 Conclusion

Innovative Improved Skip Row Sowing Technique (IISRST) is nothing but an improvement in conventionally sown soybean, by giving appearance like Broad Base and Furrow Method (BBF), to achieve all the advantages those are obtain in BBF Technology. Minimizing problems of conventionally sown sole soybean, by switching to Innovative Improved Skip Row Sowing Technique (IISRST) by maximizing benefits achieved through advantages received and converting disadvantage into advantage. The row which is skipped, reduce plant population up to certain extent but the space thus created helps to get entry in to the field, crop monitoring, timely spraying, convenience in spraying, proper aeration, even distribution of sunlight, border effect, provision for protective irrigation, in-situ rain water conservation, for provision of irrigation through furrows and irrigation by sprinkler. Thus, by minimizing problems of sole soybean crop, through adoption of Innovative Improved Skip Row Sowing Technique (IISRST), yield advantage to the extent of 15–40% can be achieved by converting disadvantage into advantage without any additional investment, that too by using resources those are available with the farmer at a source at grass root level, keeping other management practices same. Farmer need not to purchase any costly implement or he need not to use any high value externally purchased inputs or technology. Thus, by adopting low-cost technology measures, soybean cultivator can achieve sustainable yield advantage to strengthen farmers financial sustainability through adoption of available resources with the farmer at a source. Innovative Improved Skip Row Sowing Technique (IISRST) can prove to be a booster for socioeconomic sustainability of the farmer by achieving sustainable yield advantage. Thus, minimizing problems of sole soybean crop is a key to achieve sustainable yield advantage with quality produce of soybean.

References

Asawar, B. V., Pendke, M. S., & Narale, S. H. (2019). Influence of Mechanization levels & energy dynamics on production potential of Soybean under rainfed situation. *Dryland Agricultural and Research Journal, 34*(2), 71–76.

Dhakad, S. S., Ramana Rao, K. V., Bhargarva, V. K., & VermaGayatri, A. G. R. (2020). Effect of different sowing methods on growth characters and yield of soybean in vertisols—A review. *International Journal of Current Microbial Applied Science, 9*(5), 1607–1608.

Gupta, R., Gupta, P. S., Kalani, G. S., Somvanshi, S. P. S., & Besedia, A. S. (2017). Influence of broad base & furrow seed drill on growth characteristics & yield of soybean in mandsaur Dist of Madhyapradesh. *Progressive Research—An International Journal, 12*(4), 2675–2678.

Pendke, M. S., Aswar, B. V., Narale, S. H., Ravindra Chary, G., & Gopinath, K. A. (2019). Evaluation of rainwater conservation practices for reducing runoff, soil loss and enhancing moisture availability under rainfed soybean. *Indian Journal of Dryland Agriculture Research & Development, 34*(91), 38–41.

Sharma, P., Dupare, B. U., & Khandekar, N. (2019). Economic impact assessment of broad base & furrow seed drill for soybean. *Agricultural Research Journal, 9*, 392–399.

Verma, P. D., Jain, V., & Rajput, A. S. (2018). Evaluation of sowing methods of soybean in Bhataparadist of Chhattisgarh. *Journal of Krishi Vigyan, 6*(2), 109–112.

Printed in the USA
CPSIA information can be obtained
at www.ICGtesting.com
LVHW010557050324
773586LV00003B/177

9 783031 281419